Organic Materials for Non-linear Optics II

Organic Materials for Non-linear Optics II

Edited by

R.A. Hann
ICI Imagedata, Manningtree

D. Bloor
University of Durham

The Proceedings of the Second International Symposium on Organic
Materials for Non-linear Optics organised by the Applied Solid State
Chemistry Group of the Dalton Division of the Royal Society of Chemistry
held 4th–6th September 1990 in Oxford

Special Publication No. 91

ISBN 0-85186-397-3

A catalogue record of this book is available from the British
Library

© The Royal Society of Chemistry 1991

Published by The Royal Society of Chemistry,
Thomas Graham House, Science Park, Cambridge CB4 4WF

Printed in England by Redwood Press Limited, Melksham,
Wiltshire

Preface

This publication represents the proceedings of the conference on 'Organic Materials for Non-linear Optics' held in Oxford from 4-6 September 1990. OMNO 90 (as it became known) was the second in a continuing series of conferences on non-linear optics (NLO) organised by the Applied Solid State Chemistry Group. Like the first conference in the series, it was a truly international forum, with delegates and speakers from many parts of Europe, the USA, and Japan. There was also a good balance between academia, industry, and government research establishments, and all levels of approach to the subject. In the preface to the OMNO 88 proceedings, we predicted that by the time of this year's meeting, there would be credible devices based on organic materials. This prediction has been borne out in practice, as will become clear.

The first paper is an Introduction, which is intended to provide a way into the rest of the publication for those less familiar with the area; it also highlights the current areas of rapid progress and indicates the probable regions of future activity.

Each of the following Sections covers a broad area, in which the papers written by the invited speakers provide a framework for the research contributions. The division of a group of papers such as these is inevitably somewhat arbitrary, and the Section headings are only a general guide to the ground covered.

The headings are as follows:

Theory
Small Organic Molecules
Metal-organic compounds
Polymers
Devices

In editing these papers, we have been particularly impressed with some significant differences from the balance of those in the OMNO 88 publication. Firstly, there has been less emphasis on what might be termed a shot-gun approach to the synthesis and testing of new organic molecules for second order applications. This is best put in the context of some of the other trends, one of which is the greater use of figures of merit for characterising materials, and the general realisation that it is no use having a record value for the beta coefficient if the material in question absorbs strongly across most of the useful spectrum. The hunt for new materials that might give a more favourable trade-off between beta coefficient and absorption seems to have shifted towards metal-organic compounds, which are very much in an exploratory stage as far as NLO properties are concerned.

The emphasis on polymers is significantly greater than last time, and reflects the unique importance of these materials in the fabrication of devices that do not require time consuming growth of single crystals. Integrated Optical devices based on polymers have been demonstrated to have potentially useful properties, and it is likely that they will become a commercial reality in the mid term.

OMNO 92 will be held in Oxford in August 1992; the continuation of the series of meetings reflects the continuing importance of an area that is moving along the road from the laboratory to commercial exploitation. We have no doubt that there will be a further increase in the recognition of the importance of figures of merit in assessing the significance of new materials, and that more and more of the materials being developed will be polymeric. We anticipate, following the trend between OMNO 88 and OMNO 90 that there will be a larger proportion of contributions on devices, hopefully some of them showing clear advantages over those based on inorganic materials.

R A Hann
D Bloor

Contents

Introduction

Theory

Small Organic Molecules

Quadratic Enhancement of the Second Harmonic
Intensity of Interleaved Langmuir-Blodgett Films of a
Quinolinium Dye   60
    G. J. Ashwell , P. J. Martin , M. Szablewski , P. A.
    Thompson , A .T. Hewson , and S. D. Marsden

Amphiphilic Dyes for NLO in LB Films   67
    A. Laschewsky , W. Paulus , H. Ringsdorf , D. Lupo ,
    P. Ottenbreit , W. Prass , C. Bubeck , D. Neher , and
    G. Wegner

Stabilization of LB Multilayers of NLO-active Dyes by
Means of Complexation with Polymeric Counterions   75
    W. Hickel , D. Lupo , P. Ottenbreit , W. Prass , U.
    Scheunemann , J. Schneider , and H. Ringsdorf

Oriented Crystallisation and Non-linear Optics: Tools for
Studying Structural Changes of Amphiphilic Aggregates
at the Air-Solution Interface   82
    I. Weissbuch , G. Berkovic , L. Leiserowitz , and M. Lahav

Mixed-dye Langmuir-Blodgett Films as Nonlinear Optical
Materials   89
    M. E. Lippitsch , S. Draxler , and E. Koller

The Preparation of Langmuir-Blodgett Films of SDAN for
Second Harmonic Generation Application   96
    F. R. Mayers , J. O. Williams , A. Mohebati , D. West ,
    T. A. King , and G. S. Bahra

An Optical Characterisation of the Organic Nonlinear
Crystal 4-Nitro-4'-methylbenzylideneaniline (NMBA)   102
    R. T. Bailey , G. H. Bourhill , F. R. Cruickshank , D.
    Pugh , E. E. A. Shepherd , J .N. Sherwood , and G. S.
    Simpson

Second Order Non-linear Optical Properties of
4-N-Methylstilbazolium Tosylate Salts   108
    C. P. Yakymyshyn , S. R. Marder , K. R. Stewart , E. P.
    Boden , J .W. Perry , and W. P. Schaefer

Crystal Structure and SHG Activity of the New Chiral
Species 6-Nitro-2-(L-Prolinol) Quinoline   115
    K. J. Atkins , C. L. Honeybourne , K .N. Harrison , F.
    Grams , and A .G. Orpen

Metal-organic Compounds

Polymers

Devices

Introduction

Organic Materials for Non-linear Optics: Yesterday, Today, and Tomorrow

D. Bloor

APPLIED PHYSICS GROUP, SCHOOL OF ENGINEERING AND APPLIED
SCIENCE, UNIVERSITY OF DURHAM, DURHAM DH1 3LE, UK

1. INTRODUCTION

The advent of the laser is frequently identified as the catalyst for the discovery of non-linear optical effects in materials. However, the non-linearity in the optical properties of media subjected to applied electric fields have been known since the last century. The refractive index can depend either linearly (Pockels effect) or quadratically (Kerr effect) on the applied field.[1,2] The discovery of these small effects reflects the high sensitivity possible in polarization and interferometric measurements. Rayleigh, Brillouin and Raman scattering are non-linear optical phenomena that were known before the advent of the laser.[3,5] Their detection with conventional light sources was not easy and the study of these effects was confined to a few specialist laboratories. The availability of intense, coherent light sources radically changed this situation since it made the observation of these and other optically driven non-linear effects much easier. Thus the advent of the laser has resulted in a rapid development of non-linear spectroscopies and the study of the non-linear optical properties of materials. In consequence there has been an increasing interest in the technological applications of non-linear materials and a need for materials with larger non-linear optical coefficients. Though initially the focus of much of this work was on inorganic materials the field of organic non-linear optical materials has grown steadily over the last decade.[6-9]

The interest in organic compounds has been stimulated by the ability of the organic synthetic chemist to provide molecules with structures designed to optimise properties of interest. There is thus a continuous interplay between theory and experiment. Quantum mechanical calculations of the electronic properties of molecules provide a basis for the molecular design. Experimental studies of the target compounds provide data with which to refine the theoretical models. Formally the polarization of a molecule subjected to a local electric field, E^l, is given by:

$$p_i = \sum_j \alpha_{ij} E^l_j + \sum_j \sum_k \beta_{ijk} E^l_j E^l_k + \sum_j \sum_k \sum_l \gamma_{ijkl} E^l_j E^l_k E^l_l$$

where the polarization (p) and the field (E^l) are vectors and α, β and γ are tensors. The subscripts i, j, k, l ... are cartesian co-ordinate labels for the components of these quantities. β and γ are termed the first and second hyperpolarizibility tensors and characterise the second- and third- order non-linear behaviour of the molecule respectively.

When the individual molecules are assembled into a solid medium the non-linearity of the medium is expressed in terms of the external field by:

$$P_i = \sum_j \chi^{(1)}_{ij} E_j + \sum_j \sum_k \chi^{(2)}_{ijk} E_j E_k + \sum_j \sum_k \sum_l \chi^{(3)}_{ijkl} E_j E_k E_l$$

Here $\chi^{(1)}$ is related to the refractive index and $\chi^{(2)}$ and $\chi^{(3)}$ are the non-linear susceptibilities. It is these coefficients that must be maximised if useful materials are to be produced. An essential condition for finite β and $\chi^{(2)}$ is that the molecules and solids must be asymmetric. If either has a centre of symmetry then the polarization must remain unchanged in magnitude but change sign when the applied field is reversed. The sign of the even terms is unchanged on field reversal and these terms must be zero if the above condition is to be met. Thus for finite $\chi^{(2)}$ crystals must belong to a non-centrosymmetric crystal class.

The non-linear terms in the polarization are of interest since they provide frequency multiplication, ($\chi^{(2)}$; second harmonic generation, $\chi^{(3)}$; third harmonic generation) and mixing. Combinations of these processes can be used to provide coherent radiation over a wide frequency range from the ultra-violet into the infra-red from a limited number of primary laser sources. Such processes require in addition to a large non-linearity that the fundamental and output waves are phase matched, i.e. have the same phase velocity so that the output intensity can be built up over the whole of the optical path in the medium. If the phase velocities are different the phase shifts along the optical path lead to cancellation of the desired output. For phase matching to be possible the crystal should be birefringent so that combinations of extraordinary and ordinary rays satisfy the phase matching condition along specific directions within the crystals.

Low symmetry organic crystals will be birefringent but, for axial molecules, i.e. ones in which β_{zzz} dominates, there will then be special molecular orientations that will maximise $\chi^{(2)}$. Thus additional constraints are placed on the crystal packing with the molecules lying at particular angles to the symmetry axis for a particular structure.[10] This ensures efficient coupling of the molecular non-linearity to the E- and O-rays which are polarized parallel and perpendicular to the optic axis. Conversely to obtain maximum electro-optic coefficients it is necessary to align the molecular hyper-polarizibility tensors parallel to one another. Thus different molecular packing is necessary to maximise different non-linear optical coefficients.

For materials to be of practical use they must also be stable and processable. The former is a generic problem for all organic materials. The latter can be satisfied by several classes of materials with differing success. Crystals are not easy to produce either with large size or specific geometry, however, they will in general display maximum optical non-linearity. On the other hand polymers are easy to process but have complex morphology which can lead to low optical transparency due to scattering from microcrystalline regions. Glassy polymers can be used but since they are isotropic cannot possess a finite $\chi^{(2)}$. To achieve second-order non-linearity in such polymers some polar order must be imposed, usually by the orientation of polar moieties, either pendent groups or added guest molecules, by electric fields applied while the sample is heated above or near to its glass transition temperature (T_g). The imposed order is retained either by cooling well below T_g or by cross linking of the active groups during poling.[11,12] Langmuir-Blodgett (LB) films can be utilised to obtain polar structures either intrinsically or by imposing an alternating layer structure.[13] However, the production of LB film samples with a thickness sufficient to give a good macroscopic non-linearity is a time consuming process.

In the case of third-order non-linear materials since there are no structural requirements the only considerations are optical transparency and preferential alignment which will maximise the bulk non-linearity.

Progress is being made with all these classes of materials and recent work is briefly reviewed in the following sections.

2. SECOND-ORDER NON-LINEAR MATERIALS

Crystals and LB films

Theoretical models for the evaluation of β have been developed by a number of groups.[10-18] These give good agreements with measured values for known materials and provide a good guide for synthesis. The basic molecular structure required is asymmetric with electron donating and accepting groups connected by a conjugated structure. This model for molecular design leads to a trade-off between non-linearity and the wavelength range of optical transparency. Longer conjugated connecting sequences lead to larger β but also lowers the energy of the first allowed optical transition.[19] Both multiple σ-bonds and through-space interaction can also be used to provide the molecules with a collective electronic behaviour. However, while a large β can be achieved obtaining the desired crystal structure is more problematic. Chirality can be used to impose a non-centrosymmetric structure but in general the non-linearity achieved is no better than that of non-chiral molecules that form non-centrosymmetric crystals. A comprehensive survey of recent experimental studies of single crystals is given in reference 19.

Non-conjugated molecules such as urea have been known for many years and display useful non-linearity at short wavelengths.[20] Alternatively conjugated molecules can be employed with weaker donating and accepting groups. This strategy has been used to obtain optical transparency down to about 400 nm.[21] This region is of interest for frequency doubling of 800 nm region diode lasers for use with high storage density optical discs.

Derivatives of nitroaniline have been extensively studied. An example is 2-(N, N-dimethylamino)-5-nitroacetanilide (DAN) which has a noncentrosymmetric crystal structure (P2₁).[22-24] The molecular structure is shown in Figure 1. Bulk crystals of high quality have been prepared but crystal plates can be grown with a phase matching direction close to the normal so that cutting and polishing are not necessary. The intra-cavity generation of second harmonic radiation has been reported[25]. The efficiency of second harmonic generation increases quadratically with fundamental intensity. For Nd:YAG radiation at 1.06 μm there is a small resonance contribution to the SHG.[23] Resonance effects derive from the energy denominators in the quantum mechanical expressions for the non-linear coefficients. In the case of DAN SHG output at 532 nm is close to resonance with the first allowed optical absorption which has an absorption edge at a slightly shorter wavelength. Studies of SHG at high powers fit the extrapolation of the low power intra-cavity data and give efficiencies of over 25% but show saturation effects at high power levels.[26] This is primarily due to local heating of the crystals which showed good resistance to damage at power levels up to 250 MW/cm^2.

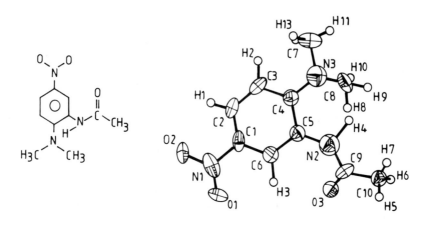

<u>Figure 1</u> Chemical structure and crystal conformation of DAN.

Introducing a longer conjugated bridge as in the merocyanine and stilbazonium type compounds results in materials opaque in the visible. However, the simultaneous observation of a large electro-optical and second and third harmonic generation effects renders these materials scientifically and technically interesting. The well known styrylpyridinium cyanine dye (SPCD) salt with a toluene sulphonate counter ion has been extensively studied.[19,27-29] However, the presence of trace impurities has dramatic effects on the observed powder SHG and THG intensities.[29] The use of salts as a route towards non-centrosymmetric crystals has been studied.[28,30] Examples are known of materials with powder SHG efficiencies over a thousand times that of a urea reference. Similar values have also been observed for compounds with relatively week donating and accepting groups, e.g. MMONS and CMONS (Figure 2).[31] These findings here resulted in speculation that there might be a contribution from collective effects in these crystals. However, it has been pointed out that both the ground state electric dipole and the optical transition moment contribute to β. These cannot be maximised simultaneously since the former requires a very asymmetric electron distribution while the latter requires a large overlap between ground and excited state electron distributions, i.e. uniform electron density over the molecule. Thus too large an asymmetry may be counter productive. The use of less asymmetric structures to maximise β has been considered by Yoshimura[32] and is discussed by Marder et al., in this volume.

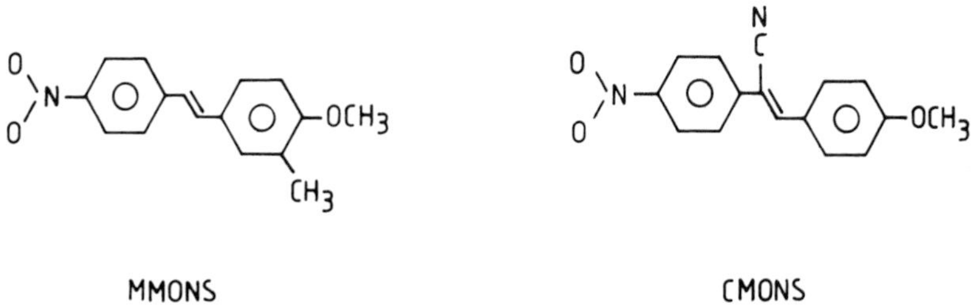

MMONS   CMONS

Figure 2 Chemical structures MMONS and CMONS.

The use of metal organic substituents has received less attention than that devoted to purely organic donors and acceptors. Despite the richer redox chemistry of metal centres the early attempts to produce metal organic nlo materials were disappointing. The first example with a substantial non-linearity to be reported was the ferrocenyl compound shown in Figure 3.[33] The close resemblance to the nitroanilines is obvious. Since then a number of other compounds showing significant SHG have been reported.[34] Chiral metal-organic moieties have been used to obtain non-centrosymmetric crystals. The compound shown in Figure 4 contains a ferrocenyl group, a stilbazonium bridge and a chiral metal group and is oriented in the crystal very close to the optimum angle for maximisation of SHG.[35] The active role of the chiral metal centres is shown by the difference in powder SHG for the isostructural Mo and W containing compounds. The former has higher activity which reflects the relative accepting power of the two substituent groups. In general these compounds are black but other more transparent metal-organic species can be employed.

Work on LB films has shown that the use of alternating layers enables SHG active films to be prepared.[13] At first the expected quadratic dependence on the number of layers deposited was not observed. However, a number of workers have reported such behaviour in optimised systems.[36] Unfortunately it is sometimes necessary to dilute the nlo active molecules with fatty acids in order to obtain good film quality. A recent notable exception is the Zwitterionic molecule related to TCNQ shown in Figure 5.[37] This material deposits as high quality Z-type (polar) layers and shows quadratic dependence on the number of layers for films several hundreds of layers thick. Recent measurements of low optical losses for

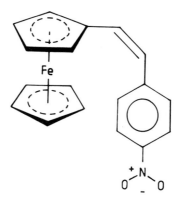

<u>Figure 3</u> Chemical structure of (cis)-[1-ferrocenyl-2-(4-nitrophenyl) ethylene]

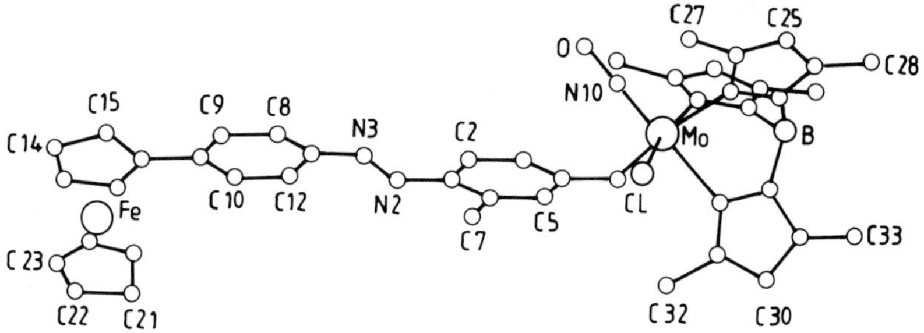

<u>Figure 4</u> Crystal conformation of the chiral metal organic compound MoNOLCl(NHC$_6$H$_4$-4-N=N-C$_6$H$_4$-4'-F$_c$) where L is HB(Me$_2$C$_3$N$_2$H)$_3$ and F$_c$ the ferrocenyl group

in-plane propagation in LB films offer prospects for their eventual use in integrated optical devices.[38]

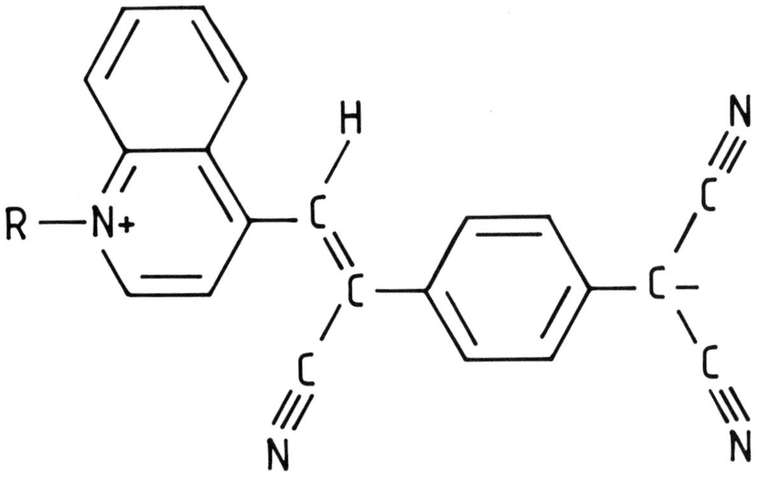

Figure 5 Chemical structure of substituted Z-β-(1-alkyl-4-quinolinium)-α-
         cyano-4-styryldicyanomethanide

Polymers

        The availability of methods to coat large areas with high quality thin
films developed for integrated circuit manufacture has been one driving
force for the development of polymers exhibiting second-order optical nonlinearity. The potential exists for the development of large electro-optical
switch arrays that would be difficult and expensive to realise in inorganic
crystalline materials. However, as noted in the introduction it is necessary
to pole glassy polymers with applied electric fields in order to introduce the
necessary non-centrosymmetry.

        Early work concentrated on glassy polymers, e.g. PMMA, loaded
with nlo active molecules.[12] In many instances the level of loading at which
the guest crystallises as a separate phase is quite low. Such composites
have been studied but even when they are well ordered scattering from the
crystallites remains a problem.[39-41] Even if a high loading is possible
without phase separation the poled host-guest system relaxes back to the
random equilibrium form on removal of the field.[42] In order to avoid this
the active groups are usually introduced as pendent groups attached to the
backbone.[42,43] This allows higher loadings to be achieved than for host-guest

systems and, by the use of post poling annealing in the presence of a field, good long term stability of the poling at reasonable operating temperatures has been achieved.

The use of pendent groups capable of producing liquid crystalline phases should in principle lead to the higher degree of orientation on poling.[44] This is because the intermolecular interactions favour a parallel alignment of the nlo active species. Thus the material is described by an Ising rather than a random orientation model giving a theoretical increase in order of a factor five. However, the scattering texture found in liquid crystals is a problem and in practise the gain from the use of a liquid crystal phase is less than the theoretical limit.

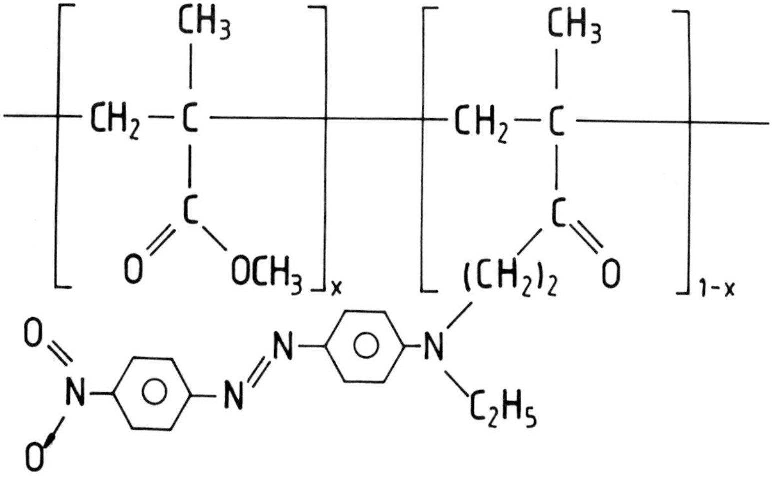

Figure 6 Chemical structure of disperse red 1 substituted polymethyl-methacrylate

A typical side chain polymer based on a methacrylate backbone is shown in Figure 6. One advantage of such polymers is that molecules with large βs, but which do not have a non-centrosymmetric crystal structure can be used. Thus the range of molecular species that can be used is much wider than is the case for crystals. Poling gives electro-optic coefficients comparable with or larger than that of lithium niobate. Most efficient poling generally occurs just below T_g where there is reasonable molecular flexibility and a high dielectric strength.[42] The decrease in dielectric strength above T_g drastically reduces poling efficiency at higher temperatures. Annealing reduces the free volume in the material and gives extended life expectancy for the poled polymer.[45,46] Even greater stability can be achieved by polymerization or crosslinking during poling[47] as discussed later in this volume by Allen et al.

Poled polymers have been used to fabricate modulation and switching devices.[48] Rapid developments are occurring as the result of collaborations between materials and device companies, e.g. as discussed by Lytel in this volume.

3. THIRD-ORDER NON-LINEAR MATERIALS

Introduction

While theory has been efficacious in aiding the molecular design of second-order materials comparable progress has not been achieved for third-order materials. One reason for this is that far more processes can contribute to the third-order non-linearity.[49] In particular multiply excited states are important. Thus though a simple two level approximation can give a reasonable estimate for β this is not the case for γ.[50,51] Since it is necessary to make some approximations to calculate the more complex terms there is a tendency for the main contributions to the third-order non-linearity to reflect the initial approximations. The situation is further complicated by experimental problems, e.g. third harmonic generation measurements are liable to be contaminated by contributions from the media adjacent to the sample.[52,53] There are also more resonance effects with one, two, three and higher-order photon resonances contributing. It is, therefore, difficult to accurately measure a truly non-resonant non-linearity for comparison with theory. Table 1 is a compilation of recent results. On the basis of the experimental details given in the literature only the first three values may be non-resonant. However, neither the contribution of higher order resonances nor the involvement of higher lying energy levels can be ruled out even in these instances.

Polymers

The general rule adopted in the search for good third-order non-linearity is to find an extended, polarizable molecular chain. Thus, one is drawn to conjugated polymers such as polyacetylene, polydiacetylene, etc. (Figure 7). The largest non-linearity has been observed for polyacetylene obtained by stretch orienting the Durham-route precursor polymer.[57] Such samples suffer from some inhomogeneity and the need for encapsulation to prevent oxidative degradation.

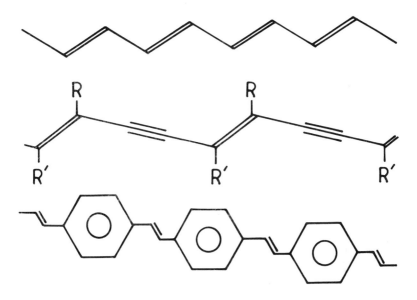

<u>Figure 7</u> Conjugated polymers with large $\chi^{(3)}$ values: top, polyacetylene; centre, polydiacetylene; bottom, polyparaphenylenevinylene.

In contrast polydiacetylenes (PDAs) are air stable and can be obtained either as large single crystals or as soluble polymers for film formation. Though the study of the non-linear optical properties was first undertaken over 15 years ago[68] it is only recently that detailed time resolved measurements have been conducted.[69] The toluene sulphonate derivative, $[CR-C\equiv C-CR]_n$ where R = - $CH_2OSO_2C_6H_4CH_3$, (PDS-TS), has been most studied because of the quality of the polymer crystals that can be obtained. $\chi^{(3)}$ varies from 4×10^{-11} esu off-resonance to 3×10^{-6} esu at the peak of the exciton absorption. Simple two level and complex exciton phase space filling models have been used to describe the results.[69,70] However, there are several sets of contradictory data, concerning the magnitude, lifetime and origin of the non-linearity and there is still a need for a consistent set of data in both the energy and time domains.[69-74] Larger nonlinearities have been reported for PDAs with conjugated sequences in the side chains directly linked to the polymer backbone.[75]

Table 1. Third-Order Nonlinearities of Polymers

Polymer[a]	Wavelength (microns)	Method[b]	$\chi^{(3)}$. $(10^{-12}$esu$)$	Condition[c]	Ref
DPATS	1.9	OptKerr	900	NR	54
PSilox	0.5	DFWM	3	NR	55
PThioMMA	?	DFWM	7	NR?	56
PAcety	1.9	THG(//)	30000	R	57
PThio	0.6	DFWM	1800	R	58
PQuinol	0.5	DFWM	1000	R	59
PMePPV	1.06	THG(//)	300	R	60
PDDThio	0.6	DFWM	300	R	61
PAzo	1.5	THG	48	R	62
P/Cyan	1.9	THG	20	R	63
PPhacet	1.06	THG	7	R	64
PPyrr	0.6	DFWM	3	R	65
PBTABQ	0.5	DFWM(s)	300000(Est)	R	66
PPtyne	0.5	DFWM(s)	0.1	R?	67

a: PDATS, polydiacetylene - toluene sulphonate; PSilox, polysiloxane; PolyThio MMA, copolymer of methyl thiophene and methyl methacrylate; PAcety, polyacetylene; PQuinol, polyquinolene; PMePPV, polydimethoxyparaphenylenevinylene; PDDThio, polydodecylthiophene; PAzo, polymethylmethacrylate with pendent azo-groups; P/Cyan, cyanine dye loaded polymers; PPhacet, polyphenylacetylene; PPyrr, polypyrrole; PBTABQ, polybisthiophenaliyl- acetoxybenzylidene copolymer; PPtyne, platinum poly-yne.

b: Opt Kerr, optical Kerr effect; DFWM, degenerate four ware mixing,(s) indicates solution measurements; THG, third harmonic generation, (//) indicates parallel to polymer chain.

c: NR, non-resonant; R, resonant

Both single crystal and cast films of soluble PDAs have been used to make optical waveguides with acceptably low loss.[76-79] Monomode guides with a variety of structures have been constructed and intensity switching between adjacent channels demonstrated.[80] The non-linearities are, however, too low to make practical all optical devices based on the optical Kerr effect, i.e. the intensity dependent refractive index.[81] Thus though high speed is possible, the response times vary from femtoseconds off-resonance to picoseconds on-resonance, power levels are unacceptably high and heating is a problem. In all measurements of electronic $\chi^{(3)}$ care must be taken to exclude heating which can produce much larger changes in refractive index than that resulting from the optical Kerr effect.

In view of the above comments it is surprising that few studies have been made of the thermo-optical effect in conjugated polymers. The intrinsic anisotropy of a chain oriented polymer such as a polydiacetylene leads to an anisotropic heating. Such an elliptical thermally induced refractive index change has an unexpected consequence.[82] This is that thermal defocussing leads to a switch from a single beam to two off-axis propagating beams in the far field. This effect has been demonstrated in PDA-TS.[83] The experimentally deduced thermo-optical coefficient is larger than that of compound semiconductors utilised in thermally driven optically bistable Fabry Perot etalons. An etalon containing the PDS-TS has been fabricated and a bistable operating characteristic with switching at a few milli-Watts incident power observed.[84]

Other polymers that have attracted interest are conjugated ladder polymers[59] and polysiloxanes.[55] The former should offer a higher π-electron density. Interest has grown in polysiloxanes as the non-linearity derives from extended σ-electron states. Because of the absence of π-electron states the polysiloxanes are stable materials with no absorption bands in the visible so that their third-order non-linearity can be observed at short wavelengths.

Crystals

Crystals of organic charge transfer (CT) salts are well known as low dimensional semiconductors and metals.[85] Thus they fit the prescription given above for useful $\chi^{(3)}$. A number of CT compounds have now been studied and shown to have large $\chi^{(3)}$ values.[86,87] However, these materials are opaque in the visible region and, in the case of the metallic salts in the infrared as well. Recent data for some selected CT compounds are given in Table 2.

Garito[50] has proposed that asymmetrically substituted polyenes should show a large $\chi^{(3)}$ since contributing transitions that are one-photon absorption forbidden in symmetric polyenes become allowed transitions for the asymmetric molecules. Such molecules should also have large β and noncentrosymmetric crystals should have large $\chi^{(2)}$ and $\chi^{(3)}$. However, very few examples of this behaviour are known.[28,29] The best example is the SPCD salt discussed above. Data for SHG and THG in crystals sufficiently thin to give quasi-phase matching are listed in Table 3. The values are given for light polarized parallel (∥) and perpendicular (⊥) to the direction of maximum absorption.

Table 2. Third-order Nonlinearities of Compounds

Material[a]	Wavelength (microns)	Method[b]	$\chi^{(3)}$ (10^{-12}esu)	Condition[b]	Ref
CTsalt	0.66	DFWM	16000	R	88
Phthal	0.6	DFWM	1800	R?	89
Deans	1.9	THG	13	NR?	90
Retinal	0.5	DFWM(s)	4	R	91

a.   CTsalt, (BEDT-FFT)$_4$ Re$_6$Se$_5$Cl$_9$; Phthal, Si phthalocyanine LB-film; Deans, diethylamino-nitrostilbene.

b: Opt Kerr, optical Kerr effect; DFWM, degenerate four ware mixing, (s) indicates solution measurements; THG, third harmonic generation.

To date the study of third-order non-linearity of crystals of low molecular weight compounds has attracted less attention than comparable work on polymers.

Table 3. Second and Third Harmonic Generation in SPCD Thin Films

	SHG[a]	SHG[a]	THG[b]	THG[b]
Thickness (microns)	//	⊥	//	⊥
4	102	0.1	2040	1
10	192	9.2	6330	38

a. SHG relative to d$_{11}$ of quartz (1.2×10^{-9}esu)
b. THG relative to $\chi^{(3)}$ of fused quartz (3×10^{-14}esu)

4. CONCLUSIONS

The study of organic materials for non-linear optics is a steadily growing area. The progress achieved by groups throughout the world between the first OMNO Conference in 1988 and the second, reported herein, is evident. Despite this much remains to be done in both the fundamental and applications areas. The search for new organic and metal organic compounds with large second-order non-linearity will continue. The shift in emphasis discussed in section 2 above from molecules with very strong electron donating and accepting groups to those with more uniformly distributed electron density will stimulate the synthesis of new materials.   Better use remains to be made of chirality to ensure noncentro-symmetric crystal structures while retaining large molecular nlo coefficients and optimal molecular orientation. Similarly there seems to be more scope for the utilisation of metal organic substituents.

At present unless new methods can be devised for the growth of thin films and fibres the use of single crystal materials will be restricted to free space propagation rather than integrated optics. With the possible exception of the efficient doubling of red semiconductor laser output into the blue this means that use as harmonic generators, optical parametric devices, etc., with high response speed will be restricted to the research laboratory. This problem may be circumvented by the development of microcrystalline composites with glassy polymers. Provided the crystallite size can be kept below about 50 nm scattering is not a problem, cf. the II-VI semiconductor microparticle in glass optical edge cutoff filters which have excellent transparency.[92] Methods for the reproducible preparation of such composites need to be developed using either polymers or sol-gel glasses as hosts.

In contrast poled polymers for electro-optical switches and modulators are being bench tested in devices. Thus there can be some optimism that the predictions of significant markets for such devices in the near future will prove to be correct. However, in the drive to produce competitive materials some of the more fundamental aspects of the materials properties and the device physics have not received adequate attention. Though the problem of depoling has largely been solved there is a remaining question of photostability. Some harmonic intensity will be produced even for low intensity fundamental power. The effect of long term exposure to short wavelength radiation needs to be investigated. However, experience in the photostabilisation of other polymers suggests that it can be prevented, e.g. by the introduction of appropriate additives.[93]

Significant improvements in third-order non-linear effects requires the synergistic interaction of theory and experiment that has been so successful for second-order effects. This demands careful experimental studies over wide wavelength and time domains using complimentary methods on very well characterized model systems. Such data is needed as a basis for the development of more realistic theoretical models which can in turn spur further materials development. At present there are no obvious candidates to produce significantly larger $\chi^{(3)}$ values than those found for oriented polyacetylene and polydiacetylene. Hence there is scope for further studies of CT complexes and other organic compounds which possess either semiconducting or metallic properties.

Thus while both scientific and technical problems remain to be solved the progress achieved since OMNO 88 suggests that the subject of organic materials for non-linear optics will remain a dynamic and expanding area of research and development for some years to come.

Acknowledgements

This brief review reflects involvement with a number of research groups under the aegis of Science and Engineering Research Council and the Department of Trade and Industry funding. Particular thanks for experimental collaboration and discussions are due to B. S. Wherrett, A. K. Kar, P. V. Kolinsky, R. J. Jones, S. R. Hall, C. J. Jones, J. N. Sherwood, R. T. Bailey, F. R. Cruickshank and P. A. Norman. In addition I wish to thank all those that I have had personal contact with and apologise to those whose work could not be included in this brief summary.

REFERENCES

1.   F. Pockels, 'Lehrbuch der Kristallphysik', Tuebner, Leipzig, 1906.
2.   J. Kerr, Phil. Mag., 1875, 1, 337.
3.   Lord Rayliegh, Phil. Mag, 1871, XLI, 274 and 447.
4.   L. Brillouin, Ann. Phys, 1922, 17, 88.
5.   C. V. Raman, Ind. J. Phys, 1928, 2, 387.
6.   'Nonlinear Optical Properties of Polymers', A. J. Heeger, J. Orenstein and D. Ulrich, Eds., Materials Res. Soc., Pittsburgh, 1988.
7.   'Molecular and Polymeric Optoelectronic Materials', A. Khanarian, Ed, SPIE Proceedings, Vol. 682, 1988.
8.   'Nonlinear Optics of Organics and Semiconductors', T. Kobayashi, Ed., Proc. in Phys., Vol. 36, Springer Verlag, Heidelberg, 1989.
9.   'Nonlinear Optical Properties of Organic Molecules and Crystals', D. S. Chemla and J. Zyss, Eds., Academic Press, New York, 1987.
10.   J. Zyss and J. L. Oudar, Phys. Rev. A., 1982, 26 2028.
11.   D. J. Williams, Angew. Chem. Intl. Ed. Eng., 1984, 23, 690.
12.   K. D. Singer, J. E. Sohn and S. J. Lalama, Appl. Phys. Lett., 1986 49, 248.
13.   I. R. Girling, N. A. Cade, P. V. Kolinsky, J. P. Earls, G. H. Cross and I. R. Petersen, Thin Sol. Films, 1985, 132, 101.
14.   S. J. Lalama and A. F. Garito, Phys. Rev. A, 1979, 20, 1179.
15.   C. C. Teng and A. F. Garito, Phys. Rev. B, 1983, 28, 6766.
16.   V. J. Docherty, D. Pugh and J. O. Morley, J. Chem. Soc. Faraday Trans.2, 1985, 81, 1179.
17.   D. Li, M. A. Ratner and T. J. Marks, J. Am. Chem. Soc., 1988, 110, 1707.
18.   C. Daniel and M. Dupuis, Chem. Phys. Lett, 1990 171, 209.
19.   J. Zyss in 'Conjugated Polymeric Materials : Opportunities in Electronics, Optoelectronics and Molecular Electronics', J. L. Bredas and R. R. Chance, Eds, Kluwer Acad. Publ., Dordrecht, 1990, p. 545.
20.   S. Kurtz and T. Perry, J. Appl. Phys., 1968, 39, 3798.
21.   Y. Kitaoka, T. Sasaki, S. Nakai, A. Yokotani, Y. Goto and M. Nakayama, Appl. Phys. Lett., 1990, 56, 2074.
22.   P. A. Norman, D. Bloor, J. S. Obhi, S. A. Karaulov, M. B. Hursthouse, P. V. Kolinsky, R. J. Jones and S. R. Hall, J. Opt. Soc. Am. B, 1987, 4, 1013.
23.   P. V. Kolinsky, R. J. Chad, R. J. Jones, S. R. Hall, P. A. Norman, D. Bloor, J. S. Obhi, Electron. Lett., 1987, 23, 791.
24.   P. Kerkoc, M. Zgonik, K. Sutter, Ch. Bossard and P. Gunter, Appl. Phys. Lett., 1989, 54, 2062.
25.   S. Ducharme, W. P. Risk, W. E. Moerner, V. Y. Lee, R. J. Twieg and G. C. Borklund, Appl. Phys. Lett, 1990, 57, 537.

26.   P. A. Norman, D. Bloor, R. T. Bailey, F. R. Cruickshank, D. Pugh, E. A. Shepherd, J. N. Sherwood, G. S. Simpson, P. V. Kolinsky, R. J. Jones, D. F. Croxall and S. R. Hall, to be published.

27.   Y. Kubota and T. Yoshimura, Appl. Phys. Letts., 1988, 53, 2579.

28.   H. Nakanishi, H. Matsuda, S. Okada and M. Kato in 'MRS. Int. Mtg on Adv. Mats', Vol. 1, Materials Res. Soc., New York, 1989, p.97.

29.   P. A. Norman, D. Bloor, P. V. Kolinsky, R. J. Jones and S. R. Hall to be published.

30.   S. R. Marder, J. W. Perry and W. P. Schaefer, in 'Molecular Electronics - Science and Technology', A. Aviram, Ed., United Eng. Trustees Inc., New York, 1989, p.55.

31.   W. Tam, B. Guerin, C. Calabresse and S. H. Stevenson, Chem. Phys. Letts., 1989, 154, 93.

32.   T. Yoshimura, Appl. Phys. Lett., 1989, 55, 534 and Phys. Rev. B, 1989, 40, 6292.

33.   M. L. H. Green, S. R. Marder, M. E. Thomspon, J. A. Bandy, D. Bloor, P. V. Kolinsky and R. J. Jones, Nature, 1987, 330, 360.

34.   B. J. Coe, C. J. Jones, J. A. McCleverty, D. Bloor, P. V. Kolinsky and R. J. Jones, J. Chem. Soc.Chem.Commun., 1989, 1485.

35.   B. J. Coe, S. Kurek, N. M. Rowley, J-D. Foulon, T. A. Hamor, M. E. Harmon, M. B. Hursthouse, C. J. Jones, J. A. McCleverty and D. Bloor, Chemtronics in the press.

36.   P. Stroeve, D. D. Saperstein and J. F. Rabolt, J. Chem. Phys., 1990, 92, 6958.

37.   G. J. Ashwell, E. J. C. Dawnay, A. P. Kuczynski, M. Szablewski, I. M. Sandy, M. R. Bryce, A. M. Grainer and M. Hasan, J. Chem. Soc. Faraday Trans., 1990, 86, 1117.

38.   Ch. Bossard, M. Kupfer, P. Gunter, C. Pasquier, S. Zahir and M. Seifert, Appl. Phys. Lett., 1990, 56, 1204.

39.   Y. Wang, Chem. Phys. Letts., 1986, 126, 209.

40.   H. Daigo, N. Okamoto and H. Fujimura, Opt. Commun., 1988, 69, 177.

41.   N. Azoz, P. D. Calvert, M. Kadim, A. J. McCaffrey and K. R. Seddon, Nature, 1990, 344, 49.

42.   M. Dubois in 'Conjugated Polymeric Materials; Opportunities in Electronics, Optoelectronics and Molecular Electronics', J. L. Bredas and R. R. Chance, Eds., Kluwer Acad. Publ., Dordrecht, 1990, p. 321.

43.   G. R. Möhlmann, Synth. Met. 1990, 37, 207.

44.   G. R. Möhlmann and C. P. J. M. van der Vorst, in 'Side Chain Liquid Crystalline Polymers', C B. McArdle, Ed., Blackie, London, 1989, p.330.

45.   K. D. Singer, M. G. Kuzyk, W. R. Holland, J. E. Sohn, S. J. Lalama, R. B. Comizzoli, H. E. Katz and M. L. Schilling, Appl. Phys. Lett., 1988, 53, 1800.

47.   D. Jungbauer, B. Reck, R. Twieg, D. Y. Yoon, C. G. Willson and J. D. Swalen, Appl. Phys. Lett., 1990, 56, 2610.

48.   G. H. Cross, A. Donaldson, R. W. Gymer, S. Mann, N. J. Parsons, D. R. Haas, H. T. Man and H. N. Yoon, SPIE Proc., 1989, 1177, 79.

49.   Y. Prior, IEEE J. Quant. Electr., 1984, QE-20, 37.

50. J. W. Wu, J. R. Heflin, R. A. Norwood, K. Y. Wong, O. Zamani-Khamiri, A.
     F. Garito, P. Kalyanaraman and J. Sounik, J. Opt. Soc. Am. B, 1989, 6, 707.
51. M. G. Kuyzk and C. W. Dirk, Phys. Rev. A, 1990, 41, 5098.
52. G. R. Meredith, B. Buchalter and C. Hanzlik, J. Chem. Phys., 1983, 78, 1533.
53. F. Kajzar and J. Messier, Phys. Rev. A, 1985, 32, 2352.
54. J. P. Hermann and P. W. Smith, J. Opt. Soc. Amer., 1980, 70, 656.
55. D. J. McGraw, A. E. Siegman, G. H. Wallraff and R. D. Miller, Appl. Phys.
     Lett., 1989, 54, 1713.
56. H. S. Nalwa, J. Phys. D : Appl. Phys., 1990, 23, 745.
57. M. R. Drury, Sol. St. Commun., 1988, 68, 417.
58. L. Yang, R. Dorsinville, Q. Z. Wang, W.K. Zhou, P. P. Ho, N. L. Yang, R.
     Alfano, R. Zamboni, R. Danieli, G. Ruani and C. Taliani, J. Opt. Soc. Am.
     B, 1989, 6, 753.
59. X. F. Cao, J. P. Jiang, D. P. Bloch, R. W. Hellworth, L. P. Yu and L. Dalton,
     J. Appl. Phys., 1989, 65, 5012.
60. J. Swiatkiewicz, P. N. Prasad, F. E. Krasz, M. A. Druy and P. Glatkowski,
     Appl. Phys. Lett., 1990, 56, 892.
61. B. P. Singh, M. Samoc, H. S. Nalwa and P. N. Prasad, J. Chem. Phys., 1990,
     92, 2756.
62. M. Amano, T. Kaino and S. Matsumoto, Chem. Phys. Lett., 1990, 170, 515.
63. S. Matsumoto, K. Kubodera, T. Kurihara and T. Kaino, Opt. Commun.,
     1990, 76, 147.
64. D. Neher, A. Wolf, C. Bubeck and G. Wegner, Chem. Phys. Lett., 1989, 163,
     116.
65. S. K. Ghoshal, Chem. Phys. Lett., 1989, 158, 65.
66. S. A. Jenekhe, W-C. Chen, S. Lo and S. R. Flom, Appl. Phys. Lett., 1990, 57,
     126.
67. S. Guha, C. C. Frazier, P. L. Porter, K. Kang and S. E. Finberg, Opt. Lett.,
     1989, 14, 952.
68. C. Sauteret, J-P. Hermann, R. Frey, F. Pradere, J. Ducuing, R. H. Baughman
     and R. R. Chance, Phys. Rev. Lett., 1976, 36, 956.
69. B. I. Greene, J. Orenstein, R. R. Millard and L. R. Williams, Phys. Rev. Lett.,
     1987, 58, 2750.
70. T. G. Harvey, W. Ji, A. K. Kar, B. S. Wherrett, D. Bloor and P. A. Norman,
     CLEO Proc., in the press
71. B. I. Greene, J. F. Mueller, J. Orenstein, D. H. Rapkine, S. Schmitt-Rink and
     M. Thaker, Phys. Rev. Lett., 1988, 61, 325.
72. M. Lequime and J. P. Hermann, Chem. Phys., 1977, 26, 431.
73. J. M. Huxley, P. Mataloni, R. W. Schoenlein, J. G. Fujimoto, E. P. Ippen and
     G. M. Carter, Appl. Phys. Lett., 1990, 56, 1600.
74. T. G. Harvey, W. Ji, J. Bolger, A. K. Kar, B. S. Wherrett and D. Bloor, SPIE
     Conference on Quantum Electronics and Laser Science, 1991, to be
     published.
75. H. Nakanishi, H. Matsuda, S. Okada and M. Kato, Polym. for Adv.
     Technol., 1990, 1, 75.

76. S. Mann, A. R. Oldroyd, D. Bloor, D. J. Ando and P. J. Wells, SPIE Proceedings, 1988, 971, 245.
77. N. E. Schlotter, J. L. Jackel, P. D. Townsend and G. L. Baker, Appl. Phys. Lett., 1990, 56, 13.
78. J. S. Patel, S-D Lee, G. L. Baker and J. A. Shelburne III, Appl. Phys. Lett., 1990, 56, 131.
79. D. M. Krol and M. Thakur, Appl. Phys. Lett., 1990, 56, 1406.
80. P. D. Townsend, J. L. Jackel, G. L. Baker, J. A. Shelburne III and S. Etemad, Appl. Phys. Lett., 1989, 55, 1829.
81. M. Thakur and D. M. Krol, Appl. Phys. Lett., 1990, 56, 1213.
82. T. G. Harvey, W. Ji, A. K. Kar and B. S. Wherrett, Optics Lett., 1990, 15, 408.
83. T. G. Harvey, W. Ji, A. K. Kar, B. S. Wherrett, D. Bloor and P. A. Norman, J. Opt. Soc. Amer., in the press.
84. T. G. Harvey, W. Ji, J. Bolger, A. K. Kar, B. S. Wherrett, D. Bloor and P. A. Norman, to be published.
85. E. Conwell, 'Semiconductors and Semimetals', Vol. 27, Academic Press, New York, 1988.
86. P. G. Huggard, W. Blau and D. Schweitzer, Appl. Phys. Lett., 1987, 51, 2183.
87. T. Gotoh, T. Kondoh, K. Egawa and K. Kubodera, J. Opt. Soc. Am. B., 1989, 6, 703.
88. B. Sipp, G. Klein, J. P. Lavoine, A. Penicaud and P. Batail, Chem. Phys., 1990, 144, 299.
89. M. K. Casstevens, M. Samoc, J. Pfieger and P. N. Prasad, J. Chem. Phys., 1990, 92, 2019.
90. T. Kurihara, H. Kobayashi, K. Kubodera and T. Kaino, Chem. Phys. Lett., 1990, 165, 171.
91. T. Sakai, Y. Kawabe, H. Ikeda and H. Kawasaki, Appl. Phys. Lett., 1990, 56, 411.
92. N. Finlayson, W. C. Banyai, C. T. Seaton, G. I. Stegeman, M. O'Neill, T. J. Cullen and C. N. Ironside, J. Opt. Soc. Am. B, 1989, 6, 675.
93. W. Schnabel, 'Polymer Degradation', Hanser Int., Vienna, 1981.

Theory

Theoretical Study of the Conjugation Ability in a Series of Hydrogen Bond Forming Molecules

J. Delhalle, M. Dory, J.G. Fripiat, and J.M. André

LABORATOIRE DE CHIMIE THÉORIQUE APPLIQUÉE, FACULTÉS UNIVERSITAIRES NOTRE-DAME DE LA PAIX, 61, RUE DE BRUXELLES B-5000 NAMUR, BELGIUM

1 INTRODUCTION

High electric responses (linear and nonlinear) observed in conjugated organic systems not only depend on the chemical nature of their delocalized networks of π-electrons but also on their configurational and conformational structure as well as packing modes in the bulk. Controlling the molecular architecture and increasing the density of active species is thus an important part of the molecular design of new materials for optoelectronics. The most important forms of organic nonlinear materials so far produced are : single crystals, solid solutions of optically nonlinear molecules in polymer matrices, Langmuir-Blodgett (LB) films, side chain polymers with active molecules and main chain polymers in which the active groups are incorporated into the polymer backbone. Presently, side chain polymers are very attractive organic materials for $\chi^{(2)}$ effects because they offer the best compromise between nonlinear optical properties and processing characteristics. In particular they are quite amenable to electric field poling to remove centrosymmetry by aligning the permanent dipoles of the side chains with the externally applied electric field.

Main chain polymers, which offer the possibility of larger concentration of active groups, preferably conjugated over large distances, into the polymer backbone are also quite interesting for $\chi^{(3)}$ effects. Scaling rules[1] predict that α, the linear molecular polarizability, should scale with the conjugation length l_π to the third power, l_π^3, whereas γ, the third-order molecular polarizability, should increase with the fifth power of that quantity, l_π^5. The molecular architecture of long chains is not easily controllable during the synthesis; defects, kinks and twists often disrupt the conjugation and spoil the large electric responses expected from these scaling laws.

Since the molecular arrangements are ultimately dictated by the specific interactions existing between molecules, it should be possible to select suitable chemical moieties, grafted to and/or incorporated in the conjugated backbone, to enforce intra- and interchain order and to some extent prevent structural defects from occuring[2,3]. Hydrogen bonds are known to yield aggregate structures and have been proposed to be used in chemistry as organizing forces[4]. In this contribution we focus on oligomers of polyenes in which hydrogen bond forming chemical moieties

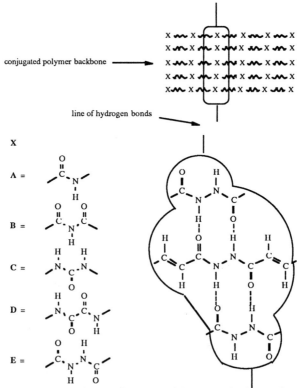

Figure 1 Schematic representations of possible links due to hydrogen bonds between polymeric chains. X represents the chemical moieties considered in this work . The circled areas delineate the size of the molecular structures used to model locally the polymer situation.

are included to improve inter- and intrachain order (Figure 1). By *ab initio*
calculations of the static electric dipole polarizability, we want to assess the relative
efficiency of hydrogen bond forming molecules as transmitters of the electronic
conjugation existing in a polyene chain.

2 COMPUTATIONAL ASPECTS AND MODEL MOLECULES

An important question about electric response calculations is the choice of the
basis set of atomic functions representing the molecular orbitals which should be
able to predict the right geometry changes and to suitably describe the reorganization
of the electron distribution in the presence of an external electric field. It must be
recalled that polarizability, and electric responses in general, are quite sensitive to
geometry and therefore it is preferable for exploratory studies that all geometries be
optimized with the same basis set. Because of the size of the systems treated in this
work and the need for comparing results obtained at the same level of description,
i.e. for the smaller and the larger systems, the minimal STO-3G basis has been
used[5] throughout. In spite of its obvious limitations, that basis predicts molecular
structures reasonably well in the sense that errors on bond distances and angles
remain fairly constant for a wide variety of molecular structures[6]. Previous works
have shown that except for the polarizability component(s) in the direction
perpendicular to the conjugation plane, the use of the STO-3G basis set yields good
qualitative estimates of the polarizability of conjugated hydrocarbons[7] as well as
systems containing classical heteroatomic linkages[2,8]. For this reason, the STO-3G
basis provides confidence in the trends to be reported in the present study and
appears as a suitable compromise between computational feasability of the study and
the required conditions for reliable results.

Full STO-3G geometry optimization has been carried out at the restricted
Hartree-Fock LCAO-SCF-MO level on the molecular systems (kept planar) using
the GAUSSIAN 86 series of programs[9]. Using these optimized geometries and the
same basis set as input data, the average static electric dipole polarizability, $\alpha = (\alpha_{xx} + \alpha_{yy} + \alpha_{zz})/3$, has been computed by the Coupled-Perturbed Hartree-Fock
method using the HONDO 7.0 program[10]. To remove the explicit dependence of the
results on the dimension of the systems, the average polarizability values are divided

by the volume V enclosed in the van der Waals spheres centered on each atoms of the molecules (the volumes due to sphere overlap are subtracted).

Figure 2 shows the various molecular arrangements considered for each of the combinations of hydrogen bonding molecules included in the reference conjugated hydrocarbon frameworks : C=C-C=C-X and C=C-X-C=C. Arrangement **(a)** corresponds to an isolated hydrogen bond forming molecule X (X = A to E), **(b)** and **(c)** are isolated conjugated oligomers representative of a possible repeat unit of an actual polymer, where they differ by the position of the hydrogen bond forming molecule in the backbone. Most often, incorporating moieties structurally and/or chemically different from the repeating structure of the conjugated chain leads to a decrease in the effective delocalization[3] which can be estimated from the polarizability changes. Thus, the purpose of considering arrangements **(b)** and **(c)** is to assess, by comparison, the extent of the interruption of the electronic conjugation due to the incorporation of a molecular fragment X at the end and in the middle of the butadienic backbone. Arrangements **(d)** and **(e)** should provide information on the influence of lateral interchain hydrogen bonds on the polarizability.

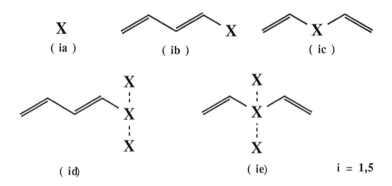

Figure 2 Schematic representation of **(i)** the five molecular arrangements **(a to e)** considered for each of the **(ii)** five systems of hydrogen bonding molecules included in the reference conjugated hydrocarbon framework : C=C-X-C=C, C=C-C=C-X.

3 RESULTS

The total energy E_T, dipole moment μ, molecular volume V, and the average polarizability divided by the molecular volume αV^{-1} are listed in Table 1.

Table 1 Total energy E_T (in au), dipole moment μ (in Debye), molecular volume V (in $\mathring{A}^3$) and the average polarizability divided by the molecular volume αV^{-1} (in au.$\mathring{A}^{-3}$) for systems (1) to (5).

	E_T	μ	V	αV^{-1}
System (1):				
(1a)	-166.68821	2.64	41.88	0.25
(1b)	-318.58177	2.44	96.60	0.40
(1c)	-318.58183	2.34	94.69	0.38
(1d)	-651.97386	9.95	179.63	0.35
(1e)	-651.97374	9.53	178.29	0.34
System (2):				
(2a)	-277.92701	4.19	60.99	0.28
(2b)	-429.82101	4.40	115.11	0.41
(2c)	-429.81923	4.26	114.53	0.37
(2d)	-985.69580	15.34	235.35	0.36
(2e)	-985.69466	15.25	236.28	0.34
System (3):				
(3a)	-221.01671	3.16	51.77	0.27
(3b)	-372.91111	3.18	107.25	0.42
(3c)	-372.91099	2.62	105.21	0.40
(3d)	-814.97200	12.94	212.11	0.37
(3e)	-814.97174	12.63	211.51	0.35
System (4):				
(4a)	-332.25215	0.0	72.59	0.29
(4b)	-484.14559	1.27	126.88	0.42
(4c)	-484.14554	0.0	125.36	0.40
(4d)	-1148.67355	1.31	266.27	0.39
(4e)	-1148.67254	0.0	265.00	0.38
System (5):				
(5a)	-332.20027	0.0	74.15	0.30
(5b)	-484.09471	1.25	126.75	0.42
(5c)	-484.09360	0.0	127.89	0.38
(5d)	-1148.53211	1.07	270.10	0.41
(5e)	-1148.53086	0.0	262.23	0.40

The notation corresponds to that introduced in Figure 2. Stability difference between the isolated isomers, $E_T(\mathbf{b})$ -$E_T(\mathbf{c})$, or between the isomers laterally connected to two hydrogen bonding molecules, $E_T(\mathbf{d})$ -$E_T(\mathbf{e})$, is negligible, about 4 kJ.mol^{-1} or less. Similarly the volume changes in comparable arrangements [(**b**), (**c**) and (**d**), (**e**)] is not significant except for system (**5**) where a 3% decrease from (**5d**) to (**5e**) is calculated.

The dipole moment, when symmetry allows it, is moderately influenced by the position of the hydrogen bond forming moiety in the backbone, but it is more dependent on the existence of interchain hydrogen bonds. For instance, in system (**1**) the dipole moment ranges from 2.64 to 2.33 D for arrangements (**a**) to (**c**), but it exceeds by more than 1.5 D the added contributions of three formamide molecules, $H\text{-}CO\text{-}NH_2$, similarly oriented. For instance, in (**e**) μ (= 9.52 D) is 1.6 D larger than the dipole moment of three formamide molecules. Similar trends are also observed in the case of systems (**2**), (**3**) and (**4**), the trend being less marked in the last two cases. The reverse situation is found for system (**5**). Note that STO-3G dipole moments are almost always underestimated, but the trends are generally correct except in situations of very small dipoles or when the electronic structure cannot be correctly described within the single determinant approximation.

The polarizability per unit volume, αV^{-1}, decreases only very slightly when the systems are laterally connected through hydrogen bonds, for instance αV^{-1} is equal to 0.40 in (**1a**) and 0.35 in (**1d**), similarly it is equal to 0.38 in (**1c**) and 0.34 in (**1e**). There is also, as expected[3], a systematic decrease in αV^{-1} when the hydrogen bond forming moiety is in the centre of the reference conjugated hydrocarbon framework : C=C-X-C=C. For the isolated chains, the largest decrease is observed for systems (**2**) and (**5**). When lateral hydrogen bonds are present as in arrangements (**d**) and (**e**), αV^{-1} decreases in all systems except for system (**5**) where $\alpha V^{-1}$is somewhat larger, 0.40, in (**5e**) than in the isolated situation (**5c**), 0.38.

The differences between the isolated molecules, [(**b**) + 2(**a**)] or [(**c**) + 2(**a**)], and those in interaction, (**d**) or (**e**), are more easily understood from the results listed in Table 2. Two properties will be of interest to us here : the stabilization energy and the variation of polarizability. The stabilization energy is an important aspect to take into account, because a larger stabilization energy is generally more propitious for the desired aggregate structure to occur.Discussion of stabilization energy is always a difficult subject due to basis superposition errors. On the basis of observations by various authors[11,12] and our own results on formic acid and formamide[3] showing that the STO-3G stabilization energies are often comparable to results obtained with larger basis sets, we use in a straightforward manner the differences indicated in

Table 2 Difference in total energy ΔE_T (in kJ.mol^{-1}), dipole moment $\Delta\mu$ (in Debye), molecular volume ΔV (in Å^3) and average polarizability $\Delta\alpha$ (in au) between the fully interacting systems, (xd) or (xe) and the sum of contributions from the isolated parts, [(xb) + 2(xa)] or [(xc) + 2(xa)], respectively (x = **1** to **5**).

	ΔE_T	$\Delta\mu$	ΔV	$\Delta\alpha$
System (**1**):				
(**1d**) - [(**1b**) + 2(**1a**)]	-41.01	2.22	-0.73	3.63
(**1e**) - [(**1c**) + 2(**1a**)]	-40.63	1.90	-0.16	3.74
System (**2**):				
(**2d**) - [(**2b**) + 2(**2a**)]	-54.46	2.55	-1.74	3.92
(**2e**) - [(**2c**) + 2(**2a**)]	-56.05	2.62	-0.23	3.61
System (**3**):				
(**3d**) - [(**3b**) + 2(**3a**)]	-72.66	3.41	+1.32	5.92
(**3e**) - [(**3c**) + 2(**3a**)]	-71.69	3.69	+2.76	5.80
System (**4**):				
(**4d**) - [(**4b**) + 2(**4a**)]	-66.80	0.04	-5.79	8.96
(**4e**) - [(**4c**) + 2(**4a**)]	-65.54	0.0	-5.54	8.35
System (**5**):				
(**5d**) - [(**5b**) + 2(**5a**)]	-96.73	-0.18	-4.95	11.48
(**5e**) - [(**5c**) + 2(**5a**)]	-96.35	0.0	-13.96	11.48

Table 2 to compare the stabilization energies of the five systems studied in this paper. In order of increasing stabilization energies one finds : (**1**) < (**2**) < (**4**) < (**3**) < (**5**), while for polarizability gains one has : (**1**) ≈ (**2**) < (**3**) < (**4**) < (**5**). It must be recalled at this point that in system (**5**) the disadvantage of incorporating a heteroatomic moiety in the C=C-C=C backbone is partially removed due to the presence of neighbouring hydrogen bonds.

4 CONCLUSION

When incorporated in the C=C-C=C backbone, the hydrogen bond forming group -CO-NH-NH-CO- presents the most favorable stabilization energy and polarizability gain of the systems considered in this communication. It should be of interest in the design of new molecules for optoelectronics when forcing organization is required. In spite of the fact that the hydrazide group is not yet part of the extensive list of hydrogen bond patterns recently published by Etter[4], several crystal determinations have been reported on hydrazide compounds. For instance, anhydrous diacetylhydrazine[13] is known to form molecular arrangements of the type shown in Figure 1. The hydrazide group could be advantageously used in works towards molecular crystals for nonlinear optics[14,15]. In the field of macromolecules, Rogers et al. have reported a quite interesting and related work on linear unsaturated polyamides and polyhydrazides[16] that should be valuable to develop in the framework of nonlinear optics.

Additional theoretical work on these compounds is in progress in our laboratory. A more extended version of this communication, where mixed hydrogen bond forming groups are considered, is in preparation[17]. In particular, molecular structure and bonding as well as first hyperpolarizability aspects will also be addressed.

ACKNOWLEDGEMENTS

The authors acknowledge the support of this work under the ESPRIT-EEC contract n°2284 on "Optoelectronics with active organic molecules". They also thank the National Fund for Scientific Research (Belgium), IBM Belgium and the Facultés Universitaires Notre-Dame de la Paix (FUNDP) for the use of the Namur Scientific Computing Facility.

REFERENCES

1. C. Flytzanis, in 'Nonlinear Optical Properties of Organic Molecules and Crystals', D.S. Chemla and J. Zyss (eds.), Academic, New York, 1987.
2. M. Dory, J. Delhalle, J.G. Fripiat and J.M. André, Int. J. Quantum Chem. Symp., 1987, 14, 85.

3.   J. Delhalle, M. Dory, J.G. Fripiat and J.M. André, in 'Nonlinear Optical Effects in Organic Polymers", J. Messier et al. (eds.), Kluwer Academic Publishers, Dordrecht, 1989.
4.   M.C. Etter, Acc. Chem. Res., 1989, 23, 120.
5.   W.J. Hehre, R.F. Stewart, J.A. Pople, J. Chem. Phys., 1969, 51, 2657.
6.   W.J. Hehre, L. Radom, P.v.R. Schleyer, J.A. Pople, 'Ab Initio Molecular Orbital Theory', Wiley, New York, 1986.
7.   A. Chablo, A. Hinchliffe, Chem. Phys. Lett., 1980, 72, 149.
8.   E. Younang, J. Delhalle, J.M. André, New J. Chem., 1987, 11, 403.
9.   M.J. Frisch, J.S. Binkley, H.B. Schlegel, K. Raghavachari, R.L. Martin, J.J.P. Stewart, F.W. Bobrowicz, D.J. DeFrees, R. Seeger, R.A. Whiteside, D.J. Fox, E.M. Fleuder, J.A. Pople, GAUSSIAN 86, release C, Carnegie-Mellon University, Pittsburgh PA 1984.
10. M. Dupuis, J.D. Watts, H.O. Villar, G.J.B. Hurst, Comp. Phys. Commun., 1989, 52, 415.
11. A.M. Sapse, L.M. Fugler, D. Cowburn, Intern. J. Quantum Chem., 1986, 29, 1241.
12. J. Sauer, P. Hobza, R. Zahradnik, J. Phys. Chem., 1980, 84, 3318.
13. R. Shintani, Acta Cryst., 1960, 13, 609.
14. T.W. Panunto, Z. Urbanczyk-Lipkowska, R. Jhonson, M.C. Etter, J. Am. Chem. Soc., 1987, 109, 7786.
15. E. Staab, L. Addadi, L. Leiserowitz, M. Lahar, Adv. Mat., 1990, 2, 40.
16. H.G. Rogers, R.A. Gaudania, J.S. Manello, R.A. Sahatjian, J. Macromol. Sci.-Chem., 1986, A23, 711.
17. M. Dory et al., to be published.

Theory of Non-linear Optical Response in Molecular Layers

R.W. Munn, S.E. Mothersdale, and M.M. Shabat

DEPARTMENT OF CHEMISTRY, UMIST, MANCHESTER M60 1QD, UK

1 INTRODUCTION

Nonlinear optics has provided a useful testing ground for the concept of molecular engineering. One seeks to design and synthesize molecules of high nonlinear optical response, and then to fabricate a molecular material in which this response is effectively expressed. In many such materials the active region is limited in thickness. Obvious examples are Langmuir-Blodgett films[1] and polymer films.[2] Less obvious are molecular crystalline thin films or crystalline cored fibres,[3] or active species diffused into the surface of an optical polymer.[4] Finally, centrosymmetric materials lacking any bulk quadratic nonlinearity may display surface nonlinearity[5] or pyroelectricity.[6]

   Molecular theories of nonlinear optical response in thin layers are not well developed. As in theories of bulk nonlinear response, a common approach has been to approximate the local electric field by the Lorentz expression. The Lorentz approximation is known to be poor for crystals of elongated molecules such as p-terphenyl[7] and the polydiacetylenes,[8] so that its reliability is questionable in Langmuir-Blodgett films, at least.   .

   A rigorous treatment of the local field in bulk nonlinear optics of molecular crystals is available, however,[9-11] and has been applied to several species.[12] Molecular theories of _linear_ optical response in thin layers were also developed some 20 years ago in connection with studies of surface excitations. Hence this paper describes a molecular theory of nonlinear optical response in ordered layers, combining the rigour

of the bulk nonlinear theory with the conceptual approach
of the thin layer linear theory.

2 LOCAL FIELD

Consider an assembly of layers labelled $g = 0,1,\ldots$.
The layers are assumed to be ordered, and for simplicity
to contain only one molecule per two-dimensional unit
cell. In a uniform applied electric field $\mathbf{E}^\circ$, a molecule
in layer g experiences a local electric field

$$\mathbf{f}_g = \mathbf{E}^\circ + \sum_{g'} \mathbf{T}_{gg'} \cdot \mathbf{p}_{g'} / \varepsilon_o v \tag{1}$$

$$= \mathbf{E}^\circ + \sum_{g'} \mathbf{T}_{gg'} \cdot \mathbf{a} \cdot \mathbf{f}_{g'}. \tag{2}$$

Here $\mathbf{T}_{gg'}$ is a dimensionless planewise dipole tensor sum,
which gives the field at a molecule in layer g due to the
induced dipole moments $\mathbf{p}_{g'}$ in layer g', where g' may
equal g; v is the volume per molecule. The induced
dipoles are related to the local field by the linear
polarizability α, leading to Eq.(2), where $\mathbf{a} = \alpha/\varepsilon_o v$. All
layers are treated as equivalent.

Now $\mathbf{T}_{gg'}$ falls off rapidly with the distance between
layers g and g'. This allows one to define a range r such
that $\mathbf{T}_{gg'}$ is negligible for $|g-g'|>r$, i.e. when r or more
layers intervene between g and g'. Then only layers g'<r
are coupled to the surface layer $g = 0$. Furthermore, such
layers couple only to further r layers deeper in the
assembly. Hence beyond layer 2r, effectively the bulk
environment is experienced, and all layers have the same
local field. If the assembly is thin enough, there is no
bulk region.

For a thin assembly of t layers, Eq. (2) is solved
directly by matrix inversion, with the result

$$\mathbf{f}_g = \sum_{g'} (\mathbf{I} - \mathbf{T} \cdot \mathbf{A})^{-1}_{gg'} \cdot \mathbf{E}^\circ \tag{3}$$

$$\equiv \sum_{g'} \mathbf{H}_{gg'} \cdot \mathbf{E}^\circ \equiv \mathbf{h}_g \cdot \mathbf{E}^\circ. \tag{4}$$

Here the matrix to be inverted is of dimension 3t x 3t,
with 3 x 3 submatrices $\mathbf{1}\delta_{gg'} - \mathbf{T}_{gg'} \cdot \mathbf{a}$. The quantity $\mathbf{h}_g$ is
the local-field tensor. This result resembles that for a

molecular crystal[15] but with the label g for a layer
replacing the label k for a molecule in the unit cell.

For a thick assembly, some layers experience the
bulk environment where the local field becomes $\mathbf{f}^b$,
independent of g. Then Eq. (2) becomes

$$\mathbf{f}^b = \mathbf{E}^\circ + \sum_{g'} \mathbf{T}_{gg'} \cdot \mathbf{a} \cdot \mathbf{f}^b = \mathbf{E}^\circ + \mathbf{T}^b \cdot \mathbf{a} \cdot \mathbf{f}^b, \qquad (5)$$

where $\mathbf{T}^b$ is the bulk dipole tensor sum

$$\mathbf{T}^b = \sum_{g'} \mathbf{T}_{gg'} \sim \sum_{g'=g-r}^{g+r} \mathbf{T}_{gg'} . \qquad (6)$$

It follows that

$$\mathbf{f}^b = (1 - \mathbf{T}^b \cdot \mathbf{a})^{-1} \cdot \mathbf{E}^\circ \equiv \mathbf{h}^b \cdot \mathbf{E}^\circ, \qquad (7)$$

with $\mathbf{h}^b$ the bulk local-field tensor.

The surface-induced deviation from the bulk local
field $\phi_g = \mathbf{f}^b - \mathbf{f}_g$ follows on subtracting Eq. (5) from Eq.
(2), to obtain

$$\phi_g = \sum_{g'} \mathbf{T}_{gg'} \cdot \mathbf{a} \cdot \phi_{g'} + \mathbf{T}_g{}^s \cdot \mathbf{a} \cdot \mathbf{f}^b. \qquad (8)$$

The quantity $\mathbf{T}_g{}^s$ is the surface-induced deviation from
the bulk dipole sum tensor, given by

$$\mathbf{T}_g{}^s = \mathbf{T}^b - \sum_{g'} \mathbf{T}_{gg'}. \qquad (9)$$

Comparison with Eq. (6) shows that since $g' \geq 0$, $\mathbf{T}_g{}^s$ is
nonzero for all $g < r$, i.e. for layers within range of the
surface. For example, the surface layer $g=0$ interacts
with r other layers on one side only, whereas a bulk
layer interacts with r layers on either side. Solution of
Eq. (8) gives ϕ_g, whence

$$\mathbf{f}_g = (1 - \sum_{g'} \mathbf{H}_{gg'} \cdot \mathbf{T}_{g'}{}^s \cdot \mathbf{a}) \cdot \mathbf{h}^b \cdot \mathbf{E}^\circ \equiv \mathbf{h}_g \cdot \mathbf{E}^\circ. \qquad (10)$$

Here $\mathbf{H} = (\mathbf{I} - \mathbf{T} \cdot \mathbf{A})^{-1}$ as before, but is now of dimension $3(2r+1) \times 3(2r+1)$, allowing for the full span of non-negligible interactions involving layers 0 to 2r.

3 MACROSCOPIC FIELD

The foregoing provides computable expressions for the linear local field in terms of the applied field $\mathbf{E}^\circ$ in the absence of the sample. However, experimental quantities such as susceptibilities relate to the macroscopic field in the presence of the sample. Although the applied field is uniform, the material response is not, and hence neither is the macroscopic field.

In layer g the macroscopic field $\mathbf{E}_g$ is given by

$$\mathbf{E}_g = \mathbf{E}^\circ - \mathbf{n}(\mathbf{n} \cdot \mathbf{p}_g)/\varepsilon_0 v, \tag{11}$$

where $\mathbf{n}$ is a unit vector normal to the layers. The normal component of the polarization in the layer $\mathbf{n} \cdot \mathbf{p}_g/v$ corresponds to a surface charge density σ contributing a field of magnitude σ/ε_0 normal to the layers. The combination $\mathbf{p}_g/\varepsilon_0 v$ can be related to $\mathbf{E}^\circ$ or $\mathbf{E}_g$ according to

$$\mathbf{p}_g / \varepsilon_0 v = \mathbf{a} \cdot \mathbf{f}_g = \mathbf{a} \cdot \mathbf{h}_g \cdot \mathbf{E}^\circ \tag{12}$$

$$= \mathbf{a} \cdot \mathbf{d}_g \cdot \mathbf{E}_g = \chi_g^{(1)} \cdot \mathbf{E}_g \tag{13}$$

Here $\mathbf{d}_g$ is the usual local-field tensor and $\chi_g^{(1)}$ is the linear susceptibility for layer g. By substituting Eqs (12) and (13) in Eq. (11) one obtains

$$\mathbf{1} + \mathbf{n}(\mathbf{n} \cdot \chi_g^{(1)}) = [\mathbf{1} - \mathbf{n}(\mathbf{n} \cdot \mathbf{a} \cdot \mathbf{h}_g)]^{-1}, \tag{14}$$

which determines $\chi_g^{(1)}$ and $\mathbf{d}_g$ since $\mathbf{h}_g$ is known from Eq. (10).

Algebraic evaluation of $\chi_g^{(1)}$ is not straightforward because of the unit vector $\mathbf{n}$. By writing

$$\mathbf{T}_{gg'} = \mathbf{L}_{gg'} - \mathbf{nn}\, \delta_{gg'}, \tag{15}$$

where $L_{gg'}$ is a planewise Lorentz-factor tensor of unit trace, one obtains

$$T^b = L^b - nn, \tag{16}$$

where L^b is the bulk Lorentz-factor tensor related to $L_{gg'}$ as T^b is to $T_{gg'}$ by Eq.(6). It is known that T^b obtained by planewise summation depends on the choice of planes in this way, leaving L^b independent of the choice.[16] Use of Eq. (15) in Eq. (1) with Eq. (11) yields

$$f_g = E_g + \sum_{g'} L_{gg'} \cdot p_{g'} / \varepsilon_o v, \tag{17}$$

showing that the $L_{gg'}$ play the same role in terms of E_g as the $T_{gg'}$ do in terms of E^o. Considerable algebra eventually leads to the result

$$\chi_g^{(1)} = \sum_{GG'} (I+k)^{-1}{}_{gG} K_{GG'} \Big/ \sum_{g'} (I+k)^{-1}{}_{gg'}. \tag{18}$$

Here $K = (A^{-1} - L)^{-1}$ and I and k are scalar matrices with elements $\delta_{GG'}$ and $n \cdot K_{GG'} \cdot n$ respectively. The local-field tensor d_g follows at once from Eq. (13).

4 OPTICAL RESPONSE

Nonlinear response can now be treated. At the molecular level, the induced dipole moment in layer g becomes

$$p_g = \alpha \cdot F_g + \beta : F_g F_g + \delta : \cdot F_g F_g F_g \tag{19}$$

Here F_g is the local field with nonlinear contributions and β and δ are the first and second hyperpolarizabilities; higher terms are neglected. At the macroscopic level the polarization in layer g becomes

$$P_g / \varepsilon_o = \chi_g^{(1)} \cdot E_g + \chi_g^{(2)} : E_g E_g + \chi_g^{(3)} : \cdot E_g E_g E_g , \tag{20}$$

where $\chi_g^{(n)}$ is the nth order susceptibility and terms beyond cubic are neglected. Since Eq. (17) still applies, substitution from Eq.(19) allows F_g to be related to E_g. Using $P_g = p_g/v$ then yields expressions for the $\chi_g^{(n)}$.

Mathematically this process resembles that for a bulk molecular crystal[11] (as for the local field). The results for the nonlinear susceptibilities are

$$\chi_g{}^{(2)} = \sum_{g'} \mathbf{b} : \cdot \mathbf{D}_{g'g} \mathbf{d}_{g'} \mathbf{d}_{g'} \tag{21}$$

$$\chi_g{}^{(3)} = \sum_{g'} [\mathbf{c} : : \mathbf{D}_{g'g} \mathbf{d}_{g'} \mathbf{d}_{g'} \mathbf{d}_{g'}$$

$$+ 2\mathbf{b} : \cdot \mathbf{D}_{g'g} \mathbf{d}_{g'} \sum_{g''} (\mathbf{D} \cdot \mathbf{L})_{g'g''} \cdot \mathbf{b} : \mathbf{d}_{g''} \mathbf{d}_{g''}], \tag{22}$$

with $\mathbf{b} = \beta/\varepsilon_o v$, $\mathbf{c} = \delta/\varepsilon_o v$, and $\mathbf{D} = (\mathbf{I} - \mathbf{L} \cdot \mathbf{A})^{-1}$.

The nonlinear susceptibilities for layer g depend not only on the local-field tensor $\mathbf{d}_g$ for that layer but also on those for other layers g', coupled by the partial local field tensor $\mathbf{D}_{g'g}$. Owing to the limited range r of the planewise sums, the evaluation of these expressions is reasonably tractable.

5 DISCUSSION

The present results provide a formal microscopic solution for the linear and nonlinear optical and electrical response in an assembly of ordered layers. Extensions to more than one molecule per two-dimensional cell and to alternating or other layer sequences are readily accommodated. The results incorporate the effect of the variation of the electric field through the surface region without invoking any explicit field-gradient or nonlocal polarizability.[17]

Numerical calculations of the nonlinear response require planewise dipole sums as input. Methods for calculating these sums have been developed and implemented previously.[13,14] In aromatic hydrocarbon crystals, interactions between adjacent layers may be only 1% of those within a layer for point molecules.[14] Our calculations for anthracene treated as three point submolecules show that interactions between adjacent layers may be 30% of those within layer, but fall off rapidly between more remote layers, so that the range r = 1. This short range implies that the surface region is very thin. Experiments at optical wavelengths in

transparent regions will then observe only an average
response over the surface region.[17]

ACKNOWLEDGEMENTS

This work is supported by SERC Grant GR/F 42195 and by
DARPA contract DAJA45-89-C-0036.

REFERENCES

1.   A. Barraud and M. Vandevyver, 'Nonlinear Optical
      Properties of Organic Molecules and Crystals', eds D.S.
      Chemla and J. Zyss, Academic, Orlando, 1987, Vol. 1, p.357.
2.   D.J. Williams, Ref. 1, Vol. 1, p.405.
3.   J. Badan, R. Hierle, A. Perigaud and P. Vidakovic, Ref 1,
      Vol. 1, p.297.
4.   R. Glenn, M.J. Goodwin and C. Trundle, J. Mol. Electron.,
      1987, 3, 59.
5.   Y.R. Shen, Nature, 1989, 337, 519.
6.   Yu. P. Piryatinsky, M.V. Kurik and S.V. Zavatsky, J. Mol.
      Electron., 1989, 5, 99.
7.   J.H. Meyling, P.J. Bounds and R.W. Munn, Chem. Phys.
      Letters, 1977, 51, 234.
8.   M. Hurst and R.W. Munn, Chem. Phys., 1988, 127, 1.
9.   J.A. Armstrong, N. Bloembergen, J. Ducuing and P.S.
      Pershan, Phys. Rev., 1962, 127, 1918.
10.   G.R. Meredith, B. Buchalter and C. Hanzlik, J. Chem. Phys.,
      1983, 78, 1533; G.R. Meredith, Proc. SPIE, 1987, 824, 126.
11.   M. Hurst and R.W. Munn, J. Mol. Electron., 1986, 2, 35.
12.   M. Hurst and R.W. Munn, J. Mol. Electron., 1986, 2, 139;
      1987, 3, 75; 'Organic Materials for Nonlinear Optics', RSC
      Special Publication No. 69, eds R.A. Hann and D. Bloor,
      Royal Society of Chemistry, London, 1989, p.3; 'Molecular
      Electronics - Science and Technology', ed A. Aviram,
      Engineering Foundation, New York, 1990, p.267; M. Hurst,
      R.W. Munn and J.O. Morley, J. Mol. Electron., 1990, 6, 15.
13.   G.D. Mahan and G. Obermair, Phys. Rev., 1969, 183, 834.
14.   M.R. Philpott, J. Chem. Phys., 1974, 61, 5306; Chem. Phys.
      Letters, 1975, 30, 387; Phys. Rev., 1975, B12, 5381.
15.   R.W. Munn, Mol. Phys., 1988, 64, 1.
16.   P.G. Cummins, D.A. Dunmur, R.W. Munn and R.J. Newham,
      Acta Cryst., 1976, A32, 847.
17.   P. Guyot-Sionnest, W. Chen and Y.R. Shen, Phys. Rev.,
      1986, B33, 8254.

The Estimation of Beta (-2w,w,w) in Pseudolinear Charge-transfer Dyes Using the Electronic Absorption Spectrum and the Solvatochromism of the Stoke's Shift: 4-(4′-Dimethylaminostyryl)-1-methylquinolinium Iodide

Colin L. Honeybourne and Karen J. Atkins

MOLECULAR ELECTRONICS AND SURFACE SCIENCE GROUP, BRISTOL
POLYTECHNIC, FRENCHAY, BRISTOL BS16 1QY, UK

INTRODUCTION.

The reasons for synthesising the title compound, 4-(4'-dimethylaminostyryl)-1-methylquinolinium iodide (DSQ), and the way in which it was accomplished, are given elsewhere in this Volume (Atkins and Honeybourne).For work on related long-alkyl species, please refer to Ashwell, Martin and Thompson(poster P12 & this Volume). Because we lack the facilities for taking EFISH (1), powder or single-crystal measurements of second harmonic generation (SHG) coefficients, we have used simple theory and, separately, spectroscopic data, to obtain two estimates for the molecular parameter relevant to SHG:i.e. $\beta(2w,w,w)$, now abbreviated as β. In both cases we used the two-level expression of Pugh and Morley (2) valid only for pseudolinear molecules that feature an intense, low-energy charge-transfer(CT) transition. Experimental values, or theoretical estimates of transition moment (M_y), change in dipole moment upon CT excitation (μ_{ge}^y) and excitation energy$(\bar{v}_{ge})$ were used to calculate $\beta(CT)$. In the foregoing we have therefore assumed that the CT excitation is confined to the long molecular (y) axis through the N(1) and C(4) atoms of the quinolinium ion.

SOLVATOCHROMISM AND DIPOLE MOMENT CHANGE.

The change in molecular dipole moment during an electronic transition is an indication of the extent of charge
transfer. Planar, conjugated molecules with strongly-
interacting donor and acceptor groups undergo charge-
separation upon excitation manifested as a large change
in dipole moment. The maximisation of μ_{ge} is essential
if a large value of ß(CT) is sought. Although the values
of $\bar{v}_{ge}$ and f_{ge}(the oscillator strength) can readily be
obtained from the spectrum, the evaluation of μ_{ge} poses
greater difficulty. However, provided that the CT-dye in
question fluoresces, it is possible to estimate μ_{ge} from
the solvent-dependence of the Stoke's Shift. This latter
is defined as the difference between the absorption and
fluoresence maxima. As emphasised by Murrell(3), at least
two solvents, preferably of very different dielectric
constant, must be used. A very detailed expression for
use in this context has been derived by Liptay(4). It
is customary to use the simplified expression, usually
termed the Lippert equation(5), shown below. When two
solvents are used, the value of the constant need not be
known, although one must assume that the "smaller terms"
can be ignored. If only one solvent is used, one is
forced to set the constant to zero, which is incorrect.

Lippert
Equation
$$\bar{v}_a - \bar{v}_f = 2P_0\,\mu_{ge}^2/(hca^3)+\text{constant}+\text{smaller terms}$$

$$P_0 = (\epsilon-1)/(2\epsilon+1) - (n^2-1)/(2n^2+1)$$

In the above equation, which is couched in the cgsesu
system of units, h and c have their usual meaning, and
the Stoke's Shift($v_a - v_f$) is in wavenumbers. The orientation polarisability, P_0, is expressed in terms of the
dielectric constant(ϵ) and refractive index(n) of the
solvent forming a cavity of radius,a, around the CT-dye.

SPECTROSCOPIC ESTIMATES OF BETA-CT(-2w,w,w)

The experimental value for the oscillator strength, f, is given by $4.319 \times 10^{-9} \times$(integrated area) in which the integrated area involves the product of the molar decadic extinction coefficient, ϵ_M, with its corresponding wave-number. Intense bands have an ϵ_M of the order of 5×10^4 and a full-width at half-height of circa $5000 cm^{-1}$. In the triangular-band-shape approximation, we see that f is of the order of unity for an intense band. In our case, the theoretical estimate of f is given as $\bar{v}_{ge} M_y^2 \times 4.701 \times 10^{29}$. We can therefore obtain M_y from the equation $M_y^2 = 2.127 \times 10^{-30} f_{ge}/\bar{v}_{ge}$ ($esu^2 cm^2$). For an intense band in the mid-optical region, M_y is circa 10^{-17} esu cm(10D).

For the Pugh and Morley CT equation we have

$$\beta_{yyy}^{CT} = 1.5 M_y^2 \, \mu_{ge} \bar{v}_{ge}^2/(h^2 c^2 (\bar{v}_{ge}^2 - 4\bar{v}^2)(\bar{v}_{ge}^2 - \bar{v}^2))$$

In the static approximation, in which the applied laser energy(i.e. $\bar{v}$ cm^{-1}) tends to zero, the equation for β_{yyy}^{CT} changes from its "laser-on" value

$$\beta_{yyy}^{CT} = 80 f_{ge} \, \mu_{ge} \bar{v}_{ge}/((\bar{v}_{ge}^2 - 4\bar{v}^2)(\bar{v}_{ge}^2 - \bar{v}^2))$$

to: $$(STATIC)\beta_{yyy}^{CT} = 80 f_{ge} \, \mu_{ge}/(\bar{v}_{ge}^3)$$

In the title compound, DSQ, the integrated area, obtained by manual integration, gave an oscillator strength of 0.77(which corresponds to a transition moment of 9.80D) for the intense band at $17053 cm^{-1}$($\epsilon_M^{max} = 4.286 \times 10^4$).

A value for μ_{ge} of 24.5D was obtained by substituting the Stoke's Shifts observed in methylene dichloride and in dimethyl sulphoxide into the Lippert equation.

The foregoing experimental quantities were used to obtain the following estimates for β_{yyy}^{CT} at laser energy, $\bar{v}$.

For $\bar{v}=0, \beta_{yyy}^{CT} = 304$; $\bar{v}=1900 cm^{-1}$, $\beta_{yyy}^{CT} = 552$; $\bar{v}=1600 cm^{-1}$, $\beta_{yyy}^{CT} = 773$. All β-values in 10^{-30} $esu^{-1} cm^5$.

THEORETICAL ESTIMATES AND VALIDATION OF BETA-CT$(-2w,w,w)$
As described elsewhere(2,6), there are only eight non-
zero tensor components of ß in a planar system(these are
itemised as column-headings in Table 1). The observable
vector average, for a system in the xy-plane, is ß-vec
given by $(\beta_X^2 + \beta_Y^2)^{\frac{1}{2}}$, in which

$$\beta_X = \beta_{xxx} + (\beta_{xyy} + \beta_{yxy} + \beta_{yyx})/3$$

$$\beta_Y = \beta_{yyy} + (\beta_{yxx} + \beta_{xyx} + \beta_{xxy})/3$$

In a perfectly linear system(y-axis), $\beta_X = 0$, and the
only component contributing to β_Y is β_{yyy}. Thus, for the
pseudolinear approximation to hold we suggest that mod.
β_Y must be greater than 0.9β-vec and that mod.β_{yyy} must
be greater than 0.9xmod.β_Y, with sign$(\beta_{yyy})=$ sign(β_Y).
In a neutral molecule, there is usually a convenient
choice of origin through which the vector joining logical
donor and acceptor passes if it is to be a candidate for
pseudolinearity. However, if this molecule is charged,
then the origin to which the ß-components refer must be
the centre of charge, although the new long-axis can be
parallel to the one originally chosen to evaluate the
atomic coordinates(see Introduction).
The various tensor components have contributions from
same-state-same-state(Diagonal) interactions and from
different-state-pair(Offdiagonal) interactions. For the
CT approximation to be valid, all Offdiagonal terms must
be negligible and only the lowest-energy state should
contribute a Diagonal term. In this work we have found
that the CT approximation holds(see Table 1).
We have used the quantum mechanical method described
elsewhere(2,6) to evaluate the diagonal and offdiagonal
contributions to the relevant components of the ß-tensor
using the nine lowest-energy states obtained by configuration interaction. The set of parameters used were those

1-METHYLQUINOLINIUM ION WITH 4-(4'DIMETHYLAMINOSTYRYL) GROUP

INDIVIDUAL STATE CONTRIBUTIONS TO BETADIAG TERMS

ST	EGY	XXXW	XXYW	XYXW	YXXW	XYYW	YXYW	YYXW	YYYW
1	2.34	2.20	14.37	14.37	12.13	86.08	75.66	75.66	441.26
2	3.93	1.34	1.15	1.15	0.76	-1.51	-1.31	-1.31	1.13
3	3.94	-0.00	0.01	0.01	0.01	0.11	0.10	0.10	0.54
4	4.05	0.72	0.97	0.97	0.76	-0.14	-0.12	-0.12	0.01
5	4.41	0.00	0.01	0.01	0.01	0.10	0.12	0.12	0.53
6	5.59	0.65	1.20	1.20	1.06	0.06	0.06	0.06	0.00
7	5.80	0.37	-0.00	-0.00	-0.04	-0.32	-0.28	-0.28	0.57
8	6.38	0.39	0.36	0.36	0.32	-0.03	-0.03	-0.03	0.00
9	6.84	0.04	0.02	0.02	0.02	-0.01	-0.01	-0.01	0.01

ST	EGY	XXX0	XXY0	XYX0	YXX0	XYY0	YXY0	YYX0	YYY0
1	2.34	1.10	6.80	6.80	6.80	39.53	39.53	39.53	220.41
2	3.93	1.09	0.83	0.83	0.83	-1.12	-1.12	-1.12	0.92
3	3.94	-0.00	0.01	0.01	0.01	0.09	0.09	0.09	0.44
4	4.05	0.59	0.74	0.74	0.74	-0.10	-0.10	-0.10	0.01
5	4.41	0.00	0.01	0.01	0.01	0.09	0.09	0.09	0.45
6	5.59	0.59	1.04	1.04	1.04	0.05	0.05	0.05	0.00
7	5.80	0.34	-0.01	-0.01	-0.01	-0.27	-0.27	-0.27	0.52
8	6.38	0.36	0.32	0.32	0.32	-0.03	-0.03	-0.03	0.00
9	6.84	0.04	0.02	0.02	0.02	-0.01	-0.01	-0.01	0.01

BETADIAG ELEMENTS IN 10*-30 cm*5 esu*-1
LASER WAVELENGTH IN nm = 1600.0

XXXW	XXYW	XYXW	YXXW	XYYW	YXYW	YYXW	YYYW
5.70	18.09	18.09	15.04	84.36	74.18	74.18	444.05

XXX0	XXY0	XYX0	YXX0	XYY0	YXY0	YYX0	YYY0
4.10	9.76	9.76	9.76	38.23	38.23	38.23	222.74

AVERAGE VALUES OF VECTOR PART OF BETA-DIAG
IN STATIC APPROXIMATION IN 10*-30 cm*5 esu*-1

BETA-X	BETA-Y	BETA-VEC
42.33	232.51	236.33

AVERAGE VALUES OF VECTOR PART OF BETA-DIAG
AT LASER WAVELENGTH AND IN 10*-30 cm*5 esu*-1

BETA-X	BETA-Y	BETA-VEC
83.27	461.12	468.58

FIRST EXCITED STATE BETA-YYY VALUES
HOMO-LUMO CONTRIBUTION = 95.8%
BETA(YYY)CT = 441.3 = 94.2% BETA-VEC
PSEUDOLINEAR CT APPROXIMATION HOLDS.

Table 1. The calculation of ß-vec(SHG) using nine excited states in both the static, "laser off" mode and at a laser wavelength of 1600nm. A value for β_{yyy}^{CT} is also presented.

QUINOLINIUM ION WITH 4-(4'-AMINOSTYRYL) GROUP

INDIVIDUAL STATE CONTRIBUTIONS TO BETADIAG TERMS

ST	EGY	XXXW	XXYW	XYXW	YXXW	XYYW	YXYW	YYXW	YYYW
1	2.45	1.87	11.94	11.94	9.99	68.22	59.82	59.82	329.99
2	3.95	0.22	0.52	0.52	0.40	-0.08	-0.07	-0.07	0.01
3	4.04	0.12	0.60	0.60	0.48	0.20	0.18	0.18	0.05
4	4.10	0.85	0.78	0.78	0.51	-1.44	-1.26	-1.26	1.47
5	4.56	0.03	0.08	0.08	0.08	0.18	0.19	0.19	0.42
6	5.72	0.62	1.30	1.30	1.17	0.09	0.08	0.08	0.00

BETADIAG ELEMENTS IN 10*-30 cm*5 esu*-1
LASER WAVELENGTH IN nm = 1600

XXXW	XXYW	XYXW	YXXW	XYYW	YXYW	YYXW	YYYW
4.48	15.65	15.65	12.99	66.75	58.56	58.56	332.65

AVERAGE VALUES OF VECTOR PART OF BETA-DIAG
AT LASER WAVELENGTH AND IN 10*-30 cm*5 esu*-1

BETA-X	BETA-Y	BETA-VEC
65.77	347.41	353.58

FIRST EXCITED STATE BETA-YYY VALUES.
HOMO-LUMO CONTRIBUTION = 96.4%
PSEUDOLINEAR AND CT APPROXIMATIONS HOLD.
BETA(yyy)CT = 329.99 = 93.33% BETA-VEC.

QUINOLINIUM ION WITH 4-(4'-DIMETHYLAMINOSTYRYL) GROUP

INDIVIDUAL STATE CONTRIBUTIONS TO BETADIAG TERMS

ST	EGY	XXXW	XXYW	XYXW	YXXW	XYYW	YXYW	YYXW	YYYW
1	2.35	2.15	14.02	14.02	11.77	83.29	73.03	73.03	421.81
2	3.94	1.17	0.96	0.96	0.60	-1.59	-1.37	-1.37	1.44
3	3.96	-0.01	0.04	0.04	0.03	0.15	0.13	0.13	0.33
4	4.04	0.81	1.03	1.03	0.81	-0.12	-0.10	-0.10	0.01
5	4.42	0.00	-0.01	-0.01	-0.02	0.06	0.08	0.08	0.57
6	5.62	0.64	1.18	1.18	1.06	0.03	0.03	0.03	0.00

BETADIAG ELEMENTS IN 10*-30 cm*5 esu*-1
LASER WAVELENGTH IN nm = 1600

XXXW	XXYW	XYXW	YXXW	XYYW	YXYW	YYXW	YYYW
5.51	17.58	17.58	14.55	81.48	71.48	71.48	424.75

AVERAGE VALUES OF VECTOR PART OF BETA-DIAG
AT LASER WAVELENGTH AND IN 10*-30 cm*5 esu*-1

BETA-X	BETA-Y	BETA-VEC
80.33	441.32	448.57

FIRST EXCITED STATE BETA-YYY VALUES.
HOMO-LUMO CONTRIBUTION = 95.8%
PSEUDOLINEAR AND CT APPROXIMATIONS HOLD.
BETA(yyy)CT = 421.81 = 94.03% BETA-VEC.

Table 2. The calculation of ß-vec(SHG) using
nine excited states(only six shown here) at a
laser wavelength of 1600nm. Values for β_{yyy}^{CT}
are also presented.

developed for dyes by Griffiths(7) for use with 9-fold
configuration interaction.

The results are shown in Table 1 for DSQ using the explicit inclusion of the methyl group in the N(+)-Me
moiety by means of the hyperconjugative model (8). The
suffix "w" refers to "laser on" and "0" refers to the
static approximation. These results show that although
the pseudolinear approximation(as defined above) still
holds, the xyy,yxy and yyx components are quite large.
Only one state contributes significantly, and this is
virtually a pure HOMO-LUMO transition. Inspection of
the molecular orbitals shows that this transition does
indeed transfer charge from the "inium" ring to the ring
of the styryl unit. Thus, the CT transition approximation
also holds. The small effect of the N+ methyl group, and
the large effect of the $N(Me)_2$ methyl groups can be seen
from the abbreviated results for related compounds in
Table 2.

In conclusion, we note that the two estimates obtained
for DSQ are similar,although the difference diverges as
the laser energy increases.

REFERENCES

1.B.F.Levine & C.G.Bethea,1975,J.Chem.Phys.,63, 2666.
2.D.Pugh & J.O.Morley, Nonlinear Optical Properties of
   Organic Molecules and Crystals,1987,Academic Press,
   New York, Edt.D.S.Chemla & J.Zyss, Vol.1, pp.193.
3.J.N.Murrell,The Theory of the Electronic Spectra of
   Organic Molecules,Chapman and Hall,1971.
4.W.Liptay,1965, Z.Naturf., 20a, 272.
5.E.Lippert,1957, Z.Electrochem., 61, 962.
6.C.L.Honeybourne, 1990, J.Appl.Phys.D, 23, 245.
7.J.Griffiths, 1982, Dyes and Pigments, 3, 211.
8.P.Lindner & O.Martensson,1967, Theoret.Chim.Acta,
   7, 352.

Small Organic Molecules

A New Tris-chromophore Non-linear Optical Oligomer for Langmuir–Blodgett Film Deposition

S. Allen and T.G. Ryan

ICI WILTON MATERIALS RESEARCH CENTRE, PO BOX 90, WILTON
MIDDLESBROUGH, CLEVELAND TS6 8JE, UK

D.P. Devonald and M.G. Hutchings

ICI COLOURS AND FINE CHEMICALS RESEARCH CENTRE, HEXAGON
HOUSE, BLACKLEY, MANCHESTER M9 3DA, UK

A.N. Burgess, E.S. Froggatt, A. Eaglesham, and R.M. Swart

ICI CORPORATE COLLOID SCIENCE GROUP, THE HEATH, RUNCORN,
CHESHIRE, UK

G.J. Ashwell and M. Malhotra

CENTRE FOR MOLECULAR ELECTRONICS, CRANFIELD INSTITUTE OF
TECHNOLOGY, CRANFIELD, BEDFORD MK43 0AL, UK

1 INTRODUCTION

There have been many reports in recent years of Langmuir Blodgett (LB) films showing high second-order nonlinear optical activity, as characterised by second harmonic generation or electro-optic (Pockels) experiments. Problems often encountered with LB films relate to their poor stability (thermal and mechanical), and poor optical quality, as shown for example by the high levels of light scattering observed in waveguide experiments. Recently there have been reports of the synthesis of polymeric LB films having improved optical and mechanical properties[1,2]. Polymers, however, in general are less tractable than their monomeric counterparts. We describe in this paper an example of an oligomeric structure which appears to retain much of the stability expected of a polymeric system along with the good film-forming properties of many monomeric LB molecules. Films of this new compound have been deposited, and characterised by a number of techniques including neutron reflectivity, optical and nonlinear optical methods.

2 FILM FORMATION

The structure of the molecule described is shown as (1) below. It is characterised by three azobenzene chromophores, attached by a diamide chain connecting the three amino donor groups. The amide group is hydrophilic and thus a potentially good head group for the amphiphile, and is also an easily synthesisable linking group for the oligomers. The azobenzene chromphores present in (1) are chemically and physically stable and have high molecular nonlinearities β when substituted with

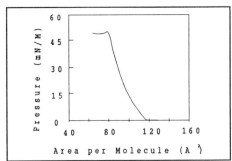

(1)

donor and acceptor groups.

A 1 mg/ml solution of (1) in chloroform (BDH Aristar) was applied to a clean water surface in a Joyce Loebl Mark IV Langmuir trough. The water was purified by Millipore-RO and Milli-Q systems and had a resistivity of greater than 18 MΩ/cm. The surface pressure (π) against area per molecule (A) isotherm is shown in Figure 1. The monolayer collapses above 50 mN/m but is stable for several hours at 8, 15 and 30 mN/m. Extrapolation from the steepest part of the isotherm gives A=103 Å². This is the area per molecule for the hypothetical state of an uncompressed, close packed layer. The area per molecule at 30 mN/m is 88.8 Å². Langmuir Blodgett deposition was successfully achieved at a pressure of 30 mN/m and a speed of 5 mm/min onto a variety of substrates. Transfer efficiencies in the range 82-100% were obtained (where 100% indicates perfect transfer), as shown for example in Table 1 below.

Figure 1. Surface pressure/area isotherm for (1) on a water subphase.

Table 1: Transfer efficiencies d for successive monolayers of (1) deposited onto a Spectrosil 'B' quartz slide.

stroke	0↓	1↑	2↓	3↑	4↓	5↑	6↓	7↑	8↓	9↑
d (%)	0	98	86	96	84	97	82	94	82	92

3 NEUTRON REFLECTIVITY

Neutron reflectivity data were collected from multilayer films of (1) deposited onto spectrosil 'B' quartz slides, with transfer efficiencies as shown in Table 1. The time of flight method was used, on the spectrometer CRISP at the Rutherford Appleton Laboratory[3]. The molecule (1) is interesting with respect to neutron reflectivity in that the aromatic section has a very different neutron scattering length density from the aliphatic chain. A Langmuir-Blodgett multilayer will consist of alternate strata of aromatic and aliphatic species, and the neutron reflectivity is very sensitive to this banding, providing an accurate measure of the aromatic and aliphatic layer thicknesses. Thus the data allow us to independently determine the orientation of the nonlinear optical chromophore and its hydrophobic tails. Figure 2 depicts the reflectivity data obtained from a nine-layer film at an incident angle of 1°. A particularly interesting feature is the Bragg peak centred at 1.7 Å which arises due to the banding described above, and which enables a detailed interpretation of the data to be made. The solid line passing through the experimental points represents the best fit to the data, and is the calculated reflectivity according to the model

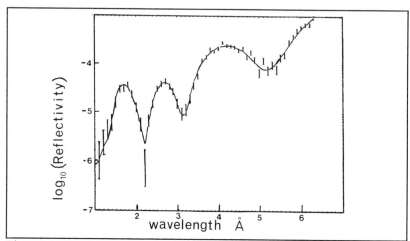

Figure 2. Neutron Reflectivity data and fitted theoretical curves for 9 layers of (1) on a Spectrosil 'B' quartz substrate.

Table 2. Monolayer properties of (1) derived from neutron scattering
and Langmuir isotherms.

	π (mN/m)	Area per molecule	Aromatic layer thickness (Å)	Tilt angle (°)
1 layer on trough	7.8	105.6	12.3	41
"	14.7	100.0	13.1	36
"	29.8	88.8	14.5	26
"	50.0	80.0	16.2(*)	0
9-layer LB film		100.4(**)	13.05	36

(*): extrapolated   (**): interpolated

discussed below.

Parallel studies have been carried out on Langmuir films of (1) on the
surface of a specially constructed Langmuir trough, at three pressures
during film compression. The results of these studies indicate that the
thickness of the aromatic band changes on compression, whilst the area
per molecule decreases (Table 2). Although the Langmuir-Blodgett films
were deposited at a pressure of 30 mN/m, where the cross-sectional area
per molecule is 88.8 Å², the neutron reflectivity results indicate an
area *in the film* of 100.4 Å² per molecule, indicating that the
molecules are relaxing during transfer. This cross-sectional area is
close to that deduced from the π/A isotherm of 103 Å², for (1) in
uncompressed but close-packed state.

The progressive increase in aromatic band thickness with
increasing pressure can be interpreted by assuming that the
chromophores in (1) are inclined increasingly towards the normal to the
trough surface. These angles of inclination can be inferred from the
differences in the thickness of the bands compared to the extended
lengths obtained from molecular modelling. The cosinusoidal
relationship between these two parameters means that for small angles
of inclination the error in the estimated angle is quite large ($\pm 10°$).

4 MOLECULAR MODELLING

In order better to understand the properties of the Langmuir-
Blodgett films of (1) it was of interest to model the conformation of
(1) in the monolayer phase. 3-D molecular modelling software (developed
by ICI in-house) has been used to aid this work. The starting point was
taken as the molecular cross-sectional area of (1), derived from the
Langmuir experiments to be 80 Å² per molecule at the highest pressure
attainable before the film collapses. It is assumed that at this point
the chromophores are aligned normal to the film trough surface. The
minimum cross-sectional area of a 2-nitro-substituted azobenzene unit

is 27.3 Å², giving a cross sectional area per molecule of 81.9 Å², gratifyingly close to the value from the monolayer measurements. The modelling problem then becomes one of accomodating the diamide linking chain in (1) within this area, and simultaneously giving a thickness of 16.2 Å for the aromatic band (as extrapolated from the neutron scattering data). The modelling presents several possibilities, one of which is shown in Figure 3. In this the chromophore and backbone thickness is 15.8 Å.

It has been necessary to assume a cis-amide conformation for the amide bond. Independent evidence for this comes from reflection absorption infra red spectroscopy (RAIRS) of multilayers of (1) coated onto aluminium coated substrates. In a 9-layer sample a carbonyl band was obtained at 1667 cm^{-1}, compared with 1639 cm^{-1} in the free (solid) state. While the latter is typical of a trans-amide, the former is close to the typical values of lactams, which are constrained to be cis, and amide dimers in solution. The model has the diamide linking chain in a "double-S" conformation (Figure 4) whilst the arylazo groups are twisted by 45° around the normal to the surface, and are close-packed. As the pressure is reduced (and in the Langmuir-Blodgett film) the chromophores tilt with respect to the surface, but this has not yet been modelled.

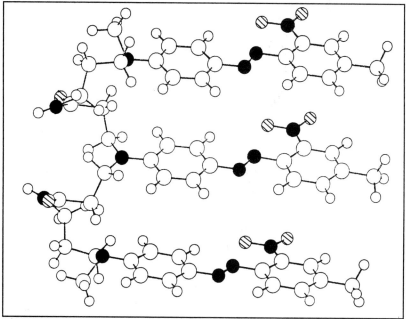

Figure 3. Side view of molecule (1) as aligned on the water surface (shown vertical), according to the model described.

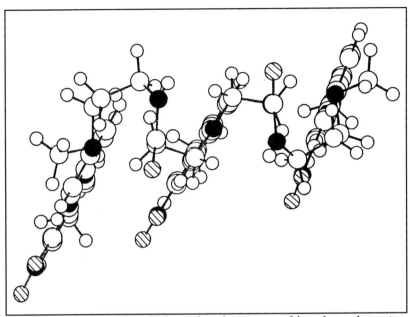

Figure 4. End on view of the molecule (1) as aligned on the water
surface, showing the conformation of the backbone.

The hydrogen bonding possibilities for the two groups in the
model (Figure 3) imply a possible four-point attachment to a
hydrophilic surface. Furthermore Figure 4 reveals that the C=O and N-H
units of the two amide groups lie at opposite corners of a rectangle,
and if this conformation carries through to the more relaxed Langmuir-
Blodgett phase they will be arranged optimally for inter-layer hydrogen
bonding between matched diamide units (Figure 3), possibly accounting
for the stability of the Y-type multilayer structures observed for (1).
The resultant thermodynamic favourability of the H-bonds more than
offsets the relative instability of cis-amide conformations compared
with trans.

The above model applies only to the Langmuir-Blodgett film phase.
High field n.m.r. spectra of (1) in solution show no nuclear Overhauser
effects between protons which are modelled to be very close, and thus
would be expected to show nOe.

5 REFRACTIVE INDICES

Refractive indices of Y-type multilayers of (1) have been measured
using surface plasmon spectroscopy[4]. The films were deposited onto

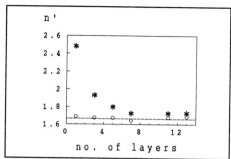

Figure 5. Refractive index (n') of films of (1) at 633 nm, with (o) and without (*) a water layer between the silver and LB layers.

glass substrates coated with a thin silver layer. The Kretschman geometry[5] is used, with a laser beam incident on this bilayer via a prism. The reflectivity of this film was observed as a function of angle of incidence, at a number of wavelengths for multilayers of different thickness. The experimental data were fitted with theoretical curves, obtained using standard Fresnel-type multilayer theory. The optical parameters (n', n'' and thickness) for the silver film were first found by fitting experimental data for an uncoated sample. The refractive indices of the Langmuir-Blodgett layers were then obtained, using in the fitting routines the film thicknesses determined by the neutron scattering experiments described above. The real part of the refractive indices n' found by this method, at a wavelength of 632.8 nm are shown in Figure 5 as a function of the number of monolayers in the film. It can be seen that the refractive index for a single layer appears to be significantly higher than that of thicker films. This effect has also been observed in other Langmuir-Blodgett films[4].

The neutron scattering data suggest that in fact there is a thin (20 Å) layer of water trapped between the surface of the slide and the first deposited monolayer, but little entrapment of water between following layers. Therefore the reflectivity data were also fitted assuming a layer of thickness 20 Å, and refractive index 1.40 (intermediate between water and the glass substrate) was present between the silver and the first monolayer. In this case a much more consistent set of refractive indices is obtained, as is also shown in Figure 5. The refractive index for thick (i.e. 13 layer) films, at 633, 820 and 1320 nm are given in Table 3, for both the above models.

Table 3. Refractive indices of thick Langmuir Blodgett films of (1)

wavelength (nm)	633	820	1320
n': no water layer	1.724	1.713	1.655
n': water layer (20Å)	1.680	1.668	1.599

6 SECOND HARMONIC GENERATION

Second harmonic generation studies were carried out on monolayers and Y-type multilayers, having an odd number of layers (1,3,5,11), using a pulsed Nd:YAG laser of fundamental wavelength 1.06 μm. The signal was constant, within experimental error, from all these samples. Since in each case there should be just a single "unpaired" layer, the constancy of the SHG is another indication of the good quality of the deposited films. The level of activity from these films was found to be superior to that from films of monomeric amphiphiles containing the same basic chromophore[6].

Alternating Y-type films were also obtained using 1-docosyl-4-methylquinolinium bromide as the counter molecule to (1). This second molecule has a low molecular β value, and therefore will not contribute significantly to the SHG. Solutions of (1) and the quinolinium salt, in Aristar grade chloroform, were spread on the pure water subphase of compartments A and B respectively of a NIMA Technology LB trough, the compartments being separated by a fixed surface barrier. Alternate layer Langmuir-Blodgett films were obtained by cycling a hydrophilically treated glass substrate via A on the downstroke and B on the upstroke. During deposition surface pressures in A and B were maintained at 30 and 35 mN/m respectively.

Films comprising up to 50 bilayers were deposited in this manner. The level of second harmonic from the film was monitored as a function of the number of layers (N), by mounting the sample at 45° to the incident laser beam. The results are shown in Figure 6, where the square root of the second harmonic intensity is plotted. It can be seen

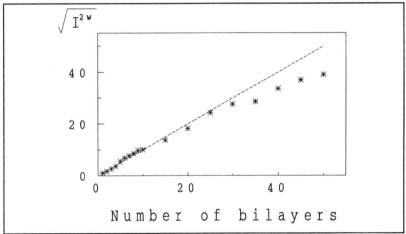

Figure 6. Square root of second harmonic intensity from AY-films of (1) and a quinolinium salt.

that up to about 25-30 layers the expected linear dependence between $\sqrt{I}$ and N is obtained, but for thicker films the level of SHG is less than expected. This is presumably due to an increased degree of disorder in the films. In these AY-type films there will be no possibility for the four-point hydrogen bond linkage between successive layers, as postulated above for the Y-type layers. This may lead to a reduction in the stability of the films. The formation of A-Y type layers of (1) with nonlinear optical layers capable of this hydrogen bonding will be reported in the future, along with a more detailed discussion of the second harmonic generation experiments.

CONCLUSION

Preliminary results have been presented on a new oligomeric nonlinear optical Langmuir-Blodgett molecule. These results are very promising, in that stable films having high levels of activity have been produced. The combination of characterisation techniques described are producing a good level of understanding of the molecular packing in these films, and have highlighted the possibility of favourable hydrogen bonding interactions. Future work will concentrate on investigating fully modifications to the basic tri-chromophore diamide structure, to improve further the properties of interest.

ACKNOWLEDGEMENTS

We are grateful to D Louden and B Cook of ICI Chemicals and Polymers for obtaining the RAIRS spectra. Financial support from the European Commission for some of the work described, through the ESPRIT programme, is acknowledged.

REFERENCES

1. R H Tredgold, M C J Young, R Jones, P Hodge, P Kolinsky and R J Jones, Electron. Lett., 1988, 24, 308.
2. W Prass, Hoechst High Chem Magazine, 1989, 7, 15.
3. J Penfold, R C Ward and W G Williams, J. Phys. E: Sci. Instrum., 1987, 20, 1411.
4. I Pockrand, J D Swalen, J G Gordon and M R Philpott, Surf. Sci., 1977, 74, 237.
5. E Kretschmann, Z. Physik, 1971, 241, 313.
6. I Ledoux, D Josse, P Vidakovic, J Zyss, R A Hann, P F Gordon, B D Bothwell, S K Gupta, S Allen, P Robin, E Chastaing and J C Dubois, Europhys. Lett., 1987, 3, 803.

Quadratic Enhancement of the Second Harmonic Intensity of Interleaved Langmuir–Blodgett Films of a Quinolinium Dye

G.J. Ashwell,* P.J. Martin, M. Szablewski, and P.A. Thompson

CENTRE FOR MOLECULAR ELECTRONICS, CRANFIELD INSTITUTE OF TECHNOLOGY, CRANFIELD, BEDFORD MK43 0AL, UK

A.T. Hewson and S.D. Marsden

DEPARTMENT OF CHEMISTRY, SHEFFIELD CITY POLYTECHNIC, SHEFFIELD S1 1WB, UK

1 INTRODUCTION

There have been many reports of second harmonic generation (SHG) from Langmuir-Blodgett (LB) film forming materials of general formula $R'R''N-C_6H_4-CH=CH-C_5H_4N^+-R$ halide, where R, R' and R" are alkyl groups[1-8] or where R is polymeric.[9-12] We now disclose the synthesis and properties of a related quinolinium dye, $(CH_3)_2N-C_6H_4-CH=CH-C_9H_6N^+-C_{18}H_{37}$ I^-, E-1-octadecyl-4-{2-(4-dimethylaminophenyl)ethenyl}quinolinium iodide, $C_{18}H_{37}QH-I$.

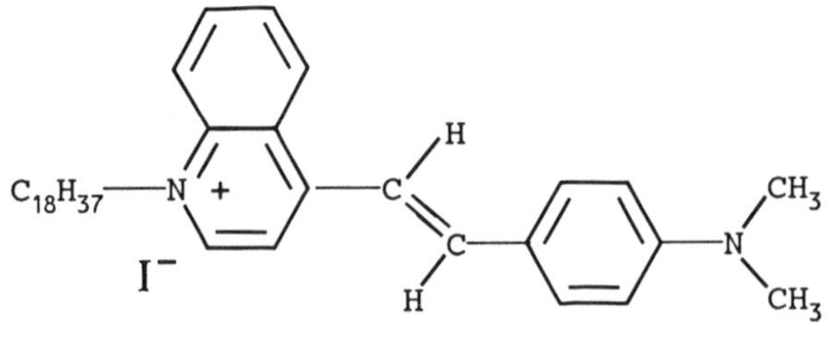

Our interest was aroused by the multifunctional behaviour of an analogue, Z-β-(1-hexadecyl-4-quinolinium)-α-cyano-

---

*To whom correspondence should be addressed.

4-styryldicyanomethanide[13,14] which has a similar, albeit zwitterionic, molecular structure and a high second-order hyperpolarisability (β). For SHG the nonlinearity is preserved as a bulk property ($X^{(2)}$) by the tendency of the zwitterions to form non-centrosymmetric Z–type structures which (i) exhibit a quadratic enhancement of the second harmonic intensity with the number of layers[15,16] and (ii) rectify when fabricated as metal|(LB monolayer)|metal devices[17]. In contrast, $C_{18}H_{37}QH$-I forms Y–type films but non-centrosymmetric structures showing enhanced SHG may be obtained by interleaving the chromophore layers with 4,4'–dioctadecyl–3,5,3',5'-tetramethyldipyrrylmethene hydrobromide (DPM).

2 SYNTHESIS

E-1-octadecyl-4-{2-(4-dimethylaminophenyl)ethenyl} quinolinium iodide, $C_{18}H_{37}QH$-I, was obtained by refluxing a solution of 1-octadecyl-4-methylquinolinium iodide (0.52 g, 1 mmol), N,N-dimethylamino-$\underline{p}$-benzaldehyde (0.13 g, 1 mmol) and piperidine in methanol for 4 h. Crystallisation of the product from ethanol gave purple microcrystals. Yield 80 %. ^{1}H NMR (DMSO): 0.9 (3H, t, J = 5.6 Hz, CH_3–); 1.3 (32H, s, $-(CH_2)_{16}-$); 2.9 (6H, s, $(CH_3)_2N-$); 4.8-5.1 (2H, m, $-CH_2N^+-$); 6.9 (2H, d, J = 5.9 Hz, Ar–H); 7.8-8.6 (9H, m, Ar–H); 9.0-9.4 (1H, m, Ar–H). UV/VIS (CH_3OH): λ_{max} 301 nm, 546 nm. Required for $C_{37}H_{55}N_2I$: C, 67.9; H, 8.4; N, 4.3 %. Found: C, 67.3; H, 8.5; N, 3.9 %.

3 LANGMUIR-BLODGETT DEPOSITION

Non-centrosymmetric films of $C_{18}H_{37}QH$-I interleaved with DPM were fabricated using a two-compartment NIMA Technology trough, the compartments being separated by a fixed surface barrier. The materials, in Aristar grade dichloromethane, were spread on the pure water subphase of separate compartments of the trough; compressed to 25 mN m^{-1} and then deposited alternately on hydrophilically treated glass slides at 60 μm/sec. $C_{18}H_{37}QH$-I was deposited initially on the first upstroke followed by DPM on the downstroke, the process being repeated for 100 layers, in each case with a transfer ratio of unity.

4 FILM SPECTRA

The u.v./visible spectrum of the alternate layer film is a simple combination of the separate spectra and has

maxima at 315, 383 and 500 nm with absorbances per bilayer
of 0.0065, 0.0051 and 0.0073 respectively (Fig. 1). The
band at 383 nm corresponds to the interleaving DPM, the
maximum being shifted from 423 nm in the homomolecular
film by the neighbouring bands of $C_{18}H_{37}QH-I$. The peak at
500 nm (cf. 504 nm in the homomolecular film) is blue
shifted from 544 nm in methanol whereas the one at 315 nm
is red shifted from 301 nm. We note that deposition of
$C_{18}H_{37}QH-I$, with a transfer ratio of unity, is maintained
for the interleaved structure. This is demonstrated by the
linear relation of the peak absorbance at 500 nm with the
number of layers (Fig. 2).

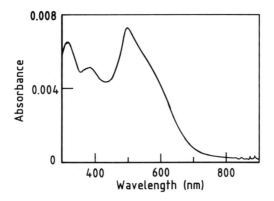

Fig. 1. U.V./visible spectrum of the alternate layer film.

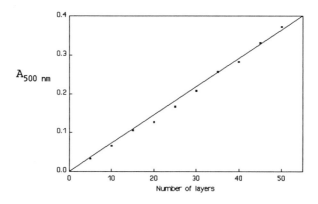

Fig. 2. Absorbance at 500 nm versus the number of layers of
$C_{18}H_{37}QH-I$ in the alternating LB film.

5 SECOND HARMONIC GENERATION

SHG measurements were carried out by irradiating the films with a Q-switched Nd:YAG laser (λ 1.064 μm; pulse width 10 ns; repetition rate 10 Hz) with the beam incident to the substrate at 45°. The SH intensity from monolayer films of $C_{18}H_{37}$QH–I on glass substrates is three to four times that of the reference hemicyanine dye, E—1—docosyl-4-{2-(4-dimethylaminophenyl)ethenyl}pyridinium bromide[1], which has a second order hyperpolarisability, β, of 200 x 10^{-30} cm^{5}esu^{-1}. The films are Y-type and SHG from multilayer films does not exceed that from the monolayer.

Non-centrosymmetric LB structures have been reported in which the active chromophore layers are interleaved by spacer layers[9-12]. In this work we had limited success by alternating with 1-docosyl-4-methylquinolinium bromide, the SHG dependence on the number of bilayers being sub-quadratic for n > 10. In contrast, ordered interleaved structures of $C_{18}H_{37}$QH–I and DPM were obtained which showed quadratic SHG enhancement to n ≈ 50 (Fig. 3). However, the normalised SH intensity, calculated from $I_{(n)}/n^2$, is only 10 % of the signal from monolayer films of $C_{18}H_{37}$QH–I on glass. It suggests that the packing within the dye layers is dependent upon the nature of the adjacent layer.

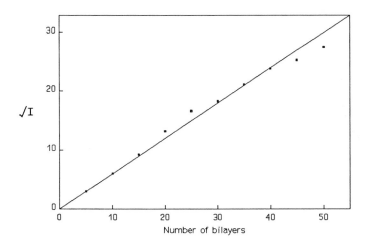

Fig. 3. Square root of the SH intensity (relative to the square root of the hemicyanine reference[1]) vs. the number of bilayers of $C_{18}H_{37}$QH-I/DPM.

Table 1. SH intensity data relative to the signal from a monolayer film of the reference hemicyanine dye[1].

Repeat	——— Relative SH intensity ———		
	Monolayer[§]	50 bilayers	100 bilayers
Quinolinium			
$C_{18}H_{37}$QH-I	3 – 4		
$C_{18}H_{37}$QH-I/DPM	0.35	753	
Pyridinium			
$C_{18}H_{37}$PH-I	2 – 4		
$C_{18}H_{37}$PH-I/DPM	1.8	4970	18,300

[§]Calculated from $I_{(n)}/n^2$ for the interleaved structures where n is the number of bilayers.

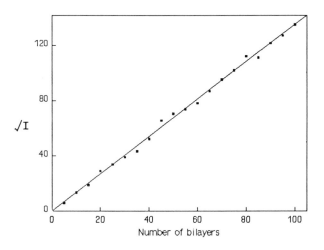

Fig. 4. Square root of the SH intensity (relative to the square root of the hemicyanine reference[1]) vs. the number of bilayers of $C_{18}H_{37}$PH-I/DPM.

DPM is also a suitable material for interleaving with the related pyridinium hemicyanine dyes, for example, E-1-octadecyl-4-{2-(4-dimethylaminophenyl)ethenyl}pyridinium iodide, $C_{18}H_{37}PH-I$, and the alternating LB film structures show quadratic SHG enhancement to ≈100 bilayers (Fig. 4). The normalised intensity, $I_{(n)}/n^2$, is similar to the value obtained for monolayers of $C_{18}H_{37}PH-I$ on glass (Table 1) whereas the intensity of SHG from 100 bilayers is **18,300** times the reference hemicyanine value. The enhancement is the highest reported for an LB film and full details will be published separately[18].

In conclusion, we have fabricated non-centrosymmetric ordered structures of two novel dyes and have achieved quadratic SHG enhancement for LB films comprising 50 layers each of DPM and $C_{18}H_{37}QH-I$ and 100 layers each of DPM and $C_{18}H_{37}PH-I$.

6 ACKNOWLEDGEMENTS

G.J.A. acknowledges the S.E.R.C. Molecular Electronics Committee for support (grants GR/E/88974, GR/F/45813 and GR/G/06596) and is grateful to the S.E.R.C. for providing studentships to P.J.M and M.S.

REFERENCES

1. I.R. Girling, N.A. Cade, P.V. Kolinski, J.D. Earls, G.H. Cross and I.R. Peterson, <u>Thin Solid Films</u>, 1985, **132**, 101.
2. D.B. Neal, M.C. Petty, G.G. Roberts, M.M. Ahmad, W.J. Feast, I.R. Girling, N.A. Cade, P.V. Kolinski and I.R. Peterson, <u>Electron Lett.</u>, 1986, **22**, 460.
3. I.R. Girling, N.A. Cade, P.V. Kolinski, R.J. Jones, I.R. Peterson, M.M. Ahmad, D.B. Neal, M.C. Petty, G.G. Roberts, W.J. Feast, <u>J. Opt. Soc. Am. B</u>, 1987, **4**, 950.
4. G.H. Cross, I.R. Girling, I.R. Peterson, N.A. Cade and J.D. Earls, <u>J. Opt. Soc. Am. B</u>, 1987, **4**, 962.
5. G.H. Cross, I.R. Peterson, I.R. Girling, N.A. Cade, M.J. Goodwin, N. Carr, R.S. Sethi, R. Marsden, G.W. Gray, D. Lacey, A.M. McRoberts, R.M. Scrowston and K.J. Toyne, <u>Thin Solid Films</u>, 1988, **156**, 39.
6. J. Schildkraut, T.L. Penner, C.S. Willand and A. Ulman, <u>Opt. Lett.</u>, 1988, **13**, 134.
7. G. Marowski, L.F. Chi, D. Moebius and R. Steinhoff, <u>Chem. Phys. Lett.</u>, 1988, **147**, 420.
8. R. Steinhoff, L.F. Chi, G. Marowski and D. Moebius, <u>J. Opt. Soc. Am. B</u>, 1989, **6**, 843.

9. B.L. Anderson, R.C. Hall, B.G. Higgins, G.A. Lindsay,
    P. Stroeve and S.T. Kowel, Synth. Met., 1989, **28**,
    D683.
10. L.M. Hayden, B.L. Anderson, J.Y.S. Lam, B.G. Higgins,
    P. Stroeve and S.T. Kowel, Thin Solid Films, 1988,
    160, 379.
11. R.C.Hall, G.A. Lindsay, B.L. Anderson, S.T. Kowel,
    B.G. Higgins and P. Stroeve, Mater. Res. Soc. Symp.
    Proc., 1989, **109**, 351.
12. R.C. Hall, G.A. Lindsay, S.T. Kowel, L.M. Hayden, B.L.
    Anderson, B.G. Higgins, P. Stroeve and M.P. Srinivasan,
    Proc. SPIE-Int. Soc. Opt. Eng., 1988, **824**, 121.
13. G.J. Ashwell, UK Patent Appl. 9007230.7, March 1990;
    European Patent Appl. 90303473.4, March 1990; Japanese
    Patent Appl. 84760/90, March 1990; US Patent Appl.,
    filed March 1990.
14. G.J. Ashwell, Thin Solid Films, 1990, **186**, 155.
15. G.J. Ashwell, E.J.C. Dawnay, A.P. Kuczynski and M.
    Szablewski, Mater. Res. Soc. Symp. Proc., 1990, **173**,
    507.
16. G.J. Ashwell, E.J.C. Dawnay, A.P. Kuczynski, M.
    Szablewski, I.M. Sandy, M.R. Bryce, A.M. Grainger and
    M. Hasan, J. Chem. Soc., Faraday Trans., 1990, **86**,
    1117.
17. G.J. Ashwell, J.R. Sambles, A.S. Martin, W.G. Parker and
    M. Szablewski, J. Chem. Soc., Chem. Commun., in press.
18. G.J. Ashwell and P.J. Martin, in preparation.

Amphiphilic Dyes for Non-linear Optics in Langmuir–Blodgett Films

A. Laschewsky, W. Paulus, and H. Ringsdorf

INSTITUTE OF ORGANIC CHEMISTRY, UNIVERSITY OF MAINZ, D-6500 MAINZ, FEDERAL REPUBLIC OF GERMANY

D. Lupo, P. Ottenbreit, and W. Praß

HOECHST AG, D-6230 FRANKFURT, FEDERAL REPUBLIC OF GERMANY

C. Bubeck, D. Neher, and G. Wegner

MPI-POLYMERE, D-6500 MAINZ, FEDERAL REPUBLIC OF GERMANY

INTRODUCTION

In respect to quadratic nonlinear optical phenomena, especially SHG, it is necessary to build up highly noncentrosymmetric structures[1,2]. The most active molecules however tend to crystallize in a centrosymmetric crystal lattice, thus it is necessary to look for alternatives to create noncentrosymmetric arrangements of the molecules.

An elegant way to build up noncentrosymmetric structures is the use of the LB-technique[3,4,5]. The monolayers are transferred one by one and thus it is possible to create metastable structures which are not accessible by other methods[6,7,8]. Unfortunately it is impossible yet to predict whether an amphiphilic molecule will form defined monolayers or not[9], especially symmetric molecules with high dipole moments. Too the transfer to solid supports may be complicated.

SYNTHESIS OF PHENYLHYDRAZONES AND STILBAZIUM SALTS

The initial studies about second-harmonic generation in Langmuir-Blodgett monolayers of stilbazium salts and phenylhydrazone dyes[10] revealed their great potential for NLO systems, because of their high second-order nonlinear optical susceptibilities $\chi^{(2)}$. Of particular interest are the nitrophenylhydrazones because these compounds do not absorb light significantly at the second-harmonic wavelength of the used Nd^{3+}:YAG Laser (1064 nm). Starting from the two known compounds 4-octadecyloxy-4´nitrophenylhydrazone (1) and N-methyl-4´(N,N-dihexadecylamino)stilbazium iodide (13) systematic variations of the molecular structure have been carried out (table 1).

TABLE 1.SYNTHESIS OF PHENYLHYDRAZONES AND STILBAZIUM SALTS

	R¹	R²	R³	R⁴	R⁵	k
1	H	$C_{18}H_{37}O$	H	H	H	0
2	H	$C_{16}H_{33}O$	$C_{16}H_{33}O$	H	H	0
3	H	$C_{18}H_{37}O$	H	OH	H	0
4	$C_{16}H_{33}O$	$C_{16}H_{33}O$	$C_{16}H_{33}O$	H	H	0
5	H	$C_{18}H_{37}S$	H	H	H	0
6	H	$(C_{16}H_{33})_2N$	H	H	H	0
7	H	$C_{16}H_{33}O$	$C_{16}H_{33}O$	H	NO_2	0
8	H	$C_{16}H_{33}O$	$C_{16}H_{33}O$	H	NO_2	1
9	$C_{16}H_{33}O$	$C_{16}H_{33}O$	$C_{16}H_{33}O$	H	NO_2	0
10	H	$(C_{16}H_{33})_2N$	H	H	NO_2	0

	R¹	R²	X⁻
11	$C_{18}H_{37}O$	CH_3	I⁻
12	$C_{18}H_{37}S$	CH_3	I⁻
13	$(C_{16}H_{33})_2N$	CH_3	I⁻
14	$(C_{16}H_{33})_2N$	$(CH_2)_3COO^-$	-
15	$(C_{16}H_{33})_2N$	$(CH_2)_6CH(COOC_2H_5)_2$	Br⁻

The intermediates for the nitrophenylhydrazones and stilbazium salts are aromatic aldehydes. They are synthesized
by etherification of the corresponding hydroxybenzaldehydes
(needed for compounds **1-4, 7,9** and **11**)[11] or nucleophilic
aromatic substitution of 4-fluorobenzaldehyde by octadecyl-
thiol (needed for compounds **5,12**). 3,4-Dihexadecyloxy-
cinnamic aldehyde was prepared by a Knoevenagel-Doebner reaction of the corresponding benzaldehyde and malonic acid.
The obtained cinnamic acid was then reduced with aluminum
lithium hydride to the alcohol and then oxidized with pyridinium chloro chromate(VI) to the desired cinnamic aldehyde
(needed for compound **8**)[11]. The synthesis of 4-(N,N-dihexa-
decyl)-aminobenzaldehyde (needed for compounds **6,10** and
13-15) has been previously described elsewhere[10].
    The phenylhydrazones (**1-10**) are obtained through con-
densation of the aldehydes with 4-nitro- or 2,4-dinitro-
phenylhydrazine[10,11]. The stilbazium salts are made by an
aldol reaction of the aldehyde component and the N-
alkylated pyridinium salts (**11-15**)[10].

PREPARATION AND CHARACTERIZATION OF MONOLAYERS

Monolayer Isotherms: All compounds were characterized by
isotherm measurements[9] at the gas-water interface. Except
the amino substituted phenylhydrazones (**6,10**) all compounds

formed defined monolayers.
The monomolecular films of the phenylhydrazones (1-5 and 7-9) exhibit only a solid analog phase in the temperature range from 20 to 30°C. At 40°C the dinitrophenyl-hydrazone derivatives (7-9) do not form stable monolayers anymore. The collapse area is varying with the number of alkyl chains of the molecule (table 2). In case of the three chain compounds (4,9), the monolayer aggregation at the collapse point is controlled by a dense packing of the hydrocarbon chains which give rise to an area of 0.60 nm²/molecule (figure 1). For the other amphiphiles the collapse area is higher than the most dense chain packing[11]. This means that the monolayer packing is controlled mainly by the head group.

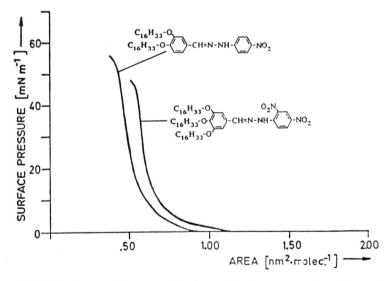

FIGURE 1. Isotherms of phenylhydrazones 2 & 9 at 20°C

All the stilbazium salts form monomolecular films. In contrast to the phenylhydrazones, the compounds with two chains (13-15) exhibit a region of coexistence and a solid analog phase (figure 2). Depending on the size and the gegenion of the hydrophilic head group the extent of the region of coexistence varies. Increasing the size of the head groups leads to larger collapse areas. While the area per molecule of the N-methylated compound (13) corresponds to a dense chain packing, the collapse area of compounds 14 and 15 is bigger (up to 0.66 nm²/molecule for 15).

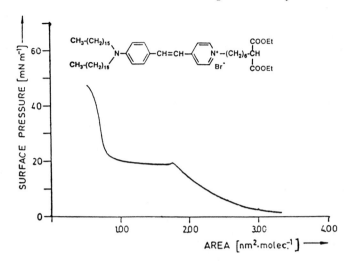

FIGURE 2. Isotherm of stilbazium salt **15** at 20°C

Langmuir-Blodgett Monolayers: Using a Lauda Langmuir
trough in a clean room, all substances were deposited during the upstroke on both sides of glass microscope slides
at 20°C. The deposition conditions were adapted to the
stiffness of the monomolecular films (table 2). Spectra of
the monolayers were recorded on a Perkin-Elmer spectrophotometer. The absorption maxima are given in table 2.
    The spectra of all phenylhydrazone derivatives (**1-5**
and **7-9**) exhibit a λ_{max} at about 420 - 430 nm. In contrast
to this the value of λ_{max} of the stilbazium salts (**11-15**)
depends strongly on their molecular structure (360 nm for
11 to 475 nm for **13**). Hence the dihexadecylamino substituted stilbazium salts (**13-15**) have a strong resonance enhanced β as their absorption maximum at 475 nm is close to the
harmonic of the Laser.
    The refractive indices n of the LB-monolayers of the
phenylhydrazones are assumed to be of the same value as the
one of **1**. The layer thickness l and refractive index of **1**
was measured by Lupo et al.[10]. Because the absorption maxima of the compounds are virtually the same, constant values for n were used. The layer thicknesses were also taken
from that work, because the length of the chromophores and
their tilt angle (measured by UV/VIS spectroscopy[12] and
SHG) is almost constant and therefore the length of the
molecules depends only on the number of carbon atoms at the
aliphatic chain.

TABLE 2. MONOLAYER & LB-FILM DATA, TRANSFER CONDITIONS

compound	I (Å)	A (Å^2)	P (mN / m)	v (cm / min)	λ_{max} (nm)
1	30	26.5	25	1.0	420
2	27.5	50	20	0.5	--
3	30	23	25	1.0	420
4	27.5	60	25	1.0	430
5	30	26	25	1.0	405
7	27.5	49	20	1.0	420
8	30	44	20	1.0	417
9	27.5	60	20	1.0	407
11	27	35	25	1.0	360
12	27	35	35	1.0	380
13	26	53	35	1.0	475
14	27	53	30	1.0	475
15	26	66	30	1.0	475

NONLINEAR OPTICAL PROPERTIES

The experimental setup for the nonlinear measurements is the same as described in the work of Lupo et al.[13]. The experiments were carried out using a Q-switched Nd^{3+}:YAG laser (λ=1064nm). The average pulse length is 35 ps and the energy of one pulse ranges from 0.3 - 0.5 mJ. The macroscopic second-order susceptibility $\chi^{(2)}$ of the LB-monolayers was determined relative to the d_{11} coefficient of quartz. The theory used to calculate $\chi^{(2)}$ is described by Sipe[14], Mizrahi et al.[15] and Neher[13,16]. The refractive indices used in the calculations are estimated 1.5 at 1064 nm and 1.7 at 532 nm for the phenylhydrazones and 1.5 at 1064 nm and 1.8 at 532 nm for the stilbazium salts.

RESULTS OF MOLECULAR VARIATION ON SHG

The results of the SHG-measurements of the phenylhydrazones are shown in table 4.

TABLE 4. $\chi^{(2)}$ AND β FOR DYE MONOLAYERS

compound	$\chi_{zzz}^{(2)}$ ($\cdot 10^{-7}$ esu)	β ($\cdot 10^{-30}$ esu)
1	3.0	57
3	4.3	66
4	0.6	44
5	4.3	81
7	1.4	60
8	2.5	62
9	1.5	64

compound	$\chi_{zzz}^{(2)}$ ($\cdot 10^{-7}$ esu)	β ($\cdot 10^{-30}$ esu)
11	3.6	42
12	7.6	65
13	14.6	304
14	12.9	197
15	6.0	180

The influence of additional alkyloxy or hydroxy substituents at the donor moiety depends on their position:
Alkyloxy substituents in meta position (**2,4**) decrease the
$\chi^{(2)}$- and β-value of the system compared to compound **1**. A
hydroxy substituent in ortho position (**3**) however increases
$\chi^{(2)}$ and β. This is probably due to the positive mesomeric
effect of the hydroxy group. In contrast to this the alkyloxy substituents in meta position only have a negative inductive effect which is responsible for the lower SHG.
     Remarkably, the thioether (**5**) has a much better donor
with respect to SHG than the alkyloxy derivative (**1**). This
is explained by the smaller electronegativity and greater
polarizability of the sulfur atom compared to oxygen atom.
Therefore the dipole moment of the ground state of **5** should
be higher than the one of **1**.
     The variation of acceptor by incorporating a second
nitro substituent (**9**) in ortho position increases β compared to the corresponding mono nitro derivative (**4**). But,
compared to compound **1** the $\chi^{(2)}$-value is lower due to the
decreased number of chromophores per area. The greater number of alkyl chains dilute the concentration of active dyes
but the chains are needed to get monolayer formation.
     The variation of the π-system (**7** compared to **8**) seems
to have no effect on β which is surprising because it is
known that the increase of conjugation leads to higher SHG
intensity[17,18].

     The influence of structural variations of the stil-
bazium salts on SHG (table 4) corresponds to the obtained

results of the phenylhydrazones. The thioether (**12**) is found to have a better donor than the alkyloxy derivative (**11**), in agreement with the explanation above. The effect of the acceptor modification (**14,15**) is unclear yet. $\chi^{(2)}$ and β are strongly decreased compared to the methyl substituted compound (**13**). This is surprising because a bigger alkyl substituent should have a smaller electronic effect. As the tilt angles of the chromophores for all stilbazium salts is virtually identical this NLO behaviour cannot be explained satisfactorily at the present.

CONCLUSION

The influence of structural variations of phenylhydrazones and stilbazium salts on monolayer behaviour and second-harmonic generation in LB-monolayers is discussed. For the phenylhydrazones, the variation of the substituents does not influence the absorption maximum of the dyes in LB-monolayers, so that the $\chi^{(2)}$-values contain almost no resonance enhancement of the nonlinearity. This allows a direct comparison of the substitution pattern on SHG.

The thioether substituted nitrophenylhydrazone exhibits the highest $\chi^{(2)}$-value measured for phenylhydrazones in LB-films and is therefore an important improvement compared to the alkyloxy substituted derivative.

A similar result is found for the stilbazium salts. The absorption maximum of 380 nm makes the thioether compound interesting, because this molecule does not absorb at the second harmonic of the Nd^{3+}:YAG laser and still has high $\chi^{(2)}$- and β-values. The highest $\chi^{(2)}$-values are still found for amino substituted hemicyanines.

References

1.   Williams,D.J. Angew.Chem., 96, 637, (1984)
2.   Chemla,D.S. Zyss,J.eds., Nonlinear Optical Properties
      of Organic Molecules and Crystals (Academic,
      New York, 1987)
3.   Langmuir,I. Trans.Faraday.Soc., 15, 62, (1920)
4.   Blodgett,K.B. J.Am.Chem.Soc., 56, 495, (1934)
5.   Kuhn,H. Möbius,D. Angew.Chem. 83, 672 (1971)
6.   Popovitz-Biro,R. et al., J.Am.Chem.Soc., 110, 2672
      (1988)
7.   Popovitz-Biro,R. et al., J.Am.Chem.Soc., 112, 2498
      (1990)

8.   Girling,I.R.et al., Opt.Commun.,55, 289 (1985)
9.   Gaines,G.L. Insoluble Monolayers at the Gas-Water
        Interface (Interscience, New York,1966)
10.   Lupo,D.et al., J.Opt.Soc.Am.B, 5, 300 (1988)
11.a)   Piancatelli,G. et al. Synthesis, 1982 245
   b)   Paulus,W. 1988 Diploma thesis, University of Mainz,
        F.R.G.
12.   Möbius,M. et al. Thin Solid Films, 132, 41 (1985)
13.   Lupo, D. et al. submitted to Adv.Mat.
14.   Sipe,J.E. J.Opt.Soc.Am.B, 4, 481 (1987)
15.   Mizrahi,V. Sipe,J.E. J.Opt.Soc.Am.B, 5, 660 (1988)
16.   Neher,D. 1990 Ph.D. thesis, MPI-Polymere, Mainz,
        F.R.G.
17.   Dulcic,A. et al. J.Chem.Phys., 74, 1559 (1981)
18.   Berkovic,G. et al. Mol.Cryst.Liq.Cryst., 150b, 607
        (1987)

Stabilization of Langmuir–Blodgett-multilayers of Non-linear Optic-active Dyes by Means of Complexation with Polymeric Counterions

W. Hickel, D. Lupo, P. Ottenbreit, W. Prass, U. Scheunemann, and J. Schneider

HOECHST AG, D-6230 FRANKFURT AM MAIN 80, FEDERAL REPUBLIC OF GERMANY

H. Ringsdorf

UNIVERSITÄT MAINZ, INSTITUT FÜR ORGANISCHE CHEMIE, J.J. BECHERWEG 18–22, D-6500 MAINZ, FEDERAL REPUBLIC OF GERMANY

1 INTRODUCTION

Langmuir-Blodgett (LB)-multilayers have gained much interest for $\chi^{(2)}$- processes in nonlinear optics because of their ability to build up highly ordered noncentro-symmetric films also from molecules with large dipole moments. Thus LB-films prepared from amphiphilic NLO-dyes exhibit very high $\chi^{(2)}$- values. So far mostly monomeric amphiphiles have been used. In the few cases where pre-formed polymers were used for film formation, the degree of order in the polymeric film was lowered.

Another way to stabilize LB-films prepared from monomeric amphiphiles is the stabilization of charged monolayers by using polymeric counterions. This has been shown already for different amphiphiles[1-3]. We now used this stabilization effect of polymeric counterions with a cationic hemicyanine amphiphile (1) which is well known to have exceptionally high $\chi^{(2)}$-values in monolayers[4]. On

(1)

the other hand the deposition of multilayers from this compound failed so far: only one double layer can successfully be transferred when pure water is the sub-phase. This is a severe problem, because for an application in nonlinear optics multilayers consisting of many single monolayers have to be used. In this paper we show that indeed the use of polymeric counterions can lead to stable multilayers without lowering the nonlinear optical properties of this compound.

2 EXPERIMENTAL DETAILS

Monolayer Isotherms

The water used for all filmbalance measurements was purified with a Millipore® water purification system (Milli-Q, 4-bowl). The salts were added to this water and, except for pure water and for Na_2SO_4, the pH was adjusted by adding a freshly prepared solution of NaOH (Riedel de Haen, p.A. grade). Na_2SO_4, Oxalic acid, and Citric acid were purchased from Riedel de Haen (p.A. grade) and used without further purification. Poly-acrylic acid and Poly-vinylsulfonic acid (Na-salt) were from Aldrich. The aqueous solutions of these polyelectrolytes were freeze-dried before use. For spreading a mixture of dichloromethane and methanol (Merck, p.A. grade) 9:1 (v/v) was used. The concentration of the amphiphile varied between 0.4 and 1.2 mg/ml.

The isotherms were measured with a commercial computer controlled filmbalance (Filmwaage 2, MGW Lauda) with sweeptimes of 10 to 90 minutes per curve. The cleanliness of the filmbalance and the purity of the water was routinely checked by measuring isotherms of a well known compound[5] (palmitic acid, reference for GC). The purity of the spreading solvent was also checked by this method.

For each isotherm the aqueous subphase was changed and the measurement was not started before the temperature of the subphase had reached equilibrium, i.e. the temperature drift had to be less than 0.1 K/min. For spreading a gas-tight syringe with a PTFE-plunger and an adjustable stopface was used (Kloehn Co., Whittier, CA, USA). The amount of solution spread (75 - 180 µl) was precisely the same for one set of measurements.

LB-film deposition

Mono- and multilayers of the hemicyanine were transferred onto clean glass slides (76x26 mm) by the vertical transfer method using the same filmbalance as above in combination with the corresponding filmlift. The glass slides were cleaned first by treatment in a freshly prepared mixture of concentrated sulfuric acid and 30% H_2O_2 1:1 (v/v) for one hour. After intense rinsing with Millipore® purified water the glass slides were ultrasonicated in an Extran® AP 11 bath (pH 9) for half an hour followed again by intense rinsing with Millipore® purified water. The freshly prepared glass slides can be stored in clean water up to one day before use. For the deposition of multilayers the glass slides were hydrophobized by treatment with hexamethyldisilazane (HMDS) vapour (10 minutes at a temperature of 70 °C). The subphase temperature for all measurements was 20 °C and the surface pressure was kept constant during deposition at values between 20 and 30 mN/m. Deposition speeds in all cases were 10 mm/min for the up- and 20 mm/min for the downstroke.

Determination of Film Properties

Thermal stability of the LB-films was measured using a setup for thermodesorption measurements as described by the group of Möhwald[6]. Eight monolayers are deposited onto an oxidized and hydrophobized Si-wafer and heated up in air at a rate of 20 °C/min while the film thickness is monitored. In addition multilayers on Si were heated up on a microscope stage at the same rate and the film was observed during the heating run.

Absorption spectra of mono- and multilayers on glass were recorded on a Perkin Elmer Lambda 9 spectrophotometer.

The apparatus for the SHG studies has already been described[7]. For the evaluation of $\chi^{(2)}$ from SHG measurements with monolayers on glass slides a model based on the formalism of Slipe[8] and described by Neher[9] has been used.

3 RESULTS AND DISCUSSION

Monolayer Isotherms

The effect of various concentrations of poly-acrylic acid in the aqueous subphase on monolayer isotherm of the hemicyanine amphiphile is shown in figure 1.

As illustrated there, at pH 6.0 subphase concentrations of poly-acrylic acid of 1 mg/l and more lead to monofilms which have a larger area requirement in the fluid phase. This is a hint for stronger headgroup interactions when polyions are added. The area requirement in the solid analogue region of the isotherms does not change up to poly-acrylic acid concentrations up to 1 g/l. Thus it can be concluded, that the order in the solid analogue film is not largely effected by the polyion-headgroup interaction. Another effect of the ionic interaction is that, even for very small polyion concentrations (0.1 mg/l) the film stability is increased as

indicated by an increased collapse pressure. These
findings are comparable to those when copolymers of
amphiphilic monomers with hydrophilic comonomers are investigated and compared with the amphiphilic monomers[10].
Thus here also the polyion complexed monofilm should have
a better deposition behaviour.

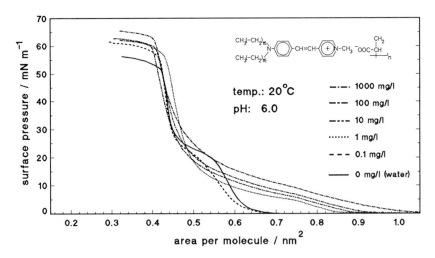

Figure 1   Effect of poly-acrylic acid in the subphase on
the surface-pressure area isotherms.

LB-Multilayers

    In deposition experiments the stabilization observed
in monofilms in fact leads to a dramatic increase of the
number of monolayers that can be transferred. As depicted
in table 1 for polymeric anions the stabilization effect
can be observed over a wide concentration and pH range
when polyions are used. For di- and trivalent anions no
change in deposition behaviour is observed.

    The thermal stability of LB-films with polyacrylic
acid as counterion was investigated as one example. Two
typical results for two different concentrations of poly-
acrylic acid are shown in figure 2. Whereas the film
thickness remains constant up to temperatures above 150
°C when the poly-acrylic acid concentration is 1 mg/l for
higher concentration of the polyion the film stability
decreases again. It could be shown that the decrease in
film thickness starting already below 100 °C is due to
the evaporation of water, which is transferred together

with the polymeric counterions from the subphase at these higher concentrations. However, for a properly chosen concentration LB-multilayers can be prepared which have similar stabilities as multilayers prepared from amphiphilic polymers.

Table 1 Effect of counterions on the deposition behaviour. Substrate material: hydrophobized glass. Amphiphilic dye: Hemicyanine **(1)**

Anion added to the subphase	Concentration of anion in the subphase	pH of subphase	number of monolayers
-	-	4	2
poly-acrylic acid	0.05	6	6
.	0.1	6	10
.	1.0	6	>20
.	1.0	7	>20
.	5.0	2-11	>20
.	10	6	>60
.	10	7	>20
.	100	6	>20
.	1000	6	2
poly-vinylsulfonic acid	10	6	>10
.	5	6	>10
.	2.5	6	>10
.	1	6	>10
sulfuric acid	5	6	2
oxalic acid	5	6	2
citric acid	5	6	2

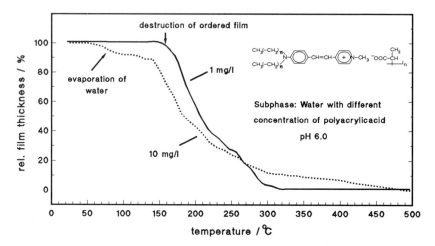

<u>Figure 2</u> Effect of polyion concentration on the thermal stability of LB-multilayers. The films (8 monolayers are heated up in air at a rate of 20 °C/min.

<u>Table 2</u> Effect of Poly-Acrylic Acid on the $\chi^{(2)}$-values (from SHG) of LB-monolayers from the hemicyanine **(1)** on glass

Subphase	$\chi^{(2)}_{zzz}$ / pm V^{-1}	$\chi^{(2)}_{zxx}$ / pm V^{-1}
Millipore $^{\textcircled{R}}$ Water	253 +/- 5	50.5 +/- 0.5
1 mg/l poly-acrylic acid, pH 6.0	270 +/- 12	36.0 +/- 0.5
5 mg/l poly-acrylic acid, pH 6.0	205 +/- 10	26.2 +/- 1.2

So far we have shown that we can obtain stable multilayers by using polyelectrolyte-monofilm complexes. For the nonlinear optical properties also the orientation

of the chromophores in the transferred layers is critical. From the isotherm measurements we concluded that the
order in the solid analogue state of the monofilms in
which the transfer is carried out is not affected by the
polymeric counterions. That the orientation of the chromophores after transfer is affected only slightly by the
polyion complexation is shown in table 2, where the $\chi^{(2)}$-
values obtained from SHG measurements with monolayers
with and without polyacrylic acid are compared. From
these values it can be concluded that the tilt angle of
the chromophore is changed only slightly to a more up-
right position after complexation with polyacrylic acid.

All these results show that this simple, but very
effective method can lead to improved LB-films with a
good stability without lowering the large hyperpolarizabilities.

Acknowledgement

We wish to thank Petra Jeckeln and Gerhard Geiß for
film characterization and Dieter Neher (MPI für Polymerforschung, Mainz, FRG) for helpful discussions. This work
was supported by a materials research program by the
Bundesminister für Forschung und Technologie under Grant
No. 03M4008.

REFERENCES

1. M. Shimomura, T. Kunitake, Thin Solid Films, 1985,
    132, 243
2. N. Higashi, T. Kunitake, Chem.Lett., 1986, 105
3. C. Erdelen, A. Laschewsky, H. Ringsdorf, J.
    Schneider and A. Schuster, Thin Solid Films, 1989,
    180, 153
4. D. Lupo, W. Prass, U. Scheunemann, A. Laschewsky, H.
    Ringsdorf, and I. Ledoux, J.Opt.Soc.Am.B, 1988, 5,
    300
5. O. Albrecht, Thin Solid Films, 1983, 99, 227
6. L.A. Laxhuber, H. Möhwald, Langmuir, 1987, 3, 837
7. C. Bubeck, A. Laschewsky, D. Lupo, D. Neher, P.
    Ottenbreit, W. Paulus, W. Prass, H. Ringsdorf, G.
    Wegner, Adv.Mat., 1990, in press
8. J.E. Slipe, J.Opt.Soc.Am.B, 1987, 4, 481
9. D. Neher, Doctoral Thesis, Univ. Mainz, 1990
10. A. Laschewsky, H. Ringsdorf, G. Schmidt, J. Schnei-
    der, J.Am.Chem.Soc., 1987, 109, 788

Oriented Crystallization and Non-linear Optics: Tools for Studying Structural Changes of Amphiphilic Aggregates at the Air-solution Interface

I. Weissbuch, G. Berkovic, L. Leiserowitz, and M. Lahav

STRUCTURAL CHEMISTRY DEPARTMENT, THE WEIZMANN INSTITUTE OF SCIENCE, REHOVOT, ISRAEL

The role played by structured aggregates as intermediates in crystal nucleation is of significant current interest. In a stereochemical approach towards understanding crystal nucleation on a molecular level, we have recently provided evidence that water soluble hydrophobic α-amino acids induce oriented nucleation of α-glycine crystals at air-solution interfaces. These amino acids accumulate at the surface of the solution forming ordered aggregates stabilized by a two-dimensional net of hydrogen bonds similar to their own crystal structure [1]. Such aggregates can induce fast oriented crystallization of suitable co-solute molecules (e.g. glycine) when there is a match between the 2-D molecular arrangement and the structure of the face from which the crystal nucleates.

For other types of amphiphiles, it may not be possible to extrapolate a unique 2-D surface structure on the basis of their 3-D crystal structure. In such cases the 2-D structure of these surface aggregates may strongly depend on the nature of the interaction between them and other solute molecules. We demonstrate here profound changes in the structure of monolayers of insoluble 4-substituted benzoic acid amphiphiles upon addition of 4-hydroxybenzoic acid (HBA) to the aqueous subphase. These structural changes have been monitored by two independent processes : the induction of oriented HBA crystals attached to the air-solution interface, and changes in the optical second harmonic generation (SHG) of the monolayer.

Our interest in these systems stems from earlier observations that addition of small amounts of 4-methoxybenzoic acid to crystallizing solutions of HBA induces fast nucleation of crystal floating at the solution surface [2]. These crystals exhibit a well developed (401) face through which the crystal attaches to the surface. This morphology is very different from that of crystals which grow as {100} plates elongated in the c direction (usually at the bottom of the crystallizing dish) in the absence of the additive.

RESULTS AND DISCUSSION

In order to elucidate the mechanism by which this nucleation occurs, we designed long chain insoluble amphiphilic molecules **1** - **6** with their hydrophilic group the X-Φ-COOH moiety (X=O,N) to be used in the form of Langmuir films.

No.	Name	Structure
1	4-(hexadecyloxy)benzoic acid	HOOC-⟨O⟩-O-$(CH_2)_{15}$-CH_3
2	4-(hexadecylamino)benzoic acid	HOOC-⟨O⟩-NH-$(CH_2)_{15}$-CH_3
3	4-(hexadecanoyloxy)benzoic acid	HOOC-⟨O⟩-OCO-$(CH_2)_{14}$-CH_3
4	4-(hexadecyloxy)phenylacetic acid	HOOC-CH_2-⟨O⟩-OCO-$(CH_2)_{14}$-CH_3
5	[4-(hexadecyloxy)phenyl] propionic acid	HOOC-$(CH_2)_2$-⟨O⟩-OCO-$(CH_2)_{14}$-CH_3
6	octadecyl-4-hydroxybenzoate	HO-⟨O⟩-COO-$(CH_2)_{17}$-CH_3

Analogous to the case of the soluble additive, 4-methoxybenzoic acid, (401) oriented nucleation of HBA crystals was also induced at the air-solution interface when amphiphiles **1**, **2** or **3** were spread on supersaturated HBA solutions.

However, amphiphilic molecules **4**, **5** and **6**, of a slightly different structure, did not induce HBA crystallization at the interface.

The different behaviour of the two groups of amphiphiles is also reflected by their π-A isotherms. All the amphiphiles, when compressed over pure water, form monolayers consisting of molecules oriented almost perpendicular to the surface as determined from the limiting area/molecule (Figure 1), and from the SHG experiments (see below).

When amphiphiles **1**, **2**, or **3** are compressed over supersaturated solutions of HBA, more expanded π-A isotherms with a larger limiting area/molecule are obtained. In contrast, the limiting area/molecule for **4**, **5**, and **6** are unchanged.

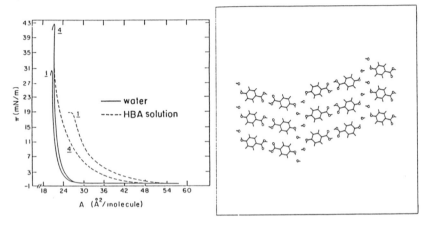

Figure 1 π-A isotherms of amphiphiles spread on water, and on 0.08M HBA solution. Isotherms for compounds **2**, **3** are similar to **1** and that of **5** and **6** are similar to **4**.

Figure 2 Packing arrangement of 4-hydroxybenzoic acid monohydrate crystal viewed perpendicular to the (401) face.

These results suggested the following mechanism: In the crystal structure [3] of the 4-hydroxybenzoic acid monohydrate (space group $P2_1/a$), the HBA molecules lie parallel to the (401) plane, forming hydrogen-bonded dimers (Figure 2). These dimers are interlinked within the (401) plane by hydrogen bonds involving the phenolic -OH groups and water molecules. The induced crystallization of HBA at the air-solution interface suggests that amphiphiles **1**, **2**, **3** form aggregates whose structure mimics that of the ($\overline{4}$01) crystalline face. This model implies that the 4-oxybenzoic acid moieties lie flat on the surface such as to form the hydrogen-bonded dimers, while their aliphatic chains emerge from the solution (Figures 3 and 4). A photograph of actual

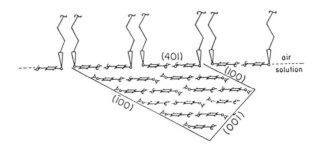

Figure 3 Packing arrangement HBA crystal with the (401) face viewed 'edge-on', delineated by the crystal faces. The layer of molecules at the interface is the Langmuir film (represented schematically).

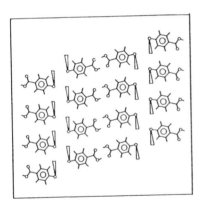 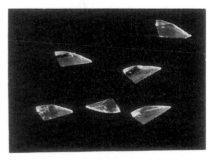

Figure 4 (left) Proposed amphiphile structure viewed perpendicular to the HBA subphase. The O(phenol)...O(phenol) distance between adjacent molecules is 4.6 Å, allowing good van der Waals contacts between chains.

Figure 5 (right) Photograph of HBA crystals.

HBA crystals grown attached to these amphiphiles is shown in Figure 5.

In order to independently verify that the specified amphiphiles change their orientation when HBA is added to the aqueous subphase we have studied these systems with second harmonic generation (SHG). This is a surface specific technique [4] which can determine the presence and orientation of surface adsorbates on centrosymmetric and isotropic substrates. Our experimental system for SHG is depicted in Figure 6.

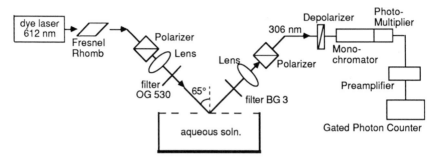

Figure 6 SHG experimental set-up. The dye laser, pumped at 10Hz by a Nd-YAG laser, produces 8ns pulses of energy 5mJ. SHG signals are typically averaged over 2000 pulses.

The second order polarization, $P^{(2)}$, induced in a medium by an optical electric field, $E(\omega)$, is described by the nonlinear susceptibility tensor, $\chi^{(2)}$:

$$P_i^{(2)} = \chi_{ijk}^{(2)} E_j(\omega) E_k(\omega) \tag{1}$$

It is readily seen from this equation that in a centrosymmetric medium $\chi_{ijk}^{(2)} = 0$, and thus dipole allowed SHG can only arise from the surface where the symmetry is broken.

The orientation of the major nonlinear optical axis (here the long axis of the phenyl group) of the amphiphile relative to the surface normal (θ) can be determined from the surface $\chi^{(2)}$ tensor. For a rod-like molecule whose azimuthal distribution in the surface plane is isotropic, there are only 2 independent $\chi^{(2)}$ components [4]:

$$\chi_{zzz}^{(2)} = N_s \ \beta \ <\cos^3\theta> \tag{2}$$

$$\chi_{yzy}^{(2)} = \chi_{zyy}^{(2)} = \chi_{xzx}^{(2)} = \chi_{zxx}^{(2)} = 1/2 \ N_s \ \beta \ <\sin^2\theta\cos\theta>$$

Here N_s is the number of monolayer molecules per unit surface area, and β is the second order nonlinearity per molecule.

The s- and p- polarized SHG signals generated by a 45° polarized input laser (I_{45-s} and I_{45-p} respectively) are related to the $\chi^{(2)}$ coefficients :

$$[I_{45-s}]^{1/2} \sim a_1 \ \chi_{yzy}^{(2)}$$

$$[I_{45-p}]^{1/2} \sim a_2 \ \chi_{yzy}^{(2)} + a_3 \ \chi_{zzz}^{(2)} \tag{3}$$

The a coefficients are evaluated [5] assuming a value of 1.5 for the refractive indices of the organic monolayer. Assuming a narrow distrubution of θ values within the monolayer, $\langle\theta\rangle$ may be evaluated from the ratio of $I_{45\text{-}s}$ to $I_{45\text{-}p}$. Magnitudes of β were evaluated from the ratio of amphiphile SHG to that of the bare water surface, for which $\chi_{yzy} = 2\text{x}10^{-17}$ e.s.u. [6].

Amphiphiles **1**, **2** and **6** spread on water all exhibited SHG signals considerably stronger than that of the bare water subphase, and $\langle\theta\rangle$ values in the range $24^{\circ} - 47^{\circ}$ (Table 1). These $\langle\theta\rangle$ values refer to the orientation of the polar XC_6H_4COO head group, but not necessarily to the orientation of the hydrocarbon chain.

Table 1 SHG signals and orientations for monolayers on water and saturated HBA solution.

Monolayer	ON WATER				ON HBA		
	$I_{45\text{-}s}$	$I_{45\text{-}p}$	$\theta(^{\circ})$	β(esu)	$I_{45\text{-}s}$	$I_{45\text{-}p}$	$\theta(^{\circ})$
Bare Subphase	1	0.6			0.4	0.8	
1	10	40	24	$2\text{x}10^{-30}$	same as subphase		90
1 80% cov.	7	30	24	$2\text{x}10^{-30}$	7	30	24
2	450	150	42	$7\text{x}10^{-30}$	same as subphase		90
6	150	40	47	$3\text{x}10^{-30}$	150	40	47

The bare HBA subphase had an SHG intensity similar to that of water. Whereas the SHG signal and $\langle\theta\rangle$ of **6** on HBA solution were the same as on water, the SHG of full monolayers of **1** and **2** on HBA were very different. The total SHG signal in these two cases was about the same as that of the bare HBA solution. This is entirely consistent with our model since the lack of SHG from the compounds **1** and **2** means, according to equation (2), that their polar moiety XC_6H_4COOH lies flat ($\theta = 90^{\circ}$) on the HBA solution surface.

The absence of SHG signals from these monolayers on HBA could in principle arise from another reason - the formation of a bilayer (at any angle to the surface) between one HBA molecule from the solution and an amphiphile molecule, leading to a cancellation of their nonlinearities. This might be possible (at least in principle) for the monolayer of <u>1</u> but can be discounted for <u>2</u> due to the large difference in β values of the 4-hydroxybenzoic and 4-aminobenzoic chromophores. Thus the SHG results can only be explained by monolayers of <u>1</u> and <u>2</u> lying flat ($\theta = 90°$) on the water surface.

At 80% of the full monolayer coverage, the $\langle\theta\rangle$ value of 1 spread on HBA is the same as on water, implying that its interaction with solute molecules at this coverage density is insufficient to change the orientation.

The fact that the HBA subphase can drastically alter the surface alignment of monolayers of <u>1</u> and <u>2</u> but not the closely related monolayer of <u>6</u> demonstrates how specific the solute-amphiphile interactions can be. We have further demonstrated this by checking the SHG of <u>1</u> spread on an aqueous solution of 4-hydoxy-phenylacetic acid (i.e. the addition of one methyl group to the HBA molecule). Over this solution, the monolayer of <u>1</u> shows exactly the same SHG signal strength and $\langle\theta\rangle$ value (24°) as on pure water, indicating the loss of the strong solute/amphiphile interaction with the small change in the solute molecule, in perfect agreement with the π-A isotherm and crystallization experiments.

In conclusion, our combination of several independent techniques, namely oriented crystallization, surface pressure-area isotherms and second harmonic generation, presents a very powerful approach for studying the structures of surface aggregates, and their interactions with solute molecules in the subphase.

REFERENCES

[1] I. Weissbuch, L. Addadi, L. Leiserowitz and M. Lahav, J. <u>Amer</u>. <u>Chem</u>. <u>Soc</u>., (1988), <u>110</u>, 561 ; I. Weissbuch, F. Frolow, L. Addadi, M. Lahav and L. Leiserowitz, J. <u>Amer</u>. <u>Chem</u>. <u>Soc</u>., (1990), in press.
[2] I. Weissbuch, G. Berkovic, L. Leiserowitz and M. Lahav, J. <u>Amer</u>. <u>Chem</u>. <u>Soc</u>., (1990), <u>112</u>, 5874.
[3] M. Colapietro, A. Domenicano and C. Marciante, <u>Acta Crystallogr</u>., (1979), B<u>35</u>, 2177.
[4] Y.R. Shen, <u>Nature</u>, (1989), <u>337</u>, 519.
[5] V. Mizrahi and J.C. Sipe, <u>J</u>. <u>Opt</u>. <u>Soc</u>. <u>Am</u>. <u>B</u>, (1988), <u>5</u>, 660.
[6] Th. Rasing, G. Berkovic, Y.R. Shen, S.G. Grubb and M.W. Kim, <u>Chem</u>. <u>Phys</u>. <u>Lett</u>., (1986), <u>130</u>, 1.

Mixed-dye Langmuir–Blodgett Films as Non-linear Optical Materials

M.E. Lippitsch and S. Draxler

INSTITUT FÜR EXPERIMENTALPHYSIK, KARL-FRANZENS-UNIVERSITÄT GRAZ, UNIVERSITÄTSPLATZ 5, A-8010 GRAZ, AUSTRIA

E. Koller

LAMBDA PROBES AND DIAGNOSTICS, GROTTENHOF-STR. 3, A-8053 GRAZ, AUSTRIA

1 INTRODUCTION

It has been recognized for several years that Langmuir-Blodgett films can have excellent second-order nonlinear optical properties. Their high degree of molecular ordering together with the possibility to produce multilayer assemblies with well-controlled structures makes them promisable for applications in various nonlinear optical devices.

For producing Langmuir-Blodgett films amphiphilic molecules consisting of hydrophilic 'heads' and hydrophobic 'tails' are used. In most work on nonlinear optical properties of these films the main contribution to the nonlinear polarizability was from the head, the tail making no significant direct contribution. On the other hand, the tails and the space between them account for more than eighty percents of the film volume. To enhance the nonlinear polarizability it would be advantageous to have a contribution also from this volume. This could be achieved by incorporating lipophilic molecules with good nonlinear properties into the film.

Within the films the molecules are closely packed and have a well-defined orientation with respect to each other. In this case it has to be expected that the electronic properties of the molecules are affected by intermolecular interactions. Girling et al.[1], Schildkraut et al.[2], Marowsky et al.[3], and Bauer et al.[4] reported significant changes of nonlinear optical properties by diluting nonlinear dyes with fatty acids. In all cases this effect has contributed to aggregation of the dye molecules and the influence of dilution on the aggregation process. These findings indicate that the ability of organic molecules in a monomolecular layer to self-organize by forming molecular aggregation or supermolecular structures and the change in optical properties brought about by this process can be of importance for nonlinear optics.

It was the aim of this study to investigate whether mixing of different nonlinear optical materials in a single Langmuir-Blodgett film can enhance the second-order nonlinear polarizability of the molecules by better filling the space between the chains using lipophilic molecules as well as by formation of hetero-molecular aggregations and electronic coupling between the different chromo-phoric systems.

2 EXPERIMENTAL PROCEDURES

The nonlinear dyes used in this investigation were chosen after an extensive study on nonlinear optical properties and suitability for building Langmuir-Blod-gett films of organic dyes. A number of dyes were found to show interesting selforganization properties under various experimental conditions. In this paper results will be reported on 4-octadecylamino-4'-nitrostilbene (OANS), N-doco-syl-4-(4-hydroxy-styryl)-pyridinium bromide (a merocyanine), and the lipophilic dye nile red.

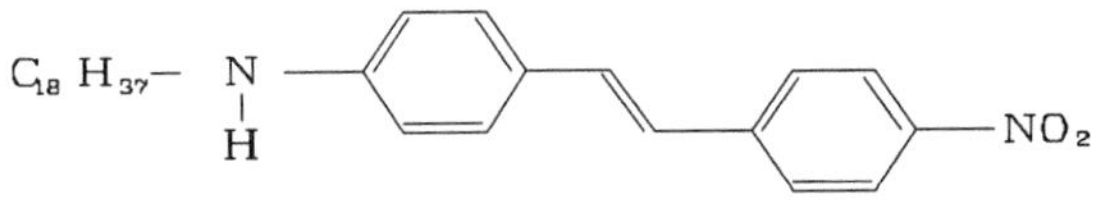

OANS

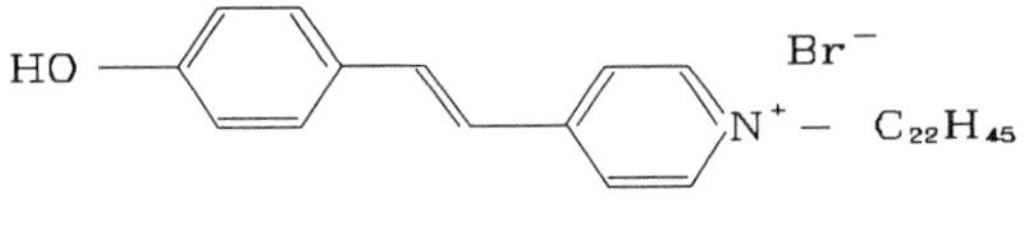

merocyanine

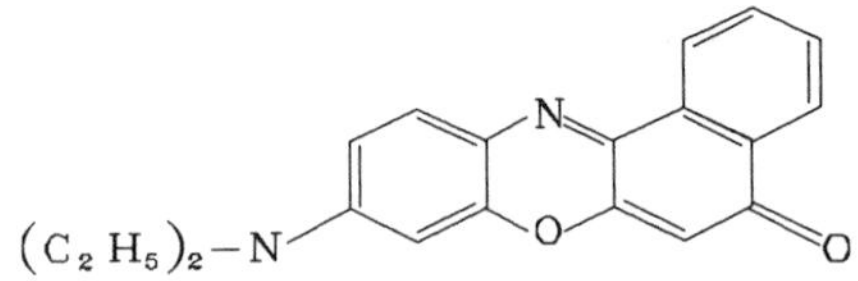

nile red

The dyes were obtained from Lambda Probes & Diagnostics, Graz, Austria (Art. nos. C-218, L-004B, and C-230, respectively). Toluene (Merck no. 8325) and chloroform (Merck no. 2445) were used as solvents. Langmuir-Blodgett films were produced on a Lauda LMF 002 Langmuir trough. The subphase was highly purified water (Barnstead NANOpure; specific resistivity 18 MΩ cm), brought to various pH values by adding nitric acid, sodium hydroxide, or borate buffer. Films were compressed to a predetermined surface pressure with a speed of 1 cm^2 s^{-1}. The films were transferred to a fused silica slide (made hydrophilic by plasma cleaning) with a speed of 0.05 mm s^{-1}.

Absorption spectra were recorded for the solutions as well as for the films on a Hewlett-Packard 8451A diode array spectrometer. Second-order polarizabilities of the molecules in solution were determined using an EFISH (electric field induced second harmonic) apparatus incorporating an actively Q-switched Nd-YAG laser, a high voltage pulse generator (Electro Optic Developments), and a two-channel boxcar integrator (Stanford Research Systems mod. SR 250). Powder measurements of the SHG (second harmonic generation) efficiency of the compounds were performed following the procedure introduced by Kurtz and Perry[5], using urea as a reference. Nonlinear polarizabilities in the Langmuir-Blodgett films were determined by measuring s- and p-polarized second harmonic radiation with s- as well as p-polarized excitation in reflection mode under various angles of incidence. Absolute values were determined by comparison with the Maker fringes from a quartz plate.

3 RESULTS AND DISCUSSION

Surface pressure / molecular area isotherms of the dyes and the mixture are given in figs. 1, 2. None of them is totally 'regular' so that interpretation is not straightforward and has to await further research. More film structure and chromophore interaction could be deduced from the absorption spectra of the dyes in solution and in Langmuir-Blodgett films (figs. 3-5). Additional information was obtained directly from the SHG measurements. In particular the tilt angle between the chromophore axes and the surface normal as well as the second-order polarizability could be determined. Values are given in tab.1,p.94.

Langmuir-Blodgett films of OANS showed interesting aggregation properties[6]. When diluted with arachidic acid (1:1), the absorption in the film was virtually the same as in solution and the second-order nonlinear polarizability β was nearly the value found in EFISH measurements. The molecules were found lying on the surface with their long axes inclined at 86° with respect to the surface normal. The pure OANS film, however, when drawn from a basic subphase, had its absorption shifted bathochromically and large domains of orientationally ordered chromophores were formed, as could be seen from considerable s-polarized SHG radiation. The direction of the chromophores was again strongly inclined to the surface normal (83°). The area occupied by a single molecule was essentially that of the hydrocarbon chain. The reason for the negligible contribution of the chromophores is not yet fully understood. Since the area consumed on the trough equalled that deposited onto the substrate bilayer formation could be excluded. The second-order nonlinear polarizability β was enhanced by a factor of 2, giving the first indication for an improvement of nonlinear optical properties due to molecular interactions.

For the merocyanine in the LB film the absorption was shifted hypsochromically compared to the solution, indicating a type of aggregation similar to that observed in the hemicyanines[1-4]. The significant s-polarized component in the SHG measurements showed that also in this case large domains with chromophores aligned in parallel were formed. The chromophoric long axes were strongly tilted with respect to the surface normal (85°). The second-order polarizability was smaller than that obtained by EFISH measurements. These findings make a molecular arrangement in the form of *H*-aggregates very probable. Different results were obtained in an uncharged merocyanine[7,8]. In that case the tilt angle was only 9°, the second-order polarizability was higher, and no indication of aggregation was found.

The third compound, nile red, is a lipophilic dye not very well suited for producing Langmuir-Blodgett films. Good films could be obtained, however, when the dye was mixed 1:1 (molar) with arachidic acid. The isotherm showed several phase transitions. In films drawn at a surface pressure of 38 mN/m, the dye molecules were obviously completely located between the chains of the arachidic acid. The SHG measurements revealed a tilt angle of 65° and no aggregation. The second-order polarizability was in reasonable agreement with results from EFISH experiments.

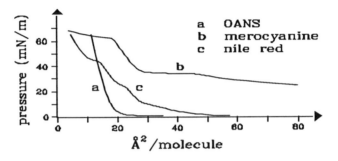

Fig. 1: Surface pressure / molecular area isotherms for the three compounds, OANS (a), merocyanine (b), and nile red mixed 1:1 with arachidic acid (c)

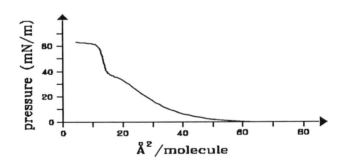

Fig. 2: Surface pressure / molecular area isotherms for the 1:1:1 mixed film

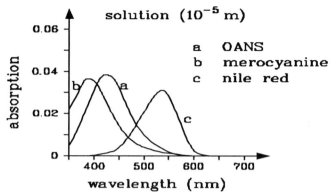

Fig. 3: Absorption spectra of the three dyes in solution

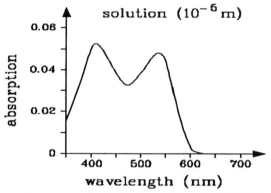

Fig. 4: Absorption spectrum of the three dyes mixed 1:1:1 in solution

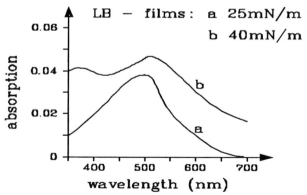

Fig. 5: Absorption spectra of mixed LB films drawn at different surface pressures

When the compounds were mixed in solution, no significant changes in the absorption spectra were observed (fig. 3). In the Langmuir-Blodgett films, however, strong molecular interactions were present as proven by dramatic changes in the absorption. The three bands merge into one and the resulting spectrum does not show any similarity with that in solution (cf. figs. 4 and 5 a). These changes occurred upon mixing of any two as well as of three compounds, indicating electronic coupling between all constituents. The pressure /area isotherm showed that the repulsive forces of the merocyanine were reduced by the presence of the other molecules. Above a surface pressure of 30 mN / m the absorption spectrum again changed significantly (fig. 5 b). The absorption bands of nile red and the merocyanine dye reappeared indicating the three dyes being uncoupled. Probably nile red is squeezed out and is now situated between the aliphatic chains rather than being between the chromophores.

The SHG measurements again yielded a significant s-polarized component, indicating large ordered domains. An analysis according to tilt angle and second-order polarizability was not reasonable now because of the contributions of three different molecular species. The SHG efficiency was two orders of magnitude higher than expected from the sum of the three compounds. The relative intensities of the polarization components of the second-harmonic radiation indicate that also in the mixed film high tilt-angles with respect to the surface normal are dominant. When an "effective second-order polarizability" is calculated from the data, the value is about one order of magnitude larger than the mean of the three components. The uncoupled dyes in the film drawn at higher surface pressure had an effective second-order polarizability much lower than the mean, indicating that the ordering of the molecules is destroyed in this case.

Table 1. Second-order polarizabilities for Langmuir-Blodgett films of the three compounds and the mixed film.

LB film composition	second-order polarizability β (10^{-30} esu)
OANS	256
merocyanine	126
nile red	190
mixed film (1:1:1)	1690*

*effective polarizability

4 CONCLUSION

Langmuir-Blodgett films are well-known to represent highly ordered supramolecular structures. It could be shown that within these films molecular self-organization may produce aggregates of strongly interacting molecules. This can also be the case if different molecules are incorporated into a single layer. The electronic interaction can be observed spectroscopically and to a certain degree can be controlled by experimental parameters like surface pressure or pH of the subphase. The interactions apparently favour specific spatial arrangements of the molecules. Moreover, they have a strong influence on the optical properties of the molecules. In aggregates of a single dye an enhancement of 2 in the second-order polarizability was observed. When incorporating three interacting dyes into a single layer, the effective second-order polarizability is one order of magnitude higher than without interaction, and the SHG efficiency is raised by a factor of 150. It is obvious that utilizing electronic interactions between nonlinear molecules in thin films could be advantageous in producing materials with even higher second-order susceptibilities.

ACKNOWLEDGEMENT

Financial support of this work by the "Forschungsförderungsfonds für die Gewerbliche Wirtschaft", grant no. 3/6421 and the "Innofinanz" is gratefully acknowledged.

REFERENCES

1   I. R. Girling, N. A. Cade, P. V. Colinsky, R. J. Jones, I. R. Peterson, M. M. Ahmad, D. B. Neal, M. C. Petty, G. G. Roberts, and W. J. Feast, *J. Opt. Soc. Am.* B 1989, 4, 950

2   J. S. Schildkraut, T. L. Penner, C. S. Willand, and A. Ulman, *Optics Lett.*, 1988, 13, 134

3   G. Marowsky and R. Steinhoff, *Optics Lett.* 1988, 13, 707

4   J. Bauer, P. Jeckeln, D. Lupo, W. Prass, and U. Scheunemann, in: '*Organic Materials for Non-Linear Optics*', R. A. Hann and D. Bloor (eds.), Royal Society of Chemistry, London 1989

5   S. K. Kurtz and T. T. Perry, *J. Appl. Phys.*, 1968, 39, 3798

6   M. E. Lippitsch, S. Draxler, and E. Koller, to be published

7   I. R. Girling, N. A. Cade, P. V. Kolinsky, and C. M. Montgomery, *Electron. Lett.*, 1985, 21, 169

8   I. R. Girling, P. V. Kolinsky, N. A. Cade, J. D. Earls, and I. R. Peterson, *Opt. Comm.*,1985, 55, 289

The Preparation of Langmuir–Blodgett Films of SDAN for Second Harmonic Generation Application

F.R. Mayers and J.O. Williams

CHEMISTRY DEPARTMENT, UMIST, P.O. BOX 88, SACKVILLE STREET, MANCHESTER M60 1QD, UK

A. Mohebati, D. West, and T.A. King

PHYSICS DEPARTMENT, UNIVERSITY OF MANCHESTER M13 9PL, UK

G.S. Bahra

RARDE, FORT HALSTEAD, SEVENOAKS, KENT, UK

1 INTRODUCTION

The recent interest in the use of organic materials for non-linear optical (NLO) applications[1-11] is sustained by their superior conversion efficiencies and damage thresholds[12] over inorganic materials[13]. The Langmuir-Blodgett (LB) technique offers obvious advantages over other methods of thin film preparation for second harmonic generation (SHG) applications in that both monomers and polymers[14] may be deposited with good control of homogeneity and thickness either as Z-type, or X-type single layers or Y-type AB layers. It does however have the disadvantage for monomeric layers that the domains formed inevitably result in scattering from the film. For frequency doubling of a Nd: YAG laser it is desirable to ensure that the process is not resonance enhanced which requires that the absorption windows around the fundamental (1064 nm) and the frequency doubled (532 nm) wavelengths are clear. This requirement is met by the para-nitroaniline derivatives which makes them ideal candidates for LB investigation.[1] 2-(N,N-dimethylamino)-5-nitroacetanilide (DAN) has demonstrated a 20 % conversion efficiency for a 2 mm thick crystal[15] as compared to 2% for a 15 mm thick potassium dideuterium phosphate (KD*P) crystal. In this paper we present results on the synthesis and characterisation of 2-(N,N-dimethylamino)-5-nitrostearylanilide (SDAN), its LB characterisation and deposition and preliminary SHG studies on the resultant films.

2 MATERIAL SYNTHESIS

SDAN was prepared by reacting 1-fluoro-5-nitroaniline with dimethylamine in ethanol. The product was then reacted with stearic acid to form the amide via the dicyclohexyl-carbodiimide (DCCI) method using dichloromethane as solvent. The reaction mixture was chromatographed on an alumina support using chloroform as eluent. The solid obtained gave satisfactory n.m.r., i.r. and microanalytical data. The resultant powdered material was crystalline as is evidenced by the X-ray diffraction (XRD) pattern recorded as compared to the starting materials (Figure 1).

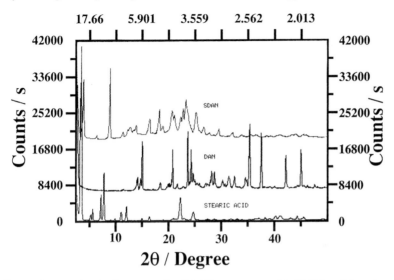

Figure 1. Powder XRD pattern for SDAN as compared to DAN and Stearic acid

The 57 XRD lines recorded could be indexed to a monoclinic unit cell with $a_0 = 34.69$ Å, $b_0 = 10.17$ Å, $c_0 = 22.62$ Å, $\alpha = 90.0^{\circ}$, $\beta = 101.3^{\circ}$ and $\gamma = 90.0^{\circ}$. Full structural determination is underway at present. Although we have not as yet determined the detailed symmetry it is apparent that the material crystallizes in a non-centrosymmetric space group as SHG activity was measured from the ungraded powder. The absorption spectrum of the material indicated transparency around both 532 and 1064 nm as is shown in Figure 2.

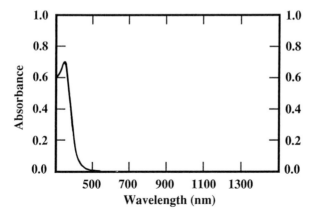

Figure 2. Absorption spectrum for SDAN in chloroform

3 LB FILM PREPARATION

The pressure/area isotherm for SDAN is shown in Figure 3 for a subphase pH of 10.8 and a temperature of 10 °C. The isotherm was recorded on one side of a Joyce Loebl AB trough using a compression rate of 2 mm/min. Isotherms were essentially invariant with pH and the area per molecule indicated in Figure 3 is surprisingly not significantly greater than that for stearic acid but this is not unknown for other systems.[16] This implies that the molecules in the film must lie almost vertical to the substrate. Langmuir films could be transferred onto hydrophobic glass such that deposition appeared to proceed only on the downstroke. The floating Langmuir film was very rigid and indeed significant displacement of the Wilhelmy plate from the vertical was usually observed. Films transferred at 25 mN/m yielded no XRD lines whereas those deposited at 35 mN/m resulted in intense XRD lines as shown in Figure 4. This would seem to agree with

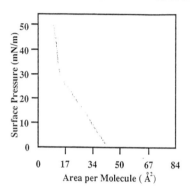

Figure 3. Pressure - Area Isotherm for SDAN

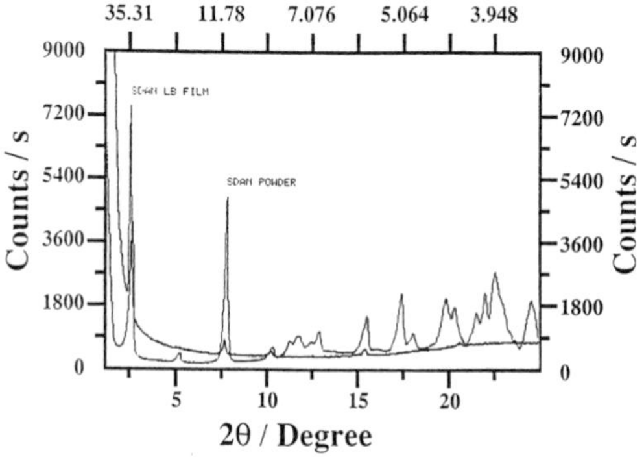

Figure 4. XRD pattern for SDAN powder as compared to the LB film

the pressure/area isotherm where the phase transition from liquid to solid appears to occur at about 30 mN/m. If the reflection corresponding to about 35 Å is second order then this would agree with the approximately vertical arrangement of the molecules indicated by the pressure area isotherm. LB films could be formed with optical transmission up to about 90 % over most of the visible spectrum but the absorption maximum is red-shifted from about 300 to 375 nm in the LB film (Figure 5).

4 SHG STUDIES

The SHG facility used for studying LB films is shown in Figure 6. Studies on the ungraded powder of SDAN held between two glass plates were performed at RARDE and the results presented in Figure 7. At the time of use the motorised rotation stage was not operational. For one downstroke and one upstroke passage of the substrate through the air/water interface at which SDAN was compressed to a surface pressure of 35 mN/m a signal strength of roughly 0.5 relative to a quartz

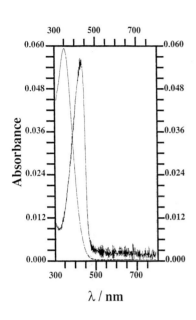

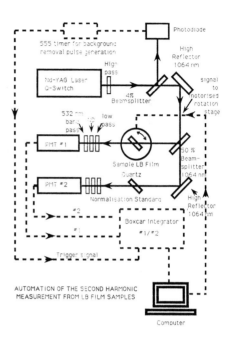

Figure 5. Absorption Spectra
for SDAN LB film and solution

Figure 6. SHG facility available

signal attenuated by an ND of 6 was obtained (Figure 8). From transfer ratio measurements it was established that transfer did not appear to occur on the upstroke which means that only one monolayer should be present on the substrate. The intensity reproducibility of the passively Q-switched Nd-YAG laser used for these measurements was monitored by regular observation of the second harmonic signal generated in transmission through a z-cut quartz plate. Reproducibility to better than ± 5 % was found. The resultant calculated relative conversion efficiency for 100 mW cm^{-2} input power at 1064 nm was 7.3 x 10^{-12} which would correspond to a susceptibility of about 3.7 x 10^{-20} CV^{-2}. This compares favourably with the value of 2.5 x 10^{-20} CV^{-2} obtained by Ledoux et al.[17] Work is in progress to secure active Q-switching of the Nd-YAG laser before a range of film thicknesses are studied.

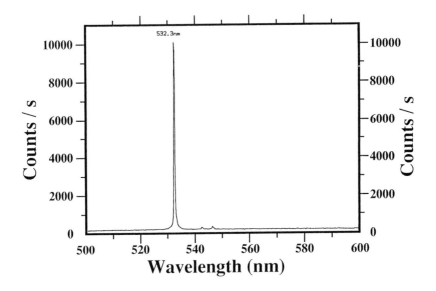

Figure 7. Frequency doubling from SDAN powder

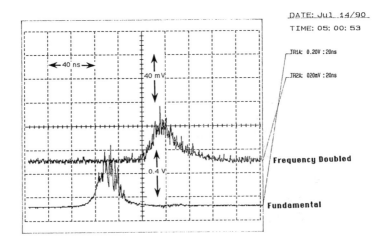

Figure 8. Frequency doubling from a monolayer of SDAN

5 ACKNOWLEDGEMENTS

The authors wish to thank the Ministry of Defence, Royal Armament Research and Development Establishment, Fort Halstead for financial support during the period of this work.

6 REFERENCES

1.   D. Pugh and J.N. Sherwood, **Chemistry in Britain** 1988 544.
2.   A.F. Garito, C.C. Teng, K.Y. Wong and O. Zammani'Kjamiri, **Mol. Cryst. Liq. Cryst.**, 1984 **106** 219.
3.   J. Zyss, **J. Non-Cryst. Solids** 1982 **47** 211.
4.   J. Zyss, D.S. Chemla and J.F. Nicoud, **J. Chem. Phys.** 1981 **74** 4800.
5.   J.L. Oudar and R. Hierle, **J. Appl. Phys.** 1977 **48** 2699.
6.   B.F. Levy, C.G. Bethes, C.D. Thurmond, R.T. Lynch and J.L. Bernstein, **J. Appl. Phys.** 1979 **50** 2523.
7.   K. Jain, J.I. Crowley, G.H. Hewig, Y.Y. Cheng and R.J. Tweig, **Opt. Laser Technol.** 1981 **13** 297.
8.   A.F. Garito and K.D. Singer, **Laser Focus** 1982 **18(2)** 5968.
9.   M. Barzoukas, D. Josse, P. Fremaux, J. Zyss, J.F. Nicoud and J.O. Morley, **J. Opt. Soc. Am. B** 1987 **4(6)** 977.
10.   J. Zyss, J.F. Nicoud and M. Coquillay, **J. Chem. Phys.** 1984 **81** 4160.
11.   I. Ledoux, J. Zyss, A. Migus, J. Etchepare, G. Grillon and A. Antonetti, **Appl. Phys. lett.** 1986 **48** 1564.
12.   R. Heirle, J. Badan and J. Zyss, **J. Cryst. Growth** 1984 **69** 545.
13.   T.A. Driscoll, H.J. Hoffman, R.E. Stone and P.E. Perkins, **J. Opt. Soc. Am. B** 1986 **3** 683.
14.   Proceedings of The Third International Conference on Langmuir-Blodgett Films (Gottingen, F.R.G.), **Thin Solid Films** 1986 **160** Nos 1 and 2 .
15.   P.A. Norman, D.A. Bloor, J.S. Obhi, S.A. Karaulov, M.B. Hursthouse, P.V. Kolinsky, R.J. Jones and S.R. Hall, **J. Opt. Soc. Am. B** 1987 **6** 1013.
16.   H. Nakahara, T. Katoh, M. Sato and K. Fukuda, **Thin Solid Films** 1988 **160 (2)** 153.
17.   I. Ledoux, D. Josse, P. Vidakovic, J. Zyss, R.A. Hann, P.F. Gordon, B.D. Bothwell, S.K. Gupta, S. Allen, P. Robin, E. Chastaing and J. Dubois, **Europhys. Lett.** 1987 **3(7)** 803.

An Optical Characterisation of the Organic Non-linear Crystal 4-Nitro-4′-methylbenzylidene Aniline (NMBA)

R.T. Bailey, G.H. Bourhill, F.R. Cruickshank, D. Pugh, E.E.A. Shepherd, J.N. Sherwood, and G.S. Simpson

THE DEPARTMENT OF PURE AND APPLIED CHEMISTRY. THE UNIVERSITY OF STRATHCLYDE. 295. CATHEDRAL STREET. GLASGOW G1 1XL, UK

ABSTRACT

Large single crystals (5x3x0.5 cm^3) of the non–centrosymmetric, monoclinic (m) form of the organic nonlinear crystal 4–nitro–4'–methylbenzylidene aniline have been prepared by seeded growth from supersaturated solutions. Optically clear specimens, with faces parallel to the (100), (010) and (001) directions were examined by the Maker fringe technique. The dielectric axes showed no dispersion between 440 and 630 nm. Refractive indices at 532 and 1064 nm are reported. The nonlinear coefficients d_{11} and d_{33} were found to be 135.5 and 0.9 pmV^{-1} respectively. Intense angle–tuned phase–matched peaks, of Types I and II, are reported.

1 INTRODUCTION

In recent years, the search for organic nonlinear compounds for optical devices has led to many potential molecular structures[1] being proposed. Primary screening of the nonlinear potential of these structures can be conveniently established by powder techniques[2]. One of the materials currently being investigated is 4–nitro–4'–methylbenzylidene aniline (NMBA).

We have established[3] means of growing large, high purity and low defect density, single crystals of these materials. As previously reported[4], NMBA crystallises in two polymorphic forms. One is triclinic, P1, but is, however, centrosymmetric. The other polymorph is monoclinic (space group Pc, point group m) and non–centrosymmetric. Recrystallisation from n–hexane solution is used to achieve the monoclinic form. Such crystals were then used to determine the linear and nonlinear optical properties.

2 EXPERIMENTAL

The purity level of NMBA, synthesised from p–nitrobenzaldehyde and p-toluidine, was 98%. This was recrystallised three times from purified n-hexane and zone refined (120 passes of a 2 cm zone at 0.1 cm per hour) in order to reduce the impurity level to 150 ppm.

Decomposition of NMBA during melt growth, using a modified Bridgman technique, resulted in strained crystals due to the incorporation of impurities. Although this effect could be minimised by careful temperature control, better quality crystals were achieved by solution growth using ethylacetate. Large single crystals, typically 5 x 3 x 0.5 cm^3, were grown by the temperature lowering procedure at 0.1 K per day.

Figure 1 shows the indexed faces of the crystal and its growth habit. In this indexing we have used the following unit cell parameters[5]: a = 0.7419, b = 1.1679, c = 0.7447 nm and β = 110° 35'

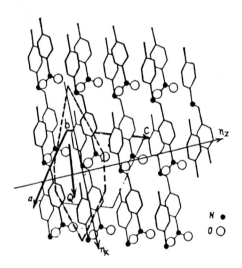

Figure 1   The indexing and habit of NMBA.   Q is the resultant molecular dipole vector and n_x, n_z are dielectric axes.

3   OPTICAL CHARACTERISATION

The z dielectric axis lies at 18° from the c crystallographic axis, in an anti-clockwise direction about b.   No dispersion (± 2%) of the dielectric axes was observed by examination under the polarising microscope, between 440 and 630 nm.

NMBA is monoclinic, point group m.   The second harmonic polarisation, P, is given by –

$$P_X = d_{11}E_X^2 + d_{12}E_Y^2 + d_{13}E_Z^2 + 2d_{15}E_XE_Z$$

$$P_Y = 2d_{24}E_YE_Z + 2d_{26}E_XE_Y$$

$$P_Z = d_{31}E_X^2 + d_{32}E_Y^2 + d_{33}E_Z^2 + 2d_{35}E_XE_Z$$

Three optical flats, indexed (100), (010) and (001) in accordance with Figure 1, were subjected to Maker fringe analysis in the equipment described previously[6], but arranged for single crystal evaluation. These samples were aligned with their dielectric planes, or axes if possible, parallel to the rotation axis. Fringes were collected for all three possible input polarisation planes (parallel, orthogonal and 45° to the rotation axis) coupled with each of the two possible output polarisations. Figure 2 shows a typical Fringe pattern obtained.

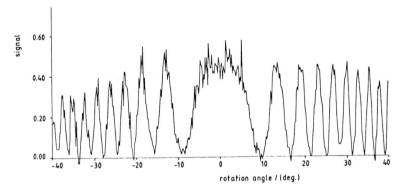

Figure 2   Fringe pattern of (010), rotated about Z axis. Input and output polarisations are parallel to the rotation axis.

Fringes from the (100) and (001) faces were obscured by angle-tuned phase-matching peaks (Table 1).

Table 1   Experimentally observed phase-matched peaks

Face	Type	Angle	Rotn. Axis	Input Polarisation	Output Polarisation	Intensity
(001)	I	+32°	Y	Para	Ortho	170,000
(001)	II	-6°	Y	45°	Para	850,000
(100)	I	-32°	Y	Para	Ortho	86,000
(100)	II	-20°	Y	45°	Ortho	20

NB   i)   input and output polarisations are relative to the rotation axis
     ii) intensity is relative to quartz, aligned on d_{11}.

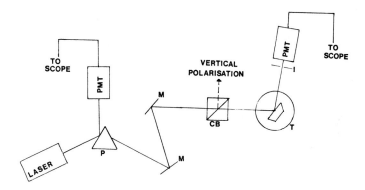

Figure 3 Brewster angle apparatus. M = mirror, I = iris, PMT = photo-multiplier tube, P = prism, CB = cubic beamsplitter, T = turntable.

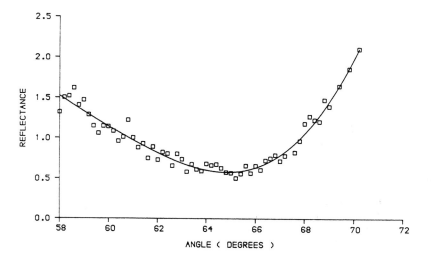

Figure 4 Brewster plot for n_x^{532}.

The principal refractive indices at 532 nm and 1064 nm (from fringe spacings) are given in Table 2. The accuracy of these refractive indices was checked by comparison of the theoretically predicted phase-matching angles with the experimentally observed values. All the values agreed well within experimental error.

Table 2   Principal refractive indices of NMBA.

Wavelength	n_x	n_y	n_z
532 nm	2.064	1.723	1.507
1064 nm	1.8172	1.6499	1.4794

From these refractive indices a phase matching locus, referred to the piezoelectric axes, was constructed (Figure 5).

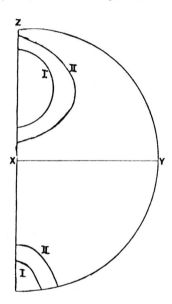

Figure 5   Phase-matching locus.

The nonlinear coefficients d_{11} and d_{33}, referenced to dielectric axes and measured relative to quartz d_{11} are presented (Table 3) together with the corresponding coherence lengths (lc) and second harmonic intensities (I_{532} nm) relative to quartz, aligned on d_{11}.

Table 3 d_{ij} and coherence lengths.

	I_{532} nm	d_{ij}	$lc/\mu m$
d_{11}	80.5	271	1.08
d_{33}	0.5	1.8	9.63

It was found that d_{11} = 135.5 pmV^{-1} and d_{33} = 0.9 pmV^{-1} based on a value[11] of 0.5 pmV^{-1} for quartz d_{11}.

4 ACKNOWLEDGEMENTS

We wish to thank the SERC for financial support.

5 REFERENCES

1. D.S. Chemla and J. Zyss, (Eds.), 'Non-Linear Optical Properties of Organic Molecules and Crystals', Acàdemic Press, N.Y., 1987, Vols. 1 and 2.
2. S.K. Kurtz and T.T. Perry, J. Appl. Phys., 1968, 39 (8), 3798.
3. B.J. McArdle and J.N. Sherwood, 'Advanced Crystal Growth', Eds. P.M. Dryburgh, B. Cockayne and K.G. Barraclough, Prentice Hall, U.K., 1987, Chapter 7, p.179.
4. R.T. Bailey, F.R. Cruickshank, S.M.G. Guthrie, B.J. McArdle, H. Morrison, D. Pugh, E.E.A. Shepherd, J.N. Sherwood, G.S. Simpson and C.S. Yoon, Mol. Cryst. Liq. Cryst., 1989, 166, 267.
5. R.T. Bailey, G.H. Bourhill, F.R. Cruickshank, S.M.G. Guthrie, G.W. McGillivray, D. Pugh, E.E.A. Shepherd, J.N. Sherwood, G.S. Simpson and C.S. Yoon, 'Materials for Non-Linear and Electro-Optics', Edited by M.H. Lyons, Institute of Physics Conference Series No. 103, 1989, Section 2.1, 119.
6. R.T. Bailey, S. Blaney, F.R. Cruickshank, S.M.G. Guthrie, D. Pugh and J.N. Sherwood, J. Appl. Phys. B., 1988, 47, 83.
7. S.P.F. Humphreys-Owen, Proc. Phys. Soc., 1960, 77, 949.
8. M. Elshazly-Zaghloul, J. Opt. Soc. Am., 1982, 72, 657.
9. C.H. Grossman and A.F. Garito, Mol. Cryst. Liq. Cryst., 1989, 168, 255.
10. 'Handbook of Lasers', Edited by R.J. Pressley, CRC, Ohio, 1971, 499.
11. S.K. Kurtz, 'Numerical Data and Functional Relationships, Group III. Crystal and Solid State Physics', Springer-Verlag, Berlin, 1969, vol. 2.

Second-order Non-linear Optical Properties of 4-N-Methylstilbazolium Tosylate Salts

Christopher P. Yakymyshyn, Kevin R. Stewart, and Eugene P. Boden

GENERAL ELECTRIC CORPORATE RESEARCH AND DEVELOPMENT, 1, RIVER ROAD, SCHENECTADY, NEW YORK 12301, USA

Seth R. Marder and Joseph W. Perry

JET PROPULSION LABORATORY, CALIFORNIA INSTITUTE OF TECHNOLOGY, 4800, OAK GROVE DRIVE, PASADENA, CALIFORNIA 91109, USA

William P. Schaefer

DIVISION OF CHEMISTRY AND CHEMICAL ENGINEERING, CALIFORNIA INSTITUTE OF TECHNOLOGY, PASADENA, CALIFORNIA 91125, USA

Introduction

Materials that show second-order nonlinear optical responses are of interest for frequency conversion and electrooptic modulation.[1] For nonlinear chromophores to exhibit non-zero bulk susceptibilities the nonlinear chromophore must reside in a noncentrosymmetric environment.[2,3] Meredith demonstrated that $(CH_3)_2NC_6H_4$-$CH=CH$-$C_5H_4NCH_3^+$ $CH_3SO_4^-$ has a large SHG efficiency, roughly 220 times that of urea.[4] He suggested that Coulombic interactions in salts could override neutral dipolar interactions which provide a strong driving force for centrosymmetric crystallization in dipolar compounds. We recently extended this approach and examined a series of over 50 organic salts in which a cationic chromophore, designed to have a large molecular hyperpolarizability, was crystallized with several counterions (anions).[5] It was our hope that perturbation of crystal structure by counter ion variation would be a general method to produce noncentrosymmetric materials. The approach in many cases led to molecules with very large macroscopic second-order optical nonlinearities as evidenced by the Kurtz powder technique.[6] One salt, Dimethyl̲A̲mino̲S̲tilbazolium T̲osylate, **10**, (DAST) had a second harmonic generation powder efficiency roughly 1000 times that of the urea reference standard.[5] X-ray studies demonstrate that, for several of the salts, the chromophore is oriented nearly along a polar axis.[5] In this paper we focus on results from our investigations on the nonlinear optical properties of 4-N-methylstilbazolium tosylate salts, electrooptic properties of DAST, and factors affecting crystallization of organic salts.

We synthesized fourteen 4-N-methylstilbazolium tosylate salts and measured their powder SHG efficiencies at 1.907 and 1.064 mm. Our results are summarized in Table 1. Of the donors and conjugation lengths explored, almost all of the tosylate salts examined had appreciable nonlinearities. Okada independently examined compounds **1**,[7] **10**,[8] as well as 4-hydroxy-4' N-methyl stilbazolium tosylate[7] and 4-cyano 4'-N-methyl stilbazolium tosylate[10] and found that each of these compounds exhibited efficient powder SHG. Of the sixteen 4-N-methyl stilbazolium tosylate salts which, to our knowlege, have been studied, 12 had powder SHG efficiencies greater than urea. Thus, the >85% incidence of noncentrosymmetric crystallization observed for these salts compares favorably to the roughly 25% incidence of noncentrosymmetric crystallization[3] in all nonchiral organic materials.

Table 1. Summary of powder SHG efficencies of $R-CH=CH-C_5H_4N-CH_3^+$ tosylate⁻ salts. The upper values were measured at 1.907 μm and the lower values at 1.064 μm.

Cmpd	R	Ref	SHG	Color
1	$4-CH_3OC_6H_4-$	5	120 100	yellow
2	$4-CH_3OC_6H_4-CH=CH-$	5	28 50	orange
3	$4-CH_3OC_6H_4-CH=CH-CH=CH-$	this work	5.9 3.3	orange
4	$4-CH_3SC_6H_4-$	5	- 1	yellow
5	$2,4-(CH_3O)_2C_6H_3-$	5	0 0.08	orange
6	$3,4-(HO)_2C_6H_3-$	this work	106 73	yellow
7	$C_{10}H_8-$(pyrenyl)	5	37 14	yellow
8	$4-(C_4H_8N)C_6H_4-$	5	0.2 0.03	red
9	$4-BrC_6H_4-$	5	1.7 5.0	pale yellow
10	$4-(CH_3)_2NC_6H_4-$	5	1000 15	red
11	$4-(CH_3)_2NC_6H_4-CH=CH-$	5	115 5	dark purple
12	$4-(CH_3)_2NC_6H_4-CH=CH-CH=CH-$	this work	8.9 2.0	dark purple
13	$C_5H_5FeC_5H_4CH=CH-CH=CH-$	9	12 -	maroon
14	$C_5H_5RuC_5H_4CH=CH-CH=CH-$	9	- 1	yellow

Crystal Structures of 4-N-Methyl Stilbazolium Tosylate Salts

For electro-optic applications a crystal comprised of highly nonlinear chromophores aligned in parallel fashion along a polar axis is desired. The crystal structure of 6 is discussed first, since it has several features in common with other stilbazolium salts.[5] Compound 6 crystallizes in the triclinic space group P1 with one cation and anion in the unit cell.[11] The exclusive translational symmetry of P1 results in exact parallel alignment of the stilbazolium cations in the crystal, an optimal situation for electro-optic applications. The cations form layers parallel to the **ab** planes. There are π–π interactions between the donor ring of a chromophore and the acceptor ring of the chromophore which lies directly below it in the stack. The interplane chromophore spacing is 3.4-3.7 Å. In between and perpendicular to the stacks of cationic chromophores lie sheets of tosylate anions. We believe that the formation of chromophore stacks separated by anion sheets facilitates the formation of macroscopically polar structures.

The general structural motif of compound **6** was observed in other stilbazolium compounds. For example, we previously reported that **10** crystallized in the monoclinic space group Cc.[5] The crystal structure of **10** is quite similar to that of **6** in that the cationic chromophores stack in a parallel manner. The perpendicular distance between chromophores within a stack is 3.5 Å. The only deviation from a completely aligned system is the 21° angle between the long axis of the molecules and the polar **a** axis of the crystal.

The X-ray crystal structures for stilbazolium compounds incorporating a tosylate counterion are summarized in Table 2. The crystals are best described as sheets of stilbazolium chromophores separated by sheets of tosylate counterions. Parallel chromophore alignment along a 'chromophore axis' is maintained within each chromophore sheet. The spacing between adjacent chromophores, within a given sheet, varies only slightly around a value of 3.4 - 3.7 Å.

As one moves through the crystal perpendicular to the chromophore sheets, one observes sheets of stilbazolium chromophores separated by sheets of perpendicular (thereby planar within their respective sheets) tosylate counterions. We also observe a consistent head (methyl) to tail (sulfonate) linear arrangement within the sheet of tosylate counterions. In previous reports, we demonstrated a strong dependence of crystal symmetry upon counterion type. The high probability of acentric crystal formation in the tosylate series reported here indicates that the tosylate moiety is an excellent choice for optimizing the intra-sheet chromophore interactions.

As one moves further through the crystal from the tosylate sheet to the next nearest sheet of stilbazolium chromophores, one observes an interesting oscillation of the chromophore axis. Viewed from above, the chromophores in neighbouring sheets form a herringbone pattern, an example of which is shown in Figure 1 for compound **10**. Minimizing the herringbone angle provides the optimum chromophore alignment for the electro-optic effect, while an optimum herringbone angle exists to maximize phase-matched second harmonic generation. In this respect, it should be noted that a large powder SHG measurement alone does not necessarily correlate with the magnitude of the electro-optic coefficient.

Growth and Characterization of DAST

Given the nearly optimal chromophore orientation in many of the tabulated compounds, extremely large electrooptic coefficients are expected. We have focussed on DAST (**10**) in an attempt to determine the linear and electro-optic properties of one member of the stilbazolium tosylate material system. In the case of DAST, if the molecular hyperpolarizability β_{zzz} (the major component of the molecular hyperpolarizability) is along the axis from the amino group to the methyl group of the pyridinium ring, then 83% of β_{zzz} is maintained along the polar **a** axis. Given the large nonresonant electronic molecular hyperpolarizability inferred from the powder SHG test, a large electro-optic coefficient is also expected in this material.

Single crystals of the monoclinic Cc phase of DAST were grown by slow evaporation from methanol, or by diffusion of ether into a methanol solution. In the former case, crystals with dimensions up to 3 x 3 x 1 mm³ resulted without the use of a seed crystal. Ether diffusion produced nearly defect-free single crystal platelets (up

Table 2 Crystallographic data for a set of R–CH=CH–C$_5$H$_4$N–CH$_3^+$ tosylate$^-$ salts. The interplane spacing (in Å) refers to spacing between adjacent stilbazolium chromophores within a sheet. The Herringbone angle is discussed in the text. V/Z is the ratio of unit cell volume to ion pairs/cell.

R	CH$_3$O-C$_6$H$_4$	CH$_3$O-C$_6$H$_4$	(OH)$_2$-C$_6$H$_3$	(CH$_3$)$_2$N-C$_6$H$_4$	OH-C$_6$H$_4$	NC-C$_6$H$_4$
Ref.	Ref. 8	Ref. 12	(6)	Ref. 5	Ref. 8	Ref. 10
Space Group	Tricl. P1	Mono. P2$_1$	Tricl. P1	Mono. Cc	Tricl. P1	Ortho. P$_{na21}$
Inter-plane spacing	(3.4)	<3.7	<3.7	<3.5	<3.5	-
Herring-bone angle	0°	49°	0°	21°	0°	≈36°
a	6.76	9.57	6.59	10.37	6.62	35.11
b	7.91	6.51	8.18	11.32	7.97	8.90
c	9.62	15.90	9.13	17.89	9.36	6.37
α	78.86°	90°	102.45°	90°	105.63°	90°
β	80.62°	91.78°	94.02°	92.24°	97.39°	90°
γ	83.49°	90°	96.62°	90°	95.4°	90°
V (Å^3)	496	990.1	474.8	2098	467	1991
Z	1	2	1	4	1	4
V/Z (Å^3)	496	495.1	474.8	524.5	467	497.9

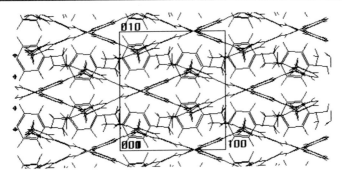

Figure 1   View of **ab** plane in compound **10**, showing the Herringbone pattern common to the 4-N-methylstilbazolium tosylate series.

to 2 x 2 x 0.1 mm^3) with mirror-like parallel facets. Single crystal X-ray orientation revealed that the platelets contained the **ab** plane (the chromophore plane), with chromophore planes stacked in the thin dimension of the platelet. Figure 2 shows the observed crystal habits and chromophore orientation. For measurements described

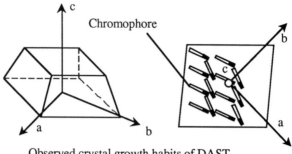

Figure 2   Observed crystal growth habits of DAST.

below, the high-quality facets and fortuitous in-plane chromophore orientation precluded the need for any additional crystal preparation. Powder SHG and powder X-ray measurements verified that the single crystals and the powdered source material were identical. DAST possesses excellent thermal properties, melting cleanly at 260 °C and remaining stable under an inert atmosphere up to 280 °C. Material stability as a melt suggests that melt processing may be a suitable method for growing large single crystals. Furthermore, crystals of DAST displayed no apparent change in optical properties after being heated to 160 °C for 200 hours in air.

Several large crystals and platelets were used to determine the linear optical and dielectric constants of DAST. The polarization dependent crystal transmission at normal incidence shown in Figure 3 displays a strong dichroism in the 600 - 700 nm range. The crystals tested showed no observable absorption dips from 750 nm to beyond 2000 nm. Refractive indices were then measured using Brewster angle reflection at 633 nm and 820 nm. At 633 nm, the crystal dichroism resulted in a large refractive index for light polarized parallel to the chromophore axis. At 820 nm, the

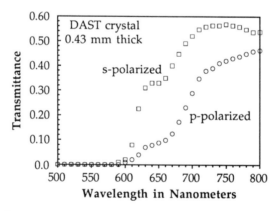

Figure 3   Transmission spectrum of a 0.43 mm thick DAST crystal:
                p-polarized, parallel to **a** axis; s-polarized, normal to **a** axis.

refractive indices are n_a= 2.75, n_b= 1.66 and n_c=1.42. The index n_a is large relative to other organic materials. This is advantageous for electro-optic applications, since the material figure of merit is given by FOM= $n_a^3 r/2$. The low-frequency (1 KHz to 100 KHz) dielectric constants ε_{ab}= 5.1 and ε_c= 3.1 indicate that low-frequency contributions to the Pockels effect are minimal in DAST. In addition, the low dielectric constant reduces the electrical drive power requirements at rf frequencies.

The electro-optic coefficient measurements used the method previously described by Yoshimura.[13] A pair of closely spaced (25 μm) electrodes supported by a glass substrate were contacted to one surface of a DAST single crystal. A properly polarized 820 nm probe beam passing between the electrode pair experienced a birefringence induced by the fringing fields between the electrodes. A polarizer and detector measured the electrically induced birefringence. With the incident beam polarized at 45° to the a axis and a 1 KHz AC electric field applied parallel to a, the measured birefringence gives $n_x^3 r_{11}$ - $n_y^3 r_{21}$. For the preliminary measurements, we assume that the r_{11} electro-optic coefficient dominates the material response.

As described by Yoshimura,[13] the crystal's inherent birefringence is first measured. The polarization rotation is then measured with an applied electric field. An electric-field induced phase delay is calculated and an electro-optic figure of merit determined using FOM=$n^3 r/2$=$\lambda \Delta D/(2\pi VL)$ where Δ is the phase delay, D is the electrode spacing, V is the applied voltage, and L is the effective crystal thickness. We assume, based on the effect of the crystal's dielectric constant on the shape of the fringing fields, that for D=25 μm, L≈10 μm, even though the crystal is actually much thicker than this. Based on this assumption, measurements on several crystals give FOM≈2100 pm/V, or r_{11}≈200 pm/V. This preliminary measurement is a lower-bound figure. An air gap of up to 100 μm between the crystal surface and the electrodes reduces the calculated field within the crystal by as much as a factor of 10. In addition, the air gap gives rise to a low-frequency (<100 Hz) roll-off of the measured birefringence, most likely caused by charge decay on the crystal surfaces. Experiments are currently under way to alleviate this problem.

Conclusions

We have described a series of 4-N-methyl stilbazolium tosylate salts which display very large second-order nonlinear susceptibilities. Powder SHG measurements demonstrated a >85% success rate in isolating acentric organic compounds. Crystal structures of six compounds have the common feature of parallel-aligned nonlinear chromophore sheets separated by sheets of 'spacer' counterions. Single crystals of one compound, DAST, were grown, and preliminary measurements of the optical, dielectric and electro-optic properties were made. The measured FOM for DAST of 2100 pm/V compares favorably with 650 pm/V determined for a related organic 4-N-methyl stilbazolium salt, styrylpyridinium cyanine dye,[12] as well as MNA (320 pm/V), the ceramic PLZT (1150 pm/V), and LiNbO$_3$ (200 pm/V). The low dielectric constant of this material bodes well for high-frequency device operation. These preliminary measurements support the tremendous potential of this material system for providing highly nonlinear, thermally stable organic materials for optical signal processing, communications and interconnect applications.

Acknowledgements

The research described in this paper was performed in part by the Jet Propulsion Laboratory, California Institute of Technology as part of its Center for Space Microelectronics Technology which is supported by the Strategic Defense Initiative Organization, Innovative Science and Technology Office through an agreement with the National Aeronautics and Space Administration (NASA). SRM thanks Professsor Robert Grubbs for access to synthetic facilities at Caltech. CPY, EPB and KRS thank the General Electric Company for permission to publish this work, and Mike DeJule for access to material processing facilities at GE CRD. We also thank Bruce Tiemann, Kelly Perry, David Fobare and Peter Phelps for expert technical assistance.

References

1.   (a) D. J. Williams, Angew. Chem. Int. Ed. Engl., 1984, 23, 690.(b) D. J. Williams, Ed., 'Nonlinear Optical Properties of Organic and Polymeric Materials', ACS Symp. Ser., 233, American Chemical Society, Washington, DC, 1983.
2.   D. S. Chemla and J. Zyss, Eds., 'Nonlinear Optical Properties of Organic Molecules and Crystals', Academic Press, Orlando, FL, 1987, Vol. 1 and 2.
3.   R. W. Twieg in 'Nonlinear Optical Properties of Organic Molecules and Crystals', D.S. Chemla and J. Zyss, Eds., Academic Press, Orlando, FL, 1987, Volume 1, p. 242.
4.   G. R. Meredith in 'Nonlinear Optical Properties of Organic and Polymeric Materials', ACS Symp. Ser., 233, D. J. Williams, Ed. American Chemical Society, Washington, DC 1983, p. 30.
5.   S. R. Marder, J. W. Perry, and W. P. Schaeffer, Science, 1989, 245, 626.
6.   S. K. Kurtz and T. T. Perry, J. Appl. Phys., 1968, 39, 3798.
7.   S. Okada, H. Matsuda, H. Nakanishi, M. Kato and R. Muramatsu, Japanese Patent 6 348 265, 1988. Chem. Abstr. 1988, 109, 219268w.
8.   T. Koike, T. Ohmi, N. Yoshikawa, S. Umegegaki, S Okada, A. Masaki, H. Matsuda and H. Nakanishi, Digest of Conference on Lasers and Electro-optics, Optical Society of America, Washington, D.C. 1990, paper CTHI31.
9.   S. R. Marder, J. W. Perry, B. G. Tiemann and W. P. Schaeffer, Organometallics, submitted.
10. T. Koike, private communication.
11. S. R. Marder, C. P. Yakymyshyn, K. R. Stewart, E.P.Boden, J. W. Perry and William P. Schaefer, Chem. Comm, to be submitted.
12. S. R. Marder, K. R. Stewart, E.P.Boden, J. W. Perry, William P. Schaefer, C. P. Yakymyshyn, L. Henling, R. E. Marsh and B. G. Tiemann, J. Amer. Chem. Soc., in preparation.
13. T. Yoshimura, J. Appl. Phys. 1987, 62, 2028.

Crystal Structure and SHG Activity of the New Chiral Species 6-Nitro-2-(L-prolinol) Quinoline

Karen J. Atkins and Colin L. Honeybourne

MOLECULAR ELECTRONICS AND SURFACE SCIENCE GROUP, BRISTOL
POLYTECHNIC, FRENCHAY, BRISTOL BS16 1QY, UK

Karl N. Harrison, Frank Grams, and A. Guy Orpen

DEPARTMENT OF INORGANIC CHEMISTRY, UNIVERSITY OF BRISTOL,
CANTOCK'S CLOSE, BRISTOL BS8 1TS, UK

INTRODUCTION

The design and synthesis of 6-nitro-2-(L-prolinol)
quinoline(PNQ) was undertaken in order to compare its
performance as a second order non-linear optical material with that of well-characterised analogues such as
N-(4-nitrophenyl)-(L)-prolinol(NPP) and 2-(L-prolinol)-
5-nitropyridine(PNP). The L-prolinol moiety has been
adopted because it proffers the favourable features of
strong electron-donating power and inherent non-centrosymmetry.

The molecular packing in crystals of PNP and NPP(1)
has been shown to be close to the optimum angle for
phase matching (54.74°). It is this optimum packing,
coupled with the efficient charge transfer, and transparency at the Nd:YAG-laser wavelength that accounts
for the exceptional second-order optical non-linearity
exhibited by crystals of NPP and PNP (Space group P2₁).
In view of these observations, it was considered that
the determination of the crystal structure of PNQ was
essential if the non-linear optical behaviour of PNQ
was properly to be explained.

EXPERIMENTAL

PNC was synthesised by a four-stage reaction sequence commencing with conversion of 6-nitroquinoline to 6-nitroquinoline-1-oxide. Acid hydrolysis of the 1-oxide yielded the 2-hydroxy derivative from which 2-chloro-6-nitroquinoline was obtained by reaction with $POCl_3$ and PCl_5. Finally, the required product was obtained by nucleophilic displacement of the chloro-group by L-prolinol (overall yield circa 3%). After purification by column chromatography, crystals were obtained by repeated recrystallisation from ethyl acetate.

A single crystal (0.5x0.4x0.5mm) was mounted on a glass fibre in air, and held in place with epoxy glue. All diffraction measurements were made at 295K on a Nicolet R3m diffractometer, using graphite-monochromated Mo-K X-radiation. Unit cell dimensions were determined from 26 centred reflections in the range $16.0 < 2\theta < 29.0°$. A total of 2753 diffracted intensities were measured in two unique octants ($\mp$h,k,l) of reciprocal space for $4.0 < 2\theta < 50.0°$ by Wyckoff scans; 2076 unique observations having $I > 2\sigma(I)$ were retained for use in the solution and refinement of the crystal structure.

The structure was solved by direct and Fourier methods and was refined in the chiral space group $P2_12_12_1$ with the molecules lying in general positions. All non-hydrogen atoms and the hydroxy-hydrogen atom, H(1), were refined without position constraints, whereas all other hydrogen atoms were constrained to idealised geometries. The chirality of the molecules in this crystal, and of the absolute structure was known since L-prolinol was used in its synthesis. This assignment was consistent with a η refinement which was of itself inconclusive ($\eta = 1.5(27))(2)$.

CRYSTAL STRUCTURE OF PNQ

The crystal structure of $PNQ(C_{14}H_{15}N_3O_3)$, of molecular weight 273.3, was found to be orthorhombic with the space group $P2_12_12_1$. The cell parameters were a = 6.877(3)Å, b = 9.741(3)Å and c = 19.811(6)Å; the cell volume was V = 1327.1(8)Å³, with four molecules per unit cell, and a calculated density of $1.368gcm^{-3}$. The molecular configuration is given in Figure 1 together with the values of observed bond lengths and bond angles.

The nitrogen atom, N(2), of the nitro-group and N(3) of the prolinol moiety lie in the same plane as the quinoline ring which is, itself, essentially planar. The oxygen atoms of the nitro-group lie in the same plane as the heterocyclic ring as is also observed in NPP; however, in PNP, these oxygen atoms lie on opposite sides of the aromatic plane.

The exocyclic bonds joining the donor and acceptor groups to the quinoline ring in PNQ are shorter than normal. The C-N bond to the nitro group is only 1.457(compared to the "norm" of 1.468) and that to the prolinol group is only 1.355(compared to the "norm" of 1.371).Such shortening of these bonds from the "norm" is also seen in NPP and PNP, and is commensurate with a contribution from a quinonoidal bonding structure to the actual ground state molecular configuration. The shortening of certain bonds in the quinoline ring also implies that there is a contribution from a quinonoidal structure.

The unit cell of PNQ is shown in Figure 2. It is of interest to note that the angle O(3)C(14)C(13) of 107.3° is significantly smaller than the corresponding angle in NPP(111.6°) and PNP(110.8°). This results from the weak hydrogen bonding in PNQ (see Fig.2) between O(3) of L-prolinol in one molecule and H(3) of the quinoline in an

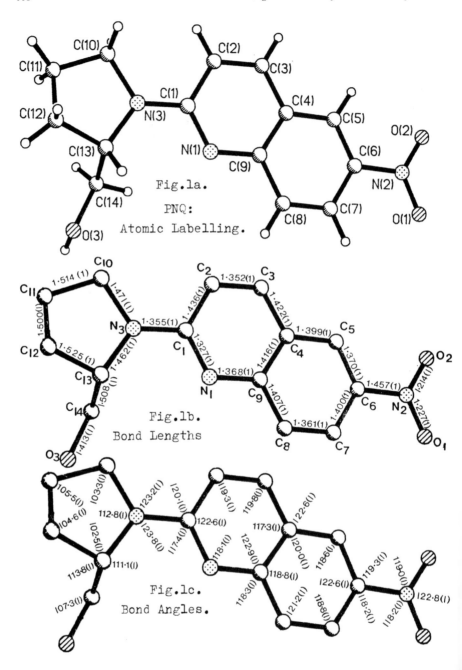

Fig.1a.
PNQ:
Atomic Labelling.

Fig.1b.
Bond Lengths

Fig.1c.
Bond Angles.

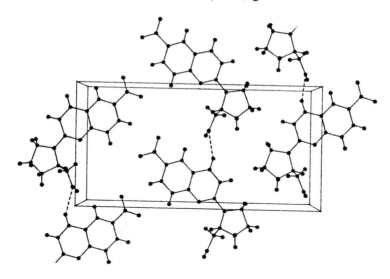

Fig.2. In-plane -OH to CH hydrogen bonding in PNQ.

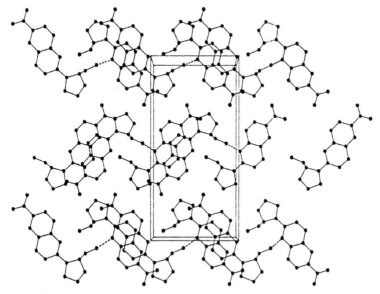

Fig.3. Out-of-plane -OH to N(quinoline) hydrogen bonding in PNQ: also note antiparallel stacking of layers.

adjacent molecule. The formation of such a hydrogen bond
in PNP is prohibited by steric crowding at the equivalent
C-H position by the nitro-group; the absence of a ring-
nitrogen precludes the formation of such a bond in NPP.
A second, stronger hydrogen bond is formed between the
hydroxyl-hydrogen of the L-prolinol group of one molecule and the heterocyclic nitrogen of a molecule in the
next layer. The intermolecular contact is 2.074Å, and a
view of this, in the b-c plane is shown in Figure 3.
Figure 3 also serves to illustrate a crucial feature of
the crystal structure of PNQ. It can clearly be seen that
molecules of QNP are packed in a quasi-antiparallel
regime, which is in sharp contrast to the quasi-parallel
alignment seen is PNP and NPP. Thus, the expected OH to
NO_2 hydrogen bond, seen in PNP and NPP, is lacking in PNQ.

CONCLUSION

In contrast to NPP and PNP, which crystallise in space
group P2₁, PNQ crystallises in group $P2_12_12_1$. Thus,
crystals of PNQ, even if molecular alignment did lead to
vector addition, can only be 19% efficient for SHG(3).
However, as the crystal structure shows, molecules of
PNQ are aligned such that vectorial subtraction of
transition moments and dipole moments will occur. The
poor performance of PNQ in SHG experiments is thus
explained. PNP and NPP can have a maximum efficiency of
38% and are aligned to facilitate vectorial addition.

REFERENCES

1. J.Zyss,J.F.Nicoud & M.Coquillay,*J.Chem.Phys.*,1984,
   81, 4160.
2. D.Rogers, *Acta Cryst.Sect.A*, 1981, 37, 734.
3. J.F.Nicoud and R.J.Twieg, *Nonlinear Optical Properties
   of Organic Molecules and Crystals*, 1987,1, 227.

Synthesis and Study of Benzo-analogues of Hemicyanines and Benzo-aza-analogues of Merocyanines with Dimethyl-aminophenyl, Hydroxyphenyl, and Ferrocenyl Substituents

Karen J. Atkins and Colin L. Honeybourne

MOLECULAR ELECTRONICS AND SURFACE SCIENCE GROUP, BRISTOL
POLYTECHNIC, FRENCHAY, BRISTOL BS16 1QY, UK

INTRODUCTION

Much of the recent synthetic programmes undertaken in
the field of second harmonic generation(SHG) have used
either the benzene nucleus or the pyridine nucleus: for
example DANS, NPP, PNP, merocyanines and hemicyanines(1).
We considered it possible that the use of a chromophore
with a larger area of conjugation, such as quinoline or
quinolinium, might enhance the molecular properties
pertinent to SHG - namely, transition moments and changes
in molecular dipole moment upon excitation. Derivatives
of quinoline have found extensive applications as dyes
and photographic sensitisers; indeed, the original definition of "cyanine dyes" was based on the quinoline
nucleus. Their usefulness as dyes arises from their
having intense intramolecular charge transfer transitions, which therefore provides one of the prerequisites
for efficient SHG performance. Zwitterionic products by
reacting quinolines with TCNQ have been obtained; these
products give large experimental and theoretical SHG
parameters(2).

THE SYNTHESIS OF QUINOLINIUM DERIVATIVES.

The synthesis of quinolinium derivatives, including one
organometallic species, was undertaken in order to
compare these donor-acceptor compounds with their cationic pyridinium(hemicyanine) analogues. In this fashion,
we intended to exploit the powerful electron-accepting
nature of the quaternised quinolinium nitrogen, particularly since polarographic studies have shown that the
quinolinium cation is a better electron acceptor than the
pyridinium cation(3). This latter feature has a marked
effect upon the reactivity of alkyl substituents of
quinoline, especially when they are at the 2- or 4-
positions of the heterocyclic ring. This reactivity
results in the facile deprotonation of the carbon atom
attached to the ring. One result of this reactivity is
that quinolinium salts of 4-alkyl-quinolines readily
condense with aldehydes with a trace of piperidine base.
Thus, benzaldehydes give styryl quinolinium salts, and
ferrocene carboxaldehyde gives ferrocenylvinyl-quinolin-
ium salts.

4-(4'-(N,N-dimethylamino)styryl)-1-methyl quinolinium
iodide.

4-methylquinoline (7.16g, 50 mM) was combined with 1-
iodomethane(7.1g, 50mM) in ethanol(20ml) and the solution
held at reflux for 1 hour. Excess solvent was removed at
reduced pressure, and the resulting yellow solid was recrystallised, first from butan-2-ol and then from methanol. 1,4-dimethylquinolinium iodide was obtained as a
yellow powder in 72% yield (10.24g, 36mM) with melting
point 172-175°C(lit.(4) 172-174°C). The quinolinium
salt(0.56g, 2mM) and 4-(N,N-dimethylamino)benzaldehyde
(0.3g, 2mM, freshly recrystallised from water) were
dissolved in dry methanol(25ml). A slight excess of

piperidine (0.26g, 3mM) was added whereupon the colour
changed from clear to magenta. The solution was held at
reflux for approximately 18 hours, and then allowed to
cool. The purple solid was collected and recrystallised
from methanol to give the required product in 56% yield
with melting point 285-288°C; this was markedly higher
than the literature(4) value. The purity and structure
of the product was determined by a detailed quantitative
and spectroscopic analysis. Calc. for $C_{20}H_{21}N_2I$ %: C,
57.70; H,5.08; N,6.73; I,30.48: Found %: C,58.03; H,
5.16; N,6.61; I,30.26.

Cognate preparations gave the new compounds 4-(4'-hexa
decyloxystyryl)-1-methylquinolinium iodide, 4-(4'-octa
decyloxystyryl)-1-methylquinolinium iodide and 4-(2-
ferrocenylvinyl)-1-methylquinolinium iodide.

SYNTHESIS OF SCHIFF'S BASE QUINOLINE DERIVATIVES.
Substitution of one of the carbon atoms in an ethylenic
bridge by a nitrogen atom provides the azomethine or
imine linkage, and compounds containing this linkage are
often referred to as Schiff's bases. The nonlinear,
optical response of some Schiff's bases has displayed
a reasonably strong powder SHG(5), in particular those
with optically active or bulky groups. The pyrrolidine
methanol(or prolinol) moiety is found in many SHG-active
materials such as N-(4-nitrophenyl)-L-prolinol(NPP) and
2-(L-prolinol)-5-nitropyridine(PNP)(1); it is also present
in a Schiff's base exhibiting a powder SHG 20 times more
efficient than urea(6). The pyrrolidine group has a much
larger electron-donating power than the dimethylamino
group and when used as an optically active resolved isomer, guarantees a non-centrosymmetric crystal structure.
The use of such a group is therefore particularly suitable in the synthesis of SHG-active materials.

4-(4'-(2-(L)-prolinol)phenyl)-iminomethylene quinoline.
The preparation of this new Schiff's base required the
prior synthesis of NPP (from l-chloro-4-nitrobenzene and
(S)-(+)-2-pyrrolidinemethanol(i.e. (L)-prolinol). The
reduction of NPP was effected by stirring under hydrogen
in dry methanol at room temperature in the presence of a
palladium-on-carbon catalyst. The product from this
reaction was not isolated, but was allowed to react with
an equimolar quantity of quinoline-4-carboxaldehyde in
dry methanol. A red solid formed almost immediately, and
was collected and recrystallised from ethanol. The
required product was obtained in 57% yield(based on
the initial quantity of NPP) and had a melting point of
191-193°C(brick-red needles). Calculated for $C_{17}H_{11}N_3$
% C, 79.36; H, 4.31; N, 16.33: Found % C, 79.70; H,
4.64: N, 16.49.
Cognate preparations gave the following 4'-derivatives
of 4-(4'-R-phenyl)iminomethylene quinoline: R = OH,N(Me)$_2$,
n-hexyloxy. These reactions are preferably performed in
toluene, using a Dean and Stark trap to remove water from
the reaction mixture by azeotropic distillation. This
prevents hydrolysis of the imine by directing the equilibrium towards the formation of a hemiaminal precursor.

COMMENTS UPON OPTICAL SPECTRA
The results from measurements of optical spectra in the
UV-visible region of the compounds shown in Figure 1 are
given below (see Table 1). Attention is drawn to the
enhanced intensity and lowered energy of the first, low-
energy transition in Ia which has the most efficient
donor-acceptor combination. The order IIc IIa IIb
arises from the same cause. On this evidence, Ia and IIc
have the larger intramolecular charge transfer.

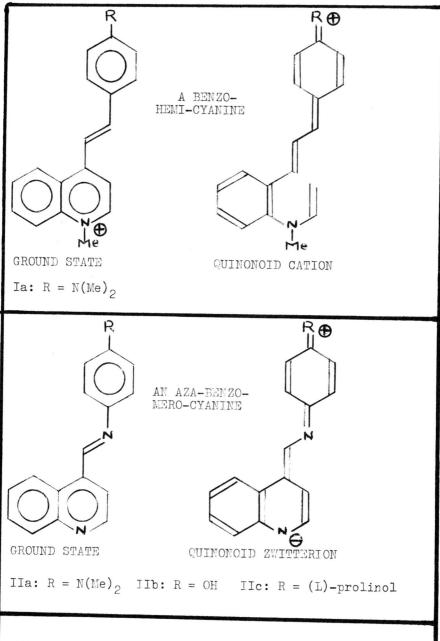

FIGURE 1. Bonding of molecules discussed in the text.

CALCULATED OPTICAL PROPERTIES

Griffiths has published a set of parameters for use in
calculating the linear optical properties of conjugated,
planar dyes. These parameters have been optimised for use
in the π-electrons-only approximation using configuration interaction between the transitions involving the
three highest filled and three lowest empty molecular
orbitals(7). The results of SCFPPP 9-CI-States calculations on two quinoline Schiff's bases are given in Table
2 in which r_x, r_y are transition moments and $\Delta r_x, \Delta r_y$
are changes in molecular(π-electron) dipole moment.
These quantities, and the state energies, can be used to
estimate the nonlinear second order(SHG) parameter
$\beta(2\omega,\omega,\omega)$ which has a vectorial average(β-vec)(2). The
calculated values of components of the β-tensor that
are of significant magnitude are given in Table 3. The
y-axis is defined by the line through the quinoline N-
and imine C-atoms.

In both molecules, only one state makes a significant
contribution to β-vec:namely, the lowest energy state.
Inspection of the dipoles in Tab.2 will confirm this,
and show that these are indeed charge transfer states.
Although the y-direction appears to dominate, the values
of β-X, and $\beta_{xyy}, \beta_{yxy}, \beta_{yyx}$ which contribute to it(2),
show that these molecules are not really pseudolinear
systems; this latter is also revealed by the disparities
between β_{yyy} and β-vec. In both molecules, the charge-
transfer state is composed of 98% of the transition from
the highest filled to the lowest empty molecular orbital.
The migration of charge in the HOMO-LUMO transition is
from the amino-phenyl to the quinoline-imine units; the
changes in charge density are widely distributed.
The results for IIa at 1064nm show the onset of a
resonance condition.

Table 1. Optical spectra: λ_{max}, nm(ϵ , $dm^3 mol^{-1} cm^{-1}$).

Ia : 587 (42,860); 309 (10,550); 244 (27,450)

IIa : 413 (16,190); 321 (8,300); 248 (22,710)

IIb : 373 (13,620); 336 (11,360); 243 (23,440)

IIc : 432 (18,830); 322 (8,990); 247 (23,430)

Table 2.Calculated properties of the five lowest
excited states(λ in nm; all dipoles in Debye).

State	λ_{max}	r_x	r_y	Δr_x	Δr_y	Molecule
1	447	-2.21	-8.20	3.81	18.26	
2	303	-0.97	0.61	-1.18	2.07	
3	294	-1.30	-1.12	-4.20	-0.61	IIa
4	272	0.46	1.46	8.80	18.76	R = $N(Me)_2$
5	246	3.83	-0.57	5.31	17.42	
1	391	-2.22	-7.50	2.59	11.56	
2	290	-0.44	-0.51	-6.67	-2.26	
3	271	-1.19	-0.13	5.10	18.88	IIb
4	249	2.15	0.21	7.20	12.83	R = OH
5	243	3.68	-0.87	6.87	17.59	

Table 3. Calculated SHG parameters(λ in nm; β in $cm^5 esu^{-1}$)

λ_{laser}	β_{xyy}	β_{yxy}	β_{yyx}	β_{yyy}	β-X	β-Y	β-vec	Molecule
0	35	35	35	150	38	160	164	
1600	57	53	53	236	62	256	263	IIa
1064	162	139	139	633	160	672	691	
0	13	13	13	58	18	73	75	
1600	20	19	19	84	25	92	96	IIb
1064	39	34	34	150	43	163	169	

REFERENCES.
1. J.F.Nicoud and R.J.Twieg, Nonlinear Optical Properties
   of Organic Molecules and Crystals, 1987, 1, 227.
2. C.L.Honeybourne,J.Phys.D, 1990, 23, 245.
3. J.B.Torrance, Accnt.Chem.Res., 1979, 12, 79.
4. V.J.Vosnyak and C.G.Savitskaya,Khim.Farm.Zh.,
   1984, 18, 951.
5. M.C.Palazzolo,Eur.Pat.G02F 1/37 C07C 119/10(03.02.87).
6. See Ref.1, pp.285.
7. J.Griffiths, Dyes and Pigments, 1982, 3, 211.

Linear and Non-linear Optical Properties of 3-Methyl-4-nitropyridine-1-oxide (POM) Crystal at Low Temperatures

P.L. Baldeck, M. Pierre, D. Block, S. Fontanell, R. Georges, and R. Romestain

LABORATOIRE DE SPECTROMÉTRIE PHYSIQUE, UNIVERSITÉ DE GRENOBLE, B.P. 87 38402 SAINT MARTIN D'HÈRES, CEDEX, FRANCE

Jo Zyss

CENTRE NATIONAL D'ETUDES DES TÉLÉCOMMUNICATIONS, 92220 BAGNEUX, FRANCE

1 INTRODUCTION

Recently, 3-Methyl-4-Nitropyridine-1-Oxyde (POM) entered the commercial market of nonlinear crystals because of its unique optical and manufacturing properties.[1] POM single crystal exhibits high-nonlinearities in the infrared and visible region because of its intense internal charge transfer in the excited state.[2-7] Applications of a nonlinear material are limited by its absorption bands and damage threshold. POM crystal has a cut-off frequency at 450 nm which prevents the efficient second-harmonic generation of green and blue frequencies at room temperature.

In this paper, we present measurements of linear and nonlinear optical properties of POM crystals at liquid-helium and liquid-nitrogen temperatures. Our motivation for this work was threefold: 1) characterization of the strong internal charge-transfer transition responsible for the nonlinear properties and the cut-off frequency, 2) identification of absorption bands limiting the performance of POM crystal in the visible, and 3) investigation of temperature effects on absorption bands, and on the second-harmonic generation properties.

2 INTERNAL CHARGE TRANSFER TRANSITION $\pi \rightarrow \pi^*$ AT 378 NM

Near ultra-violet internal charge transfer transition $\pi \rightarrow \pi^*$ is well known for the molecule in solution but not yet measured in the crystal (Fig.1).[8-10] Figure 2 shows the imaginary part ε'' of the dielectric constant of a POM crystal at 1.5 K for polarizations parallel to

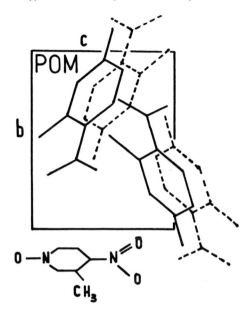

Fig. 1 Molecular and crystallographic (projection <1, 0, 0 >) data.

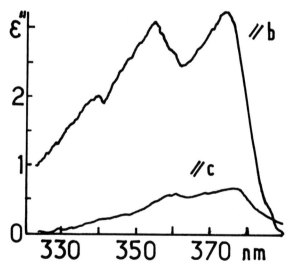

Fig. 2 Spectral data (imaginary part ε'' of the dielectric constant) characterizing the strong internal charge transfer transition $\pi \rightarrow \pi^*$ of a POM crystal at 1.5 K for polarizations along the b and c axis.

the b and c optical axis. The complex dielectric constant was calculated from reflectivity measurements because the high optical density of samples precluded absorption measurements.[11] The absorption peak at 375 nm (Fig. 2) corresponds to the charge transfer transition. Compared to room temperature solutions, this transition is red-shifted by about 20 nm. The absorption peaks at 355 nm and 340 nm correspond to vibronic states linked to the strong (IR and Raman active) stretching vibrations of the $-NO_2$ substituent.[12]

3   TRANSITION $n \rightarrow \pi^*$ AT 440 NM

At liquid-helium temperature, transmission measurements show a structured spectrum at about 440 nm (Fig. 3). A similar transition in NPO crystal was attributed to a $n \rightarrow \pi^* B_2$ transition involving one of the $-NO_2$ lone pair.[11] The presence of this transition, very close to the main charge transfer system, induces a predominant relaxation channel for the internal charge transfer state. This $n \rightarrow \pi^*$ transition also enhances the absorption tail of the charge transfer transition, which limits applications of POM crystals in the green-blue part of the spectrum.

4   TRANSMISSION SPECTRA IN THE VISIBLE REGION

Figure 4 displays b- and c-polarized transmission loss (optical density) spectra of POM crystals at 300 K and 1.5 K in the visible. Figure 4-a corresponds to a 3-mm sample freshly cleaved from an ingot grown from an acetonitrile solution. Figure 4-b corresponds to a 1-mm sample (natural faces) obtained by slow evaporation of an acetone solution.

Figures 4-a and 4-b present three salient features: 1) a blue cut-off frequency at about 450 nm determined by the tail of the internal charge transfer transition. (notice that for a 1-mm sample, an optical density of 1 corresponds to $\varepsilon'' \approx 10^{-4}$ compared to $\varepsilon'' = 3.2$ at the maximum of the charge transfer transition), 2) a transparency region ranging from the near infrared to about 450 nm. In this spectral region, losses are mainly due to reflection, more important in the b-polarization than in the c-polarization ($n_b > n_c$). 3) several impurity absorption bands are visible in the b-polarized spectrum (Fig. 4-a).

These measurements demonstrate the possibility of using low-temperatures to blue shift the cut-off frequency of POM crystal by a maximum of 10 nm. However, they are not sufficiently accurate to quantify the residual absorption (<5%) and temperature effects in the transparency range. The presence of impurity absorption bands in

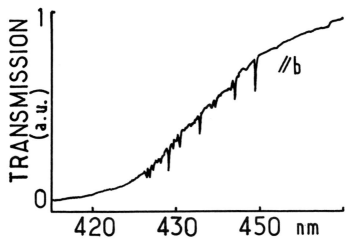

Fig. 3   b-polarized transmission spectrum of a 250 μm-thick POM crystal showing the n→π* transition near 440 nm.

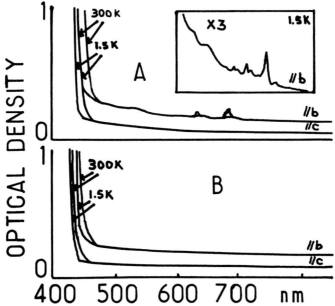

Fig. 4   Polarized transmission loss spectra of POM crystals at room-temperature and low-temperature. a) 3-mm thick sample (CNET) grown from an acetonitrile solution of POM powder. b) 1-mm thick sample grown from an acetone solution of POM powder.

crystals grown from acetonitrile solutions indicates that, in order to obtain large POM crystals suitable for the second-harmonic generation of visible wavelengths, the growing conditions have to be carefully controlled.

5   TEMPERATURE EFFECTS ON CW SECOND-HARMONIC GENERATION AT 532 NM

The efficient doubling of CW Nd:YAG lasers is one of the major potential application of nonlinear organic crystals. In the last part of this paper, we present results of preliminary measurements to test the possibility of using POM crystals for the second-harmonic generation of CW laser light at 532 nm.

POM crystals, grown from acetone solutions, were cut (1-2 mm thick) to satisfy the phase-matching conditions for the fundamental beam normal to the incident surface. A 20-cm focal length lens was used to focus the CW Nd:YAG laser to a 50 μm beam waist. Within these experimental conditions the power of second-harmonic signal was kept in the microwatt range.

Figure 5 displays the power of the second harmonic as a function of the incident infrared power. At room temperature, an incident power of 450 mW damaged the sample. In addition, the phase-matching angle had to be adjusted continuously for optimum second-harmonic generation. Low temperature measurements were done using nitrogen liquid to cool POM samples in a cryostat. At 77 K (crystal immersed in liquid nitrogen) second-harmonic generation was obtained with a phase-matching angle of 48° as compared to 54°18' at 300 K. However, nitrogen gas bubbles were generated in the cryostat by heating the sample and prevented from reaching the damage threshold. When the sample was maintain in a cold nitrogen gas, up to 800 mW of CW infrared power could be focused in the POM crystal without damage. However, phase-matching conditions were again unstable due to thermal effects. As seen in Fig. 5, a dramatic improvement was obtained at room temperature using a strong air flow to stabilize the sample temperature. More than 4W of polarized infrared power (maximum available) could be focused in the sample without damage and change of phase matching angle.

6   SUMMARY

In this work, we investigated absorption bands which limit nonlinear applications of POM crystals in the visible. The strong internal charge transfer transition $\pi \rightarrow \pi^*$ is characterized at 1.5 K. In

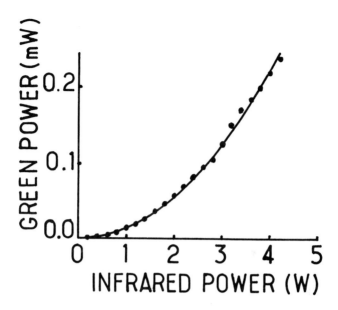

Fig. 5 Second-harmonic generation at 532 nm from a CW Nd:YAG laser. The temperature of the POM crystal was stabilized with a strong air flow (at room temperature).

comparison to measurements in solution, the transition is red-shifted by 20 nm. A n→π* transition, identified at about 440 nm, limits the transparency range of POM crystal in the visible. The cut-off frequency of POM crystal is blue-shifted by 10 nm at 1.5 K. Impurity absorption bands are found in the transparency range of crystals grown from acetonitrile solution. Finally, preliminary measurements of second harmonic generation of CW Nd:YAG laser at 77 K and 300 K show important thermal effects on damage threshold and phase-matching conditions. More than 4W of polarized infrared power could be safely focused to a 50 μm beam waist in a temperature stabilized POM crystal.

REFERENCES

1. Quartz & Silice, Cedex 27   92096 Paris-La-Defense.
2. J. Zyss and D. S. Chemla, eds, *Nonlinear Optical Properties of Organic Molecules and Crystals*, (Academic Press, New York, 1987), 2 volumes.
3. J. Zyss, D. S. Chemla, and J. F. Nicoud, J. Chem. Phys. **74**, 4800-4811 (1981).
4. M. Sigelle, J. Zyss, and R. Hierle, J. Noncryst. Solids **47**, 287-290 (1982).
5. R. Hierle, J. Badan, and J. Zyss, J. Cryst. Growth **69**, 545-554 (1984).
6. J. Zyss, I. Ledoux, R. B. Hierle, R. K. Raj, and J. L. Oudar, IEEE J. of Quantum Electron. **QE-21**, 1290-1295 (1985).
7. D. Josse, R. Hierle, I. Ledoux, and J. Zyss, Appl. Phys. Lett. **53**, 2251-2253 (1988).
8. M. Yamakawa, T. Kubota, K. Ezumi, and Y. Mizuno, Spectrochimica Acta 30, 2103-2119 (1974).
9. M. Shino, M. Yamakawa, and T. Kubota, Act. Cryst. B **33**, 1549-1556 (1977).
10. G. V. Kulkarni, A. Ray, and C. C. Patel, Journal of Molecular Structure 71, 253-262 (1971).
11. M. Pierre, P.L. Baldeck, D. Block, R. Georges, H.P. Tromsdorff, and J. Zyss, "NPO Crystal Electronic Spectra," to be published.
12. M. Joyeux, G. Menard, N.Q. Dao, J. of Raman Spect. 19, 499-502 (1988).

Non-linear Optics in Poled Nematic Matrices: New Effects in Second Harmonic Generation and Molecular Organization

Garry Berkovic, Valeri Krongauz, and Shlomo Yitzchaik

DEPARTMENT OF STRUCTURAL CHEMISTRY, THE WEIZMANN INSTITUTE
OF SCIENCE, REHOVOT, ISRAEL 76100

Poling a polymeric film containing dopant chromophore molecules or side chain groups with high molecular hyperpolarizabilities (β) enables the preparation of films exhibiting large second order nonlinear optical effects such as second harmonic generation (SHG) and electro-optic modulation [1]. At a temperature above the glass transition temperature (T_g), a strong external electric field preferentially aligns dipolar chromophore groups along the field direction (designated as the x direction); this alignment may be preserved if the polymer is cooled to below T_g before removing the field. This process leads to the formation of a medium where the major component of $\chi^{(2)}$, the nonlinear susceptibility tensor, is $\chi^{(2)}_{xxx}$. Other nonzero components are $\chi^{(2)}_{xyy} = \chi^{(2)}_{xzz}$, and $\chi^{(2)}_{yxy} = \chi^{(2)}_{zxz}$. All other $\chi^{(2)}$ components are zero [1].

In the usual method of sample preparation, a thin (several μm) planar film is poled by a field applied perpendicular to the plane of the film. This field may be applied by sandwiching the film between electrodes, or by a corona discharge. In either case, DC fields as high as 10^6 V/cm may be applied to the film, which have resulted in $\chi^{(2)}$ coefficients as high as 10^{-7} esu [2,3].

The present work describes the observation of some surprising, and potentially very useful nonlinear optical phenomena on thin films which are poled parallel to the plane of the film (see Figure 1). We have observed [4,5] that poling certain polymer films in this geometry can lead to more nonzero $\chi^{(2)}$ coefficients than obtained by the usual poling procedure, and that the nonlinearity **PERPENDICULAR** to the poling direction is much stronger than that parallel to it. In particular, the $\chi^{(2)}_{zzz}$ component can be much stronger that the $\chi^{(2)}_{xxx}$ component (x is the direction of poling in the film plane, and z is normal to the film). The "cross term" $\chi^{(2)}_{zxx}$ can also be stronger than $\chi^{(2)}_{xxx}$. The largest coefficient we have observed from these type of films is $\chi^{(2)}_{zzz} = 6 \times 10^{-8}$ esu for a sample poled at 5×10^4 V/cm [5] i.e. a nonlinearity comparable to the best results from films poled in the "conventional" manner by much stronger DC fields.

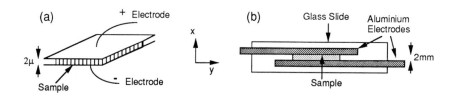

Figure 1. (a) Sandwich geometry and (b) in-plane geometry for electric field poling of organic polymer films. The poling field direction is designated as the x-axis in both cases. The film thickness, in the x- and z- directions for (a) and (b) respectively, is typically a few microns.

The systems producing this effect are nematic polymer glasses containing the high β molecules merocyanine (MC) and dimethylamino-nitrostilbene (DANS). This effect has been observed when the nematic glass is a liquid crystal polyacrylate containing the high β molecules either as a co-monomer or blended with it, as well as in a special nematic-like glass (known as a "quasi liquid crystal") of merocyanine in equilibrium with its structural isomer, spiropyran.

EXPERIMENTAL

We have synthesized the various polymers shown in Figure 2. In order to incorporate the merocyanine group in the side chain of a liquid crystal polymer, our strategy [4] was to synthesize a photochromic liquid crystal polymer (PLCP) with a spiropyran side group, and convert this group to merocyanine by UV irradiation after polymerization.

The unique properties of the so-called quasi liquid crystals (QLC) and their preparation are discussed elsewhere [6]. An equilibrium mixture of the merocyanine and spiropyran forms of this low molecular weight material can form a nematic like glass.

DANS molecules have also been blended with other QLCs and PLCPs, and also with a liquid crystal polymer (LCP) which contains no high β groups.

The preparation of films according to Figure 1(b) and our set up for SHG measurements are described in an earlier publication [4].

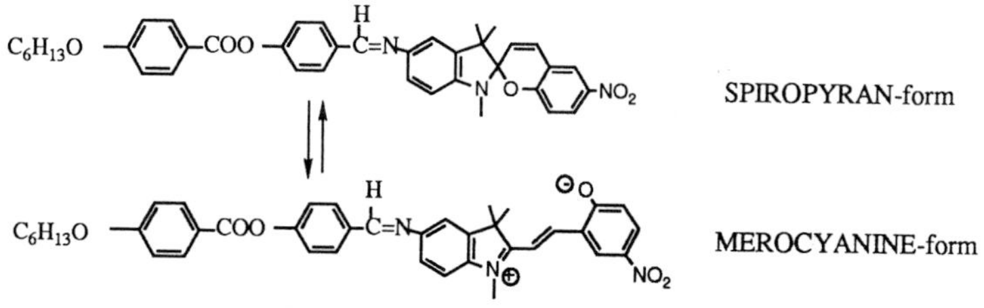

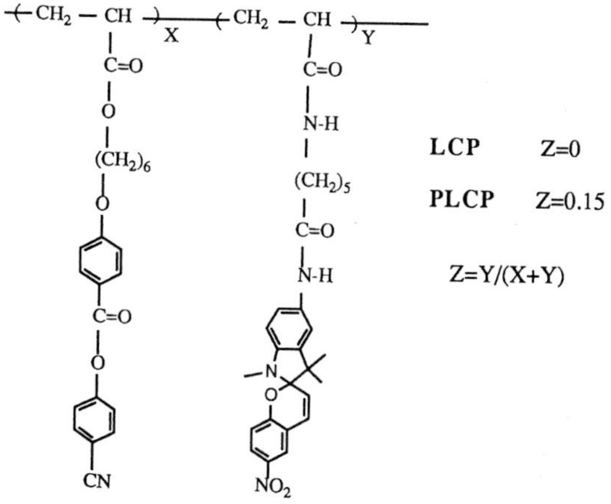

Figure 2.

RESULTS AND DISCUSSION

Second harmonic generation from these films was
measured several (4-7) days after the original
preparation and poling procedure. Measurements are
performed while a DC electric field is re-applied at
room temperature. SHG signals are measured as relative
intensities I_{v-v}, I_{h-h} and I_{v-h}, where the first
subscript (vertical or horizontal) represents the
polarization of the input laser, and the second
subscript represents the polarization of the detected
SHG signal. The vertical polarization direction
coincides with the sample x-axis, while the horizontal
polarization direction contains components along the
sample y and z directions (see Figure 3a). Results are
presented in Table 1, using light incident on the
sample at an external angle of 45°.

For a polymer in which poling breaks the symmetry
along the field direction (x) only, we would expect to
observe an I_{v-v} SHG signal due to the $\chi^{(2)}_{xxx}$ component. A
signal I_{h-h} only arises via $\chi^{(2)}$ components with all
three subscripts either y or z, and these components
are all expected to be zero (see above) in such a case.
Similarly, no I_{v-h} signal would be expected, since $\chi^{(2)}_{yxx}$
and $\chi^{(2)}_{zxx}$ should be zero.

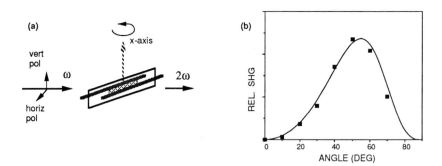

Figure 3. (a) Experimental geometry for SHG
experiments. (b) Dependence of the I_{h-h} signal of
2% DANS in LCP as the sample is rotated about its
x-axis. The angle is the external angle of
incidence, measured from the normal.

Table 1 Relative SHG signals in the presence of a 10 kV/cm electric field for the various combinations of input and output polarizations. The I_{h-v} signals (not shown) were always weaker than the other three.

Sample	I_{v-v}	I_{h-h}	I_{v-h}
QLC ; fresh	20	600	100
QLC ; 3 weeks old	5	400	40
PLCP ; fresh	10	60	
PLCP ; 3 weeks old	2	70	6
2% DANS in LCP	70	1×10^3	
2% DANS in QLC-PLCP (1:4) in 50kV/cm field	8×10^3	5×10^5	1×10^4

The results of Table 1 show that the above assumptions are clearly invalid in the case of our poled polymers. The appearance of I_{h-h} and I_{v-h} signals indicate that there is also asymmetry along either the y or the z direction. By measuring the I_{h-h} signal as the sample is rotated about the x axis, thus changing the projection of the horizontal polarization onto the y and z axes, we could determine which of these directions is responsible for the unexpected SHG signals. In Figure 2b we see that at normal incidence (horizontal polarization coincident with the y axis) the I_{h-h} signal vanishes, indicating that the y direction in the film is symmetric and thus not responsible for the SHG. Increasing the angle of incidence, and thus the projection of h onto z, increases the I_{h-h} SHG signal, indicating the fact that asymmetry along z (i.e.normal to the plane of the thin film) is causing these SHG signals. The signal again decreases for angles approaching grazing incidence, due to the large reflection losses of the incident light, as well as to the enlargement of the laser spot size on the sample.

Closer examination of the dependence of I_{h-h} on the angle of incidence of the laser reveals the contributions of the y-symmetric $\chi^{(2)}$ components to the I_{h-h} signal. Taking into account the angular dependent components of the laser field along the y and z directions inside the film, we could fit the data of Figure 2b using the ratio $\chi^{(2)}_{yzy}$ / $\chi^{(2)}_{zzz}$ as an adjustable parameter (The Kleinman condition, $\chi^{(2)}_{yzy} = \chi^{(2)}_{zyy}$ is assumed). The observed data in Figure 2(b) are fitted very well by $\chi^{(2)}_{yzy} = (0.2 \pm 0.1)\chi^{(2)}_{zzz}$.

Knowing that $\chi^{(2)}_{zzz}$ makes the dominant contribution to the observed I_{h-h} signals, we may estimate the magnitude of this coefficient in the various samples by comparing their SHG to that of a standard material of known $\chi^{(2)}$. We used the Maker fringe maximum SHG of crystalline quartz, and found that for the strongest signal, DANS in QLC/PLCP at 50 kV/cm, $\chi^{(2)}_{zzz} \approx 6 \times 10^{-8}$ esu/cm^3. In this sample, as well in the other compositions, the dependence of the SHG signals on the DC field [4,5] increases more rapidly than the usual quadratic dependence [3].

When the DC electric field was turned off, the I_{h-h} signal decayed on a time scale varying from minutes to hours. This decay was usually faster that the decay of the I_{v-v} signal. However, we have recently reported [5] two blends where the I_{h-h} signal decays much more slowly. In these cases the strong I_{h-h} signals were observed, without the need to re-apply the DC field, one week after poling.

The source of the $\chi^{(2)}_{zzz}$ nonlinearity in these samples which have been poled in the x direction was investigated. Spurious effects arising from $\chi^{(3)}$ nonlinearity, electric field components along z, or polarization rotations have been ruled out [4]. We concluded that the SHG originates mainly from dimers or higher aggregates of the high β species because [4]:

(i) Dyes including MC and DANS are known to aggregate readily in solution and amorphous and liquid crystalline polymers. The existence of aligned aggregates has been verified for these samples by absorption spectroscopy and linear dichroism.

(ii) Merocyanine containing samples which have been left in the dark for several weeks contain spiropyran monomers and merocyanine aggregates only, since merocyanine monomers decay to spiropyran [4]. These aged samples exhibited similar I_{h-h} signals to fresh samples, and since spiropyran has only very weak optical nonlinearity, we can conclude that the SHG arises mainly from MC aggregates.

In order that the aggregates of the high β species will have a non-isotropic arrangement in the z-direction, addition interactions must be invoked. The source of such a symmetry breaking interaction is not entirely clear. Two possibilities are long range interactions between the glass substrate and the nematic medium, or some charge transfer asymmetry induced when the strong poling field is applied, and some current inevitably flows through the system. It should be noted that in our samples the electrode thickness (several hundred $\mathring{A}$) is less than that of the polymer (1-2 μm) and so any trapped charges in the system will not be uniformly distributed along the z direction. If the latter explanation is correct, this effect could also arise in amorphous polymer glasses. Further investigation of this possibility are underway.

In conclusion, we have demonstrated an intriguing phenomenon, whereby poled polymers exhibit much stronger nonlinearity perpendicular to the poling direction than parallel to it. The $\chi^{(2)}$ coefficients obtained may be extremely high. Poled polymers films with such "two dimensional" nonlinearity could find applications in electro-optic devices, where this special property could be exploited for phase matched SHG, and the ability to perform electro-optic modulation on light of different polarizations.

REFERENCES

1. D.J. Williams, Angew. Chem. Int. Ed., 1984, 23, 690.
2. K.D. Singer, M.G. Kuzyk, W.R. Holland, J.E. Sohn, S.J. Lalama, R.B. Comizzoli, H.E. Katz and M.L. Schilling, Appl. Phys. Lett., 1988, 53, 1800.
3. S. Esselin, P. Le Barny, P. Robin, D. Broussoux, J.C. Dubois, J. Raffy and J.P. Pocholle, SPIE Conf. Proc., 1988, 971, 120.
4. S. Yitzchaik, G. Berkovic and V. Krongauz, Chem. Mater., 1990, 2, 162.
5. S. Yitzchaik, G. Berkovic and V. Krongauz, Macromolecules 1990, 23, 3539.
6. F.P. Shvartsman and V.A. Krongauz, Nature, 1984, 309, 608.

Laser Induced Gratings and Non-linear Optics in Organic Materials

S. Graham, M. Thoma, and C. Klingshirn

FB PHYSIK DER UNIVERSITÄT, 6750 KAISERSLAUTERN, FEDERAL REPUBLIC OF GERMANY

M. Eyal, D. Brusilovsky, and R. Reisfeld

CHEMISTRY DEPARTMENT, HEBREW UNIVERSITY, 91904 JERUSALEM, ISRAEL

S.V. Gaponenko, V.Yu. Lebed, and L.G. Zimin

INSTITUTE OF PHYSICS, BYELORUSSIAN ACADEMY OF SCIENCES, LENINSKY PR. 70, MINSK 220602, USSR

1. INTRODUCTION

We have studied the nonlinear optical properties of two different organic dyes, methyl orange and acridine orange, in different glass matrices, namely in a composite glass matrix prepared by the sol-gel method[1] and in a lead-tin (heavy) glass matrix. In the following section we describe the experiments with laser induced gratings (LIG) on the composite glasses with which we determined the origin and the value of the optical nonlinearity. Section 3 explains the measurements concerning the decay times of the luminescence of composite and heavy glasses and section 4 covers the one and two beam experiments performed on acridine orange in heavy glass with which the bleaching effects are investigated.

2. EXPERIMENTS WITH LASER INDUCED GRATINGS

Transient, reversible laser induced gratings measurements were performed on methyl and acridine orange in composite glass. The grating is created by crossing two coherent chopped beams of equal intensity of a cw Ar^+ - laser (λ_{exc} = 514.5 nm) under a small angle of 0.42° inside the sample. On both materials self-diffraction is observed. By measuring the diffracted intensity as a function of the incident one we can evaluate the grating efficiency and deduce the value of the optical nonlinearity. Further we can determine whether the optical nonlinearity is due to a change in the optical refractive index and/or in the absorption coefficient.

In figure 1 we have plotted the intensity of the first diffracted order I_1 versus the exciting intensity I for methyl and acridine orange in composite glass. The grating constant was Λ = 35 μm.

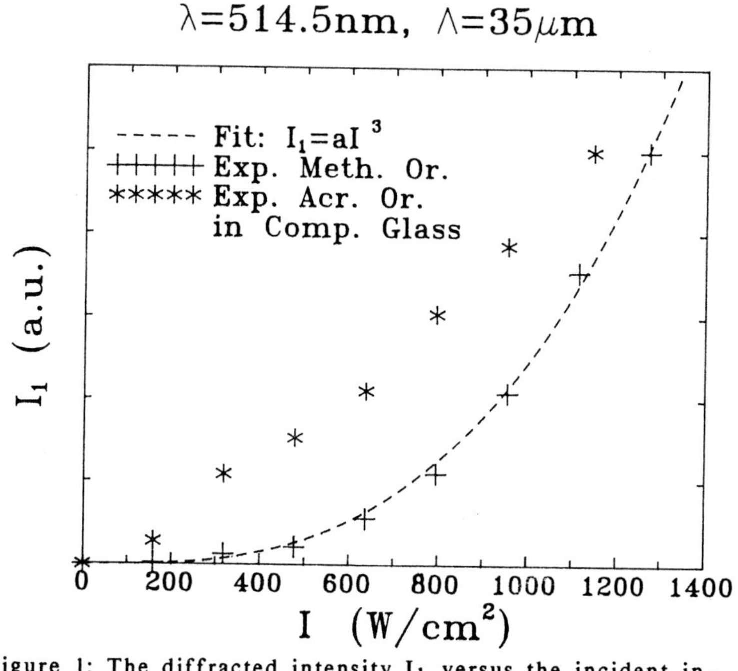

$$\lambda=514.5nm, \quad \Lambda=35\mu m$$

Figure 1: The diffracted intensity I_1 versus the incident intensity I; (***) experimental points for acridine orange and (+++) methyl orange in composite glass, (- - -) a calculated power-three dependence $I_1 \propto I$

In the case of the methyl orange doped glass the intensity of the first diffracted order shows a cubic dependence on the exciting intensity, which speaks for the assumption that the optical nonlinearity is mainly due to changes in the refractive index[2]. The grating efficiency reaches 0.02% at a peak intensity of the laser of 1.3 kW/cm^2. In the case of the acridine orange doped glass we do not have such a simple $I_1(I)$ dependence. Here the grating efficiency reaches a value of 0.33% at a peak intensity of the laser of 1.2 kW/cm^2.

In a cw pump- and probe beam experiment, the laser beams were not chopped, we detected a change in the absorption of about 10% at an excitation intensity (averaged) of 100 W/cm^2. From this finding we can conclude that in the acridine orange composite glass the non-linear optical change is of absorptive as well as of dispersive origin, but compared to the methyl orange doped glass the absorptive part is more dominant, because we do not observe a power-three dependence. Further details of this experiment and of the decay times of the LIG can be found in [3]. The origin of the LIG is for both dyes thermal.

3. MEASUREMENTS OF THE DECAY TIMES OF THE LUMINESCENCE

The time-resolved experiments on the nanosecond luminescence ($S_1 \rightarrow S_0$ transition) of the organic dyes in different glass matrices were performed by exciting the samples with a pulse lamp which operates with a frequency of 15 kHz and has a pulse duration of several nanoseconds. In figure 2 we see the curves of the normalized luminescence intensity ($\cdots$) and of the normalized excitation intensity (---) in a logarithmic scale versus the time in nanoseconds for acridine orange in composite glass. The excitation wavelength was 350 nm, the luminescence was detected at 575 nm. We see that the decay of the luminescence is nonexponential. Only parts of the experimental points can be fitted by exponentials which yield for acridine orange in composite glass decay times of the luminescence of τ = 2.4 ns (the luminescence has decayed to 30.3%) up to τ = 13 ns in the time range of 40-70 ns. The same experiment was performed at a different excitation wavelength λ_{exc} = 450 nm. Here the luminescence decay times range from τ = 2.5 ns (43.9%) to τ = 6.2 ns (time range 30-50 ns).

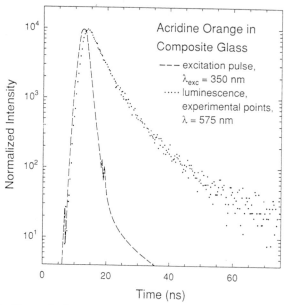

Figure 2: Plot of the normalized luminescence intensity ($\cdots$) and of the excitation intensity (---) in dependence of the time for acridine orange in composite glass

When incorporated into a different matrix, namely into a heavy glass matrix, acridine orange has decay times of τ = 1.2 ns (67%) to τ = 13 ns (time range 40-60 ns) at an excitation wavelength of 337 nm and λ_{lum} = 580 nm. The same experiment was performed on acridine orange in an ethanol solution in two different concentrations. Solution No.1 had approximately the same concentration as the heavy glass containing this dye, solution No.2 had a ten times lower concentration. In both solutions we observe an exponential decay of the luminescence with, contrary to acridine orange in the glass matrices, only one time constant. The decay time increases with the decrease of the concentration of the dye solution. We evaluated a decay time of 1.7 ns for solution No.1 and of 3.8 ns for solution No.2. Besides the nanosecond luminescence in acridine orange in heavy glass we also observe a delayed luminescence in the millisecond range, which is created by the population of the lowest lying triplet state; the decay times have been published in [4].

In methyl orange in composite glass the decay times of the luminescence vary from $\tau < 0.5$ ns to 8 ns (time range 20-35 ns). The excitation wavelength was 337 nm and $\lambda_{lum} = 590$ nm.

4. BLEACHING EXPERIMENTS IN AN ACRIDINE DOPED HEAVY GLASS

On acridine orange in heavy glass we performed one and two beam experiments with tunable dye lasers as excitation sources in order to investigate the bleaching effects due to phase-space filling. In figure 3a we see the results of the two beam experiment, we have plotted the optical density versus the wavelength of the probe beam. The solid curve was measured without laser excitation. Under excitation with $\lambda_{exc} = 485$ nm and $I_{exc} = 10$ MW/cm^2 (T=300 K) we observe a bleaching in the range of 460-540 nm (xxx in figure 3a). The value of the absorption change increases when exciting the sample at a wavelength of $\lambda_{exc} = 515$ nm in the wavelength range of 460-475 nm and between 495 and 535 nm ($\cdots$ in figure 3a). These measurements complement pump- and probe beam experiments which have been performed at longer wavelengths, where we detected an induced absorption between 550 and 680 nm and bleaching between 535 and 545 nm at an excitation wavelength of 500 nm and excitation intensities between 6 and 15 MW/cm^2. Measurements at an excitation intensity of 30 MW/cm^2 have been already described in detail in [5].

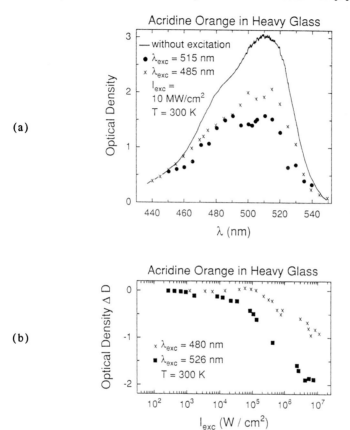

(a)

(b)

Figure 3: (a) Two beam experiment and (b) One beam experiment on acridine orange in heavy glass (further explanations in the text)

Figure 3b shows the change of the optical density ΔD as a function of the excitation intensity measured in a one beam experiment for two different wavelengths. We see that for $\lambda = 480$ nm (xxx) a bleaching is observed only for excitation intensities above 10^5 W/cm^2. For λ = 526 nm (squares) the bleaching effect is detected already at lower intensities, that is above 10^4 W/cm^2.

In conclusion we state that these results are mainly important for basic research and that we do not see a possibility for application in data-processing, waveguide development or related fields in the near future which would be compatible with the semiconductors.

5. REFERENCES

1. R. Reisfeld, "Optical Behaviour of Molecules in Glasses prepared by the Sol-Gel Method", Proc. Winter School on Glasses and Ceramics from Gels, Sol-Gel Science and Technology, Brazil, August 1989, Eds. M.A. Aegerter, M. Jafelicci Jr., D.F. Souza and E.D. Zanotto, World Scientific, Singapore, New Jersey, London, Hong Kong, pps. 323-345

2. H.J. Eichler, P. Günter, D.W. Pohl, "Laser-Induced Dynamic Gratings", Springer Series in Optical Sciences Vol. 50, Berlin, Heidelberg, 1986

3. S. Graham, M. Eyal, M. Thoma, D. Brusilovsky, R. Reisfeld and C. Klingshirn, J. Lumin., to be published, Proc. of the Intern. Conf. on Luminescence 1990, Lisboa

4. R. Reisfeld, "Luminescence and Nonradiative Processes in Porous Glasses", Intern. School on "Advances in Nonradiative Processes in Solids", Erice 1989, to be published in NATO ASI Series, Ed. B. Di Bartolo

5. S. Graham, R. Renner, C. Klingshirn, W. Schrepp, R. Reisfeld, D. Brusilovsky and M. Eyal, Proc. Intern. Conf. Materials for Non-Linear and Electro-Optics, Cambridge (U.K.), 1989, Inst. Phys. Conf. Ser. No.103, Ed. M.H. Lyons, pps. 157-162

Second-order Non-linear Optical Properties of Donor–Acceptor Acetylene Compounds

A.E. Stiegman and Joseph W. Perry

JET PROPULSION LABORATORY, CALIFORNIA INSTITUTE OF TECHNOLOGY
4800, OAK GROVE DR., PASADENA, CALIFORNIA 91109, USA

Lap-Tak Cheng

CENTRAL RESEARCH AND DEVELOPMENT DEPARTMENT, E. I. DU PONT DE
NEMOURS AND CO., INC., EXPERIMENTAL STATION, P.O. BOX 80356,
WILMINGTON, DELAWARE 19880-0356, USA

We present here a detailed study of the relationship between the electronic structure and the microscopic second-order nonlinear polarizability for the series of donor-acceptor phenylacetylene compounds having the general structure shown below (where D and A are an electronic donor and acceptor group, respectively).[1]

For this series, systematic changes in the conjugation length and in the donor-acceptor group can be readily accomplished with the effect of these changes on the molecular second-order hyperpolarizability (β) correlated. From this, an understanding of which changes result in the most significant enhancement of the hyperpolarizability can be deduced.[2]

This series of compounds was also screened for SHG (second-harmonic generation) activity. A number of the materials proved to be SHG active with several showing high conversion efficiencies relative to urea.

The second-order molecular hyperpolarizability, β, is determined by the measurement of the electric field induced second harmonic generation (EFISH).[3] This measurement yields a scalar quantity which is the projection of the vector invariant of the second-order hyperpolarizability tensor onto the ground state dipole moment of the molecule, $\beta \cdot \mu$. If the molecular axis is taken as the z axis, then dividing out the dipole moment yields the z projection of the tensor (eq.1).[4]

$$\beta_z = \beta_{zzz} + 1/3 \; (\beta_{zyy} + \beta_{zxx} + 2\beta_{xxz} + 2\beta_{yyz}) \qquad (1)$$

For linear molecules in which the dipole moment lies along the molecular axis, which is the case for the donor-acceptor acetylenes, the contribution from the off-diagonal tensor elements is minimal yielding $\beta_z \approx \beta_{zzz}$.

The magnitude of β can be related to the spectroscopic and photophysical properties of a molecule through the familiar two-state second-order perturbation expression (eq. 2).[5]

$$\beta = \frac{3\Delta\mu_1 m_1^2}{2h^2} \frac{\omega_1^2}{\left(\omega_1^2 - 4\omega^2\right)\left(\omega_1^2 - \omega^2\right)} \qquad (2)$$

From this expression, the magnitude of β is a function of ω_1 and ω (the energy of the transition and the incident (laser) radiation, respectively) and is proportional to m_1 (the transition moment integral) to $\Delta\mu_1$ (the dipole moment change). It is the relative importance of these spectroscopic and photophysical factors on the magnitude of β that emerges from this investigation.

The acetylene bridged donor-acceptor molecules have characteristic spectroscopic and excited state properties that are dependent on the *donor-acceptor strength* and the *conjugation length*.[6]   These molecules all show low-energy intermolecular charge-transfer (ICT) transitions which are characterized by high oscillator strengths and large excited-state dipole moments. For this series of molecules the energy of the ICT band, the oscillator strength (f), the transition moment integral (m_1) and the dipole moment change ($\Delta\mu$) are given in Table I along with the value of β obtained from the EFISH experiment.[7]

The dependence of β of donor-acceptor strength.

The energy of the ICT band varies as the donor and acceptor functional groups change, giving a relative ordering scheme for the different groups.   The ordering of donor groups based on a fixed acceptor (NO_2) is: $(R)_2N > (R)(H)N > H_2N > RS > RO$.   The ordering for acceptor groups, based on a fixed donor (H_2N) is: $NO_2 > CN > SO_2CH_3 \approx C(O)OCH_3 \approx C(O)CH_3$. The oscillator strength and transition moment integral for the ICT band is, however, relatively invariant with

Table I. Spectroscopic and Nonlinear Optical Properties of Donor-Acceptor DiphenylAcetylenes:

n	A	D	$\lambda_{max}(\varepsilon \times 10^{-4})$	f	$|m_1|^2 (\times 10^{-35}\ esu)$	$\mu(\times 10^{-18}\ esu)$	$\Delta\mu(\times 10^{-18}\ esu)$	$\beta(\times 10^{-30}\ esu)$
1	SO_2Me	NH_2	338 (3.3)	.53	3.8	6.5	—	13
1	CO_2Me	NH_2	337 (2.3)	.40	2.9	3.8	10.6	15
1	C(O)Me	SMe	332 (3.1)	.47	3.3	3.7	4.6	9.8±2
1	C(O)Me	NH_2	336 (2.4)	.42	3.0	3.3	12.6	12
1	CN	SMe	326 (3.1)	.52	3.6	4.0	8.0	15
1	CN	NH_2	343 (2.9)	.49	3.6	5.2	4.3	20
1	CN	NMe_2	373 (2.0)	.28	2.2	6.1	3.1	29
1	NO_2	OMe	347 (2.0)	.42	3.1	4.4	6.2	14
1	NO_2	SMe	358 (2.2)	.50	3.8	4.0	5.4	20
1	NO_2	NH_2	379 (2.1)	.47	3.8	5.5	9.5	24
1	NO_2	NMe_2	416 (1.8)	.37	3.3	6.1	5.7	46
2	NO_2	SMe	373 (2.1)	.54	3.1	3.9	—	17
2	NO_2	NH_2	384 (1.8)	.37	3.0	6.3	8.5	28±2

a) chloroform solvent

changes in donor-acceptor group. Overall, the transition moment integral varies only by a factor of two over the entire series. In general the dipole moment changes, $\Delta\mu$, do not correlate with the energy or intensity of the ICT band with large dipole moment changes occurring for relatively weak donor-acceptor pairs. What is also significant is that, while an order of magnitude range is observed in the extreme the value of $\Delta\mu$, most of the donor-acceptor pairs vary by only a factor of three.

Consistent with these observations, the magnitude of β is expected to be dominated by the *energy* of the ICT transition with the product of the dipole moment change and transition moment integral being relatively constant. If this is correct then eq. 2 predicts (assuming $\omega_1 >> \omega$) that β should scale with $1/\omega^2$ of the transition energy. Figure 1 is a plot of β vs $1/\omega^2$ of the ICT transition for the whole series of donor-acceptor diphenylacetylenes.

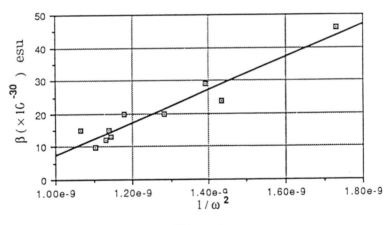

Figure 1

This plot is highly linear indicating that, at least for this class of molecules, the energy of the ICT transition does dominate the magnitude of β while variables such as the transition moment and the dipole moment change play a less significant role.

The dependence of β of conjugation length.

As the number of acetylenes increases the energy of the ICT band remains relatively unchanged with a small drop in the oscillator strength being observed. Optical transitions falling at higher energy, which are nominally "$\pi \rightarrow \pi^*$" transitions, decrease dramatically in energy with increasing conjugation length until, for the p-nitro-p'-aminodiphenylhexatriyne molecule, they overlap the ICT band.

Increasing the length of the molecules, however, does not result in an increase in the excited-state dipole moment or in the dipole moment change. In fact, the dipole moment change actually decreases in going from one to two acetylenes. This may be due to a greater delocalization of charge over the molecule as the π bonding molecular orbitals move to lower energy with increasing conjugation length.

As a result, the magnitude of β is relatively invariant as the number of acetylenes increases from one to two for all of the donor-acceptor pairs. For the amino-nitro series of molecules three different conjugation lengths corresponding to 1,2 and 3 acetylenes are available for comparison. To compare the entire series (Table II) it is necessary to compare values of the product $\beta \cdot \mu$ as the relative insolubility of the p-amino-p'-nitrodiphenylhexatriyne molecule required the use of a polar solvent (N-Methylpyrrolidone) which prevented an accurate determination of the ground state dipole moment. The values of the product $\beta \cdot \mu$ are relatively constant for the diphenylacetylene and the diphenylbutadiyne molecules while for the diphenylhexatriyne $\beta \cdot \mu$ increases sharply. As the ground-state dipole moment varies only slightly with increasing conjugation length the large increase seen in the $\beta \cdot \mu$ product is due primarily to an increase in β itself.

Table II. EFISH Measurements of the Nonlinear-Optical Properties of Donor-Acceptor Diphenylacetylenes in N-Methylpyrrolidone

Donor	Acceptor	# acetylenes	$\mu\beta$ (esu)
NH_2	NO_2	1	22×10^{-47}
NH_2	NO_2	2	24×10^{-47}
NH_2	NO_2	3	41×10^{-47}

We attribute the large increase in the value of β observed for the p-nitro-p'-aminodiphenylhexatriyne molecule to the superposition of several essentially isoenergetic transitions, all contributing to the magnitude of β. This superposition occurs when high energy "$\pi \rightarrow \pi^*$" bands move to lower energy with increasing conjugation length. In the case of the amino-nitro-diphenylhexatriyne molecule the electronic spectrum is comprised of a large intense transition with an oscillator strength of f = 2.0, suggesting that it is composed of a number of transitions including the ICT band (which appears as a shoulder). It is the sum of all of these transitions that determine the value of the second-order hyperpolarizability.

Second harmonic generation.

The series of donor-acceptor acetylene molecules were screened for their SHG activity using the Kurtz powder method.[8] The efficiencies (relative to a urea standard) of those found to be active are shown in Table III. Several of the materials proved to have relatively high SHG efficiencies and one compound, p-amino-p'methylcarboxylatediphenylacetylene, proved to be phase matchable.

Table III Second Harmonic Generation Efficiencies for Donor-Acceptor Acetylene Compounds.

Donor	Acceptor	n	SHG Efficiency (relative to urea)	
			1064 nm	1907 nm
NH_2	NO_2	1	0.2	0.4
NH_2	NO_2	2	0.05	0.07
NH_2	NO_2	3	0.03	0.7
NH_2	CN	1	0.2	—
NH_2	CN	2	0.4	0.3
NH_2	CO_2CH_3	1	120	120
NH_2	CO_2CH_3	2	16	34
CH_3S	NO_2	1	65	200
CH_3S	NO_2	2	0.2	0.6
CH_3S	CN	2	0.7	—

Summary and conclusions.

In general, a number of conclusions about the effect of structural changes on the magnitude of the second-order molecular hyperpolarizability for this class of molecules can be drawn. Overwhelmingly, the value of β is dominated by the energy of the ICT band with strong donor-acceptor pairs producing large values of β primarily because they have low energy absorption bands. The effect of the transition moment and dipole moment change on the value of β were less important and less predictable.

As the conjugation length increased the dipole moment change actually decreased and the value of β varied only slightly. One particularly interesting result from this study was the origin of the large value of β achieved for the amino-nitrodiphenylhexatriyne molecule. The degree to which the superposition of electronic states in long conjugated molecules contributes to large values of β is worth further investigation.

Acknowledgement

This work was performed by the Jet Propulsion Laboratory, California Institute of Technology, as part of its Center for Space Microelectronics Technology, which is supported by the Strategic Defense Initiative Organization, Innovative Science and Technology Office through an agreement with the National Aeronautics and Space Administration.

REFERENCES

1. (a) Perry, J. W., Stiegman, A. E.,Marder, S. R., Coulter, D. R. in *Organic Materials for Non-linear Optics*,; Hann, R. A., Bloor, D., ed. Royal Chem. Soc. (b) Stiegman, A. E. *J. Am. Chem.* Soc. **1989**, 111, 8771

2. *Nonlinear Optical Properties of Organic and Polymeric Materials*, Williams, D. J., ed., ACS Symp. Ser 233, American Chemical Society, Washington, D. C., **1983**.

3. (a) Levine, B. F., Bethea, C. G. *J. Chem. Phys.* **1975**, 63, 2666 (b) Oudar, J. L. *J. Chem. Phys.* **1977**, 67, 446

4. (a) Jerphagnon, J. *Phys. Rev (B)* **1970**, 2, 1091 (b) Ulman, A. *J. Phys. Chem.* **1988**, 92, 2385

5. Oudar, J. L., Zyss, J. *Phys. Rev.* (A) **1982**, 26, 2016.

6. Stiegman, A. E., Miskowski, V. M., Perry, J. W., Coulter, D. R. *J. Am. Chem. Soc.* **1987**, 109, 5884.

7. See, for example, Mataga, N., Kubota, T. *Molecular Interactions and Electronic Spectra* **1970**, Marcel Dekker, New York.

8. Kurtz, S. K., Perry, T. T. *J. Appl. Phys.* **1968**, 39, 3798.

Non-linear Optical Properties of 2-Adamantylamino-5-nitro-pyridine (AANP) Crystal

S. Tomaru, T. Kurihara, H. Suzuki, N. Ooba, and T. Kaino

NTT OPTO-ELECTRONICS LABORATORIES, NIPPON TELEGRAPH AND TELEPHONE CORPORATION, TOKAI-MURA, NAKA-GUN IBARAKI-KEN 319-11, JAPAN

S. Matsumoto

NTT APPLIED-ELECTRONICS LABORATORIES, NIPPON TELEGRAPH AND TELEPHONE CORPORATION, MIDORI-MACHI, MUSASHINO-SHI, TOKYO-TO 180, JAPAN

1 INTRODUCTION

Some organic materials have attracted much attention as they show promise for application in the field of nonlinear optics.[1,2] In particular, some crystals, for example, NPP[3] and DAN[4], are expected to be utilized in frequency converters for the near IR to visible wavelength region.

It is well known that these crystals should have a noncentrosymmetric structure to realize second-order optical nonlinearity. An effective method for achieving this structure is to add the bulky group to nitroaniline derivatives.[5,6]

We have developed several kinds of organic crystals with a bulky donor group to obtain second order optical nonlinearity. This paper reports the second-order nonlinear optical properties of 2-adamantylamino-5-nitropyridine(AANP) which shows very large nonlinearity among the materials reported to date.

2 SYNTHESIS AND SHG EXPERIMENTS

The materials were prepared using the following reaction:

$$X = CH \text{ or } N \qquad R = \text{(adamantyl)} \text{ or } \text{(norbornyl)}$$

The obtained crude product was purified by column chromatography and subsequent recrystallization from

methanol. Finally, the crystal was purified by the sub-
limation technique. The purified materials were analyzed
by means of NMR and IR spectroscopy.

The SHG efficiency of all the sythesized materials
was evaluated by the powder technique developed by Kurts
and Perry. A Q-switched Nd:YAG laser with a wavelength of
1.064 μm was used as a fundamental light source. Table 1
shows the relative SHG intensity ratio to urea[7] of AANP
and other materials. DAN[4] was also used as a reference
material. All the sythesized materials have a bulky amino
group as a donor group. It was found that almost all the
materials in table 1 had a noncentrosymmetric crystal
structure. Among these, the SHG intensity of AANP was the
largest, 300 times larger than that of urea, and about 3
times larger than that of DAN. Therefore, we have made a
detailed investigation of the second order optical nonlinearity of the AANP single crystal.

3 CRYSTAL GROWTH OF AANP

Crystal growth of AANP from the melt, the solution and the
vapour, have been investigated. It was found to be difficult to grow large single AANP crystals from the vapour.
It was easy to grow large AANP crystals from dimethyl for-
mamido solution, however, it was difficult to obtain good
optical quality. So, we grew AANP crystals from the melt
using a modified vertical Bridgeman technique. The melting
point of AANP is 167°C. At temperatures higher than mp,
AANP crystals decompose gradually due to residual
impurities. Therefore, carefully purified material needs
to be used and we grew the crystals using the Bridgeman
technique in a N_2 atmosphere. The crystal growth rate was
about 1mm/hr. AANP single crystals with typical dimensions
of 5x5x1 mm^3 were obtained. Figure 1 shows a photograph of
an AANP crystal grown from the melt.

Table 1.   Relative SHG intensity ratio to urea of some developed
organic materials with a bulky donor group.

compound	SHG intensity ratio	compound	SHG intensity ratio
NO_2—⟨O⟩—N(H)— (bicyclic)	~0	NO_2—⟨O⟩(N)—N(H)— (bicyclic)	12
NO_2—⟨O⟩—N(H)— (adamantyl)	30	NO_2—⟨O⟩(N)—N(H)— (adamantyl) (AANP)	300
urea[7]	1	DAN[4]	115

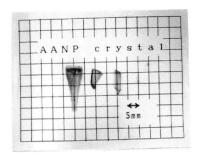

Figure 1.   Photograph of AANP crystal grown from the melt
using a modified vertical Bridgeman technique.

4 CRYSTAL STRUCTURE OF AANP

The crystal structure of AANP was determined by X-ray
diffraction.   AANP crystals belong to the orthorombic system with space group Pna2$_1$ and point group. mm2.   The lattice parameters are ; a=7.992 Å, b= 26.313 Å, c=6.601Å.[9]
There are four molecules per unit cell, i.e., Z=4.   Figure
2 shows the structure of AANP crystals with unit cell axes
viewed along the c axis.   The cleaved plane was found to be
a (010) plane, that is an ac plane.

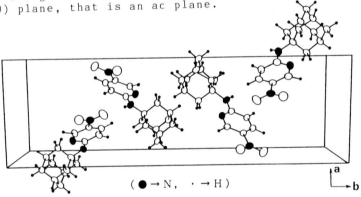

$(\bullet \rightarrow N, \quad \cdot \rightarrow H)$

Figure 2.   Packing diagram viewed along the c axis.

Zyss and Oudar have reported the optimum molecular orientations for efficient phase-matching, and have concluded that
the charge transfer axis should lie at an angle of $\theta = 54.74$[9]
in a molecular orientation with the point group mm2.   The
orientation of the molecular charge transfer axis with
respect to the crystallographic symmetry axis could be estimated from the result of the X-ray diffraction.   For the
AANP molecular orientation, it was found that the angle(θ)
is about 58°.

5 OPTICAL PROPERTIES

Figure 3 shows the absorption spectrum of a 250 μm thick-AANP crystal in the ultra-violet and visible wavelength region. The cut off wavelength was about 460 nm.

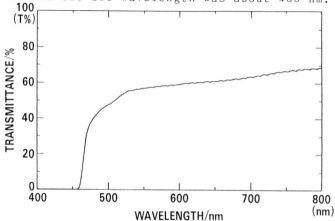

Figure 3. Absorption spectrum of a 250 μm-thick AANP crystal in the visible wavelength region.

Table 2 shows the refractive indices of AANP. The refractive indices at 0.53 μm were measured directly by using the Becke line method.[10] Those at 1.06 μm were estimated by measuring several refractive indices in the visible wavelength region.

6 NONLINEAR OPTICAL PROPERTIES OF AANP CRYSTALS

For crystals belonging to point group mm2, the second harmonic polarization P is expressed by[5]

$$P_1 = 2\epsilon_0 d_{15} E_1 E_3$$

$$P_2 = 2\epsilon_0 d_{24} E_2 E_3$$

$$P_3 = \epsilon_0 (d_{31} E_1^2 + d_{32} E_2^2 + d_{33} E_3^2)$$

Table 2. The refractive indices of AANP crystals.

refractive index	* $\lambda = 0.53 \mu$m	** $\lambda = 1.06 \mu$m	
n_a	1.77	1.67	
n_b	1.61	1.59	* measured values
n_c	1.86	1.71	**estimated values

where d_{ij} are tensor components of the second-order non-linear optical coefficients and E_1, E_2 and E_3 are the electric field components of the fundamental wave along the crystallographic axis a,b and c, respectively.

The coefficients of the AANP crystal were measured using the Maker fringe method.[11] A Q-switched Nd:YAG laser with a 1.064 μm wavelength, a repetition rate of 10 Hz and a pulse duration of 8ns was used as a fundamental light source. As b-plates of AANP ,that is cleaved planes, were used for the measurement, the coefficients d_{31} and d_{33} could be estimated. The fundamental light polarized along the rotation axis was incident on the ac plane. Maker fringes for d_{31} and d_{33} were observed with the rotation axis parallel to the a axis and c axis, respectively. The generated second-harmonic light for d_{31} was polarized vertically to the rotation axis and that for d_{33} was polarized horizontally. The observed Maker fringe pattern for d_{31} is shown in Fig.4. Coefficients d_{31} and d_{33} were estimated by comparing the SHG intensity with that of a quartz plate for which d_{11}=0.5 pm/V.[12] Table 3 shows d_{31} and d_{33} of the AANP crystal. These values were comparable with the coefficients of MBANP[13] and NPP[3].

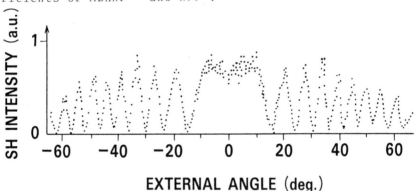

EXTERNAL ANGLE (deg.)

Figure 4.   Maker fringes for d_{31} of a 250 μm thick AANP crystal plate (a cleaved plane).

Table 3.   d coefficient and coherent length of AANP crystals.

d coefficient		coherent length; lc	
d_{31}	~80 pm/V	l_c^{31}	~2.0 μm
d_{33}	~60 pm/V	l_c^{33}	~2.1 μm

7 PHASE MATCHED SHG

Figure 5 shows the relationship between the particle size of AANP crystal powder and its relative SHG intensity ratio to urea. These relative intensities increase monotonously dependent on increases in particle size.

The results of the powder SHG measurements and refractive indices measurements indicate the possibility of phase-matched SHG in AANP crystals. Therefore, angle-tuned phase-matched second harmonic generation was attempted by rotating a 1mm thick AANP crystal.

A cw Nd:YAG laser with a 1.064 μm wavelength was used as a fundamental light source. SHG intensity in AANP crystals exhibited angular dependencies. When the rotating angle was about 55° to the cleaved plane, the SHG signal was enhanced. SHG efficiency was estimated to be about $2 \times 10^{-3} (W^{-1})$. We believe that the efficiency can be increased by improving the optical quality of AANP crystal and developing the polishing technique for organic crystals.

From the refractive indices measurements, the possibility of realizing SHG in the 1.3-1.55 μm fundamental wavelength region was also indicated.[14]

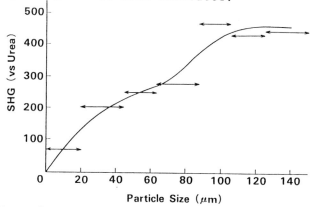

Figure 5. Relationship between particle size and relative SHG intensity ratio to urea in AANP crystals.

8 CONCLUSION

By adding a bulky donor group to a nitrobenzene or nitropyridine ring, some organic materials for second order optical nonlinearity were developed. Among these materials, it was found that 2-adamantylamino-5-nitropyridine (AANP) crystals had one of the largest second-order nonlinear optical susceptibilities reported so far.

Coefficients d_{31} and d_{33} were estimated using the Maker
fringe method, and were found to be comparable to that of
NPP. Angle-tuned second harmonic generation of Nd:YAG
laser radiation at 1.064 μm was demonstrated using a 1 mm
thick AANP crystal and the conversion efficiency was found
to be about $2 \times 10^{-3} (W^{-1})$.

Since the new organic crystal, AANP, has large second-
order optical nonlinearity, it is a promising material for
use in frequency converters in the visible to near IR
wavelength region.

9 ACKNOWLEDGEMENT

The authors would like to thank H.Takara for his helpful
discussions and K.Murase and H.Hiratsuka for their continuous support and encouragement.

10 REFERENCES

1. D.J.Williams,ed., 'Nonlinear Optical Properties of Organic
   and Polymeric Materials', Am.Chem.Soc.Symp.Ser. 233, American
   Chemical Society, Washington,D.C., 1983.
2. D.J.Williams, Angew.Chem.Int.Ed.Engl., 1984, 23, 690.
3. J.Zyss, J.F.Nicoud and M.Coquillay, J.Chem.Phys.,1984,81,
   4160
4. J.-C.Baumert,R.J.Twieg,G.C.Bjorklund,J.A.Logon, and C.W.Dirk,
   Appl.Phys.Lett., 1987, 51, 1484.
5. P.Gunter,Ch.Bosshard,K.Sutter,G.Chapuis,,R.J.Twieg and
   D.Dobrowolski, Appl.Phys.Lett., 1987, 50, 486.
6. J.F.Nicoud, Mol.Cryst.Liq.Cryst., 1988, 156, 257.
7. J.A.Morrell,A.C.Albrecht,K.H.Levin and C.L.Tang,J.Chem.Phys.,
   1979, 71, 5063.
8. N.Ooba,H.Suzuki,S.Tomaru and T.Kaino, to be submitted.
9. J.Zyss and J.L.Oudar, Phys.Rev., 1982, A26, 2028.
10. N.H.Hartshorne and A.Stuart,'Crystals and the Polarizing
    Microscope', Edward Arnold, London, 1960.
11. J.Jerphagonon and S.K.Kurts, J.Appl.Phys., 1970, 41, 1667
12. R.Bechmann and S.K.Kurtz, in Landolt-Bornstein, 'Numerical
    Data and Functional Relationships', Group 3. Crystal and
    Solid State Physics, Vol.2., Springer-Verlag, Berlin, 1969
13. R.T.Bailey, F.R.Cruickshank, S.M.G.Guthrie, B.J.Mcardle,
    H.Morrison,D.Pugh, E.A.Shepherd, J.N.Sherwood, C.S.Yoon,
    R.Kashyap, B.K.Nayar and K.I.White, Opt.Commun., 1988, 65,
    229.
14. H.Takara, private communication

Metal-organic Compounds

Structure/Property Relationships for Organic and Organometallic Materials with Second-order Optical Non-linearities

Seth R. Marder,* David N. Beratan,* Bruce G. Tiemann

JET PROPULSION LABORATORY, CALIFORNIA INSTITUTE OF TECHNOLOGY, 4800 OAK GROVE DRIVE, PASADENA, CALIFORNIA 91109, USA

Lap-Tak Cheng* and Wilson Tam

CENTRAL RESEARCH AND DEVELOPMENT DEPARTMENT, E. I. DU PONT DE NEMOURS AND CO., INC., EXPERIMENTAL STATION, P.O. BOX 80356, WILMINGTON, DELAWARE 19880-0356, USA

Second-order nonlinear optical coefficients arise in molecules lacking a center of symmetry.[1-3] Conjugated organic molecules with electron donating and accepting moieties can exhibit large electronic second-order nonlinearities (i.e. first hyperpolarizabilities), β. Generally, β increases with increasing donor and acceptor strength (defined approximately by the ionization potential and electron affinity, respectively) and increasing separation assuming strong electronic coupling through the bridge at all separations. A perturbation theory expression for β based on a molecular orbital description of the electronic states is consistent with experiments.[4] For donor-acceptor substituted benzenes such as 4-nitroaniline, a two-state model for β is adequate - the term associated with the charge transfer (CT) excited state in the perturbation theory expression for β dominates. The two-state expression for the dominant component of the β tensor is proportional to[4,5]:

$$(\mu_{ee} - \mu_{gg}) \frac{\mu_{ge}^2}{E_{ge}^2}$$

where g is the index of the ground state, e is the index of the CT excited state, and μ is the dipole matrix element between the two subscripted states. Two-state expressions account reasonably well for the observed β values obtained by solution DC electric field induced second harmonic generation (EFISH) measurements on many organic molecules.[5,6] EFISH provides direct information on the vectorial projection of β along the molecular dipole (μ_{gg})

direction. For organic compounds, structure-property trends concerning donor-acceptor strength and the conjugated intervening bridge structure have been the topics of many studies.[7,8]

In this paper we will show that by using the two state approximation for β, one can find a combination of donor and acceptor strengths for a given separation and bridge structure that maximizes $|\beta|$. We propose and implement a molecular design strategy that should enable one to optimize β for a given bridge structure and number of intervening bonds. In the second part of the paper we show that metallocenes can exhibit substantial β.

THE ROLE OF THE BRIDGING π SYSTEMS IN SECOND-ORDER NONLINEAR OPTICS

Manipulation of the chemical properties of a chromophore can substantially alter the first hyperpolarizability, and these changes can be understood from Eq. 1. The terms in this equation are (1) the change in dipole moment between the two states, (2) the square of the transition matrix element (proportional to the oscillator strength), and (3) the inverse square of the transition energy. Maximizing the product of these factors requires compromise; the individual terms each have their own characteristic maxima as a function of the chemical details of the molecule. More specifically, changing electron donor and acceptor strength for a given conjugated bridge varies these three factors substantially. To gain insight into the interrelationships among $\mu_{ee} - \mu_{gg}$, μ^2_{ge}, the energy of the CT transition and the dominant component of β we performed a simple series of Hückel molecular orbital calculations on a four orbital model system consisting of donor, acceptor, and bridge.[9,10] The coulomb energy difference between the donor (α_D) and the acceptor (α_A) was varied in 100 uniform increments from zero to 3t (where t is the exchange matrix element between the bridge orbitals). Molecular orbitals were calculated for each case as well as $\mu_{ee} - \mu_{gg}$, μ^2_{ge}, the energy difference between the highest occupied molecular orbital (HOMO) and the lowest unoccupied molecular orbital (LUMO), and β using Eq. 1. The results are shown in Fig. 1.

Several qualitative features emerge from the analysis of the two-level model for β in bridged donor acceptor molecules. Both μ^2_{ge} and $1/E_{ge}^2$ are maximized when the molecular asymmetry is zero ($\alpha_D - \alpha_A = 0$, i.e. equal coulomb energies for D and A orbitals) and decrease with increasing asymmetry, consistent with the behavior of cyanine chromophores. The change in dipole moment, $\mu_{ee} - \mu_{gg}$, is zero when $\alpha_D - \alpha_A$ is zero, increases to a maximum value, and then decreases as $\alpha_D - \alpha_A$ becomes large. When $\alpha_D - \alpha_A$ is zero, the charge distribution is symmetrical as is the case in a symmetrical cyanine dye. As $\alpha_D - \alpha_A$ increases, the induced asymmetry in the electron population of the HOMO and LUMO increases and, therefore, $\mu_{ee} - \mu_{gg}$ increases. This is the case for a somewhat unsymmetrical cyanine dye. As $\alpha_D - \alpha_A$ becomes very large, the bridge mediated donor-acceptor orbital mixing decreases. Thus, for molecules such as stilbenes substituted with weak donors and acceptors, the HOMO-LUMO transition, which is dominated by bridging atoms, looses charge-transfer character. Since β_{xxx} (the component of β along the charge-transfer axis) is the product of these three peaked functions, it also has a maximum that is on the low $\alpha_D - \alpha_A$ side of the $\mu_{ee} - \mu_{gg}$ peak. This result suggests that β_{xxx} is not maximized when the two states have the largest degree of charge transfer.

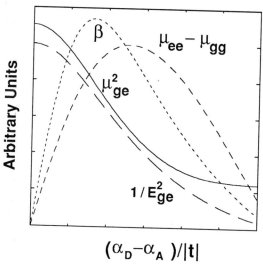

Figure 1. Plots illustrating the dependence of $\mu_{ee} - \mu_{gg}$ μ^2, $1/E_{ge}^2$ and β on $\alpha_D - \alpha_A$ derived from molecular orbital calculations on a four orbital system consisting of a donor, an acceptor and two bridge orbitals.

The relative energies of the limiting resonance structures of a donor-acceptor molecule are influenced by two factors. First, a Coulombic factor can be positive, negative, or zero depending on whether charge separation is created, annihilated, or simply shifted. The second factor, a resonance energy, depends on the aromaticity of the π electron donor and acceptor in the two limiting cases. The majority of systems that have been explored for second-order nonlinear optics fall into the categories of substituted benzenes, biphenyls, stilbenes, tolanes, and chalcones. These compounds all have aromatic ground states. The corresponding electron transfer states involve quinoidal ring structures. Electronic polarization arising from the applied electric field results in an increase net quinoidal character and therefore to a net loss of resonance stabilization energy (Figure 2a). Thus, the aromatic nature of the ground state should impede electronic polarization in an applied field. As a result, the effective donor and acceptor strength of a given pair connected by an aromatic bridge will be lower than when they are attached to a degenerate π-electron system. Compounds that have quinoidal ground states and aromaticity in the excited state can be synthesized and are predicted to assist charge polarization (Figure 2b). A degenerate π-electron bridge has identical resonance stabilization in both the ground and polarized (CT) states (Figure 2c and 2d).

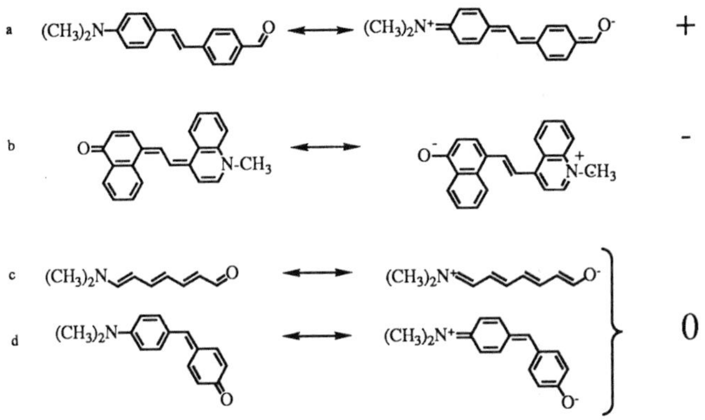

Figure 2. Change in resonance energy between neutral and dipolar states for (a) an aromatic, (b) quinoidal, (c) polyene and (d) aromatic/quinoidal π systems. In these molecules the neutral resonance structure contributes more to the ground state and the dipolar structure contributes more to the excited state.

For all of the donor-acceptor substituted benzenes, biphenyls, stilbenes, tolanes, and chalcones that have been studied, β increases with increasing donor and acceptor strength (i.e. decreasing $\alpha_D - \alpha_A$ in Figure 1). Indeed, one of the challenges identified for achieving large β in these molecules has been the search for stronger donors and acceptors. *We suggest that the choice of an aromatic ground state is the predominant factor that has limited the magnitude of β for a given chain length.* Molecules that have ground states degenerate in both charge transfer configurations, i.e., have partially quinoidal ground states, will not lose aromaticity upon charge transfer. This will greatly diminish the *effective* $\alpha_D - \alpha_A$ and may allow one to optimize β. An odd polyene donor-acceptor molecule (Figure 2c and d) should be inherently more polarizable than an aromatic system of comparable length (Figure 2a). The drawback of donor-acceptor polyene molecules is their instability with respect to decomposition reactions, even for relatively short chain lengths. We therefore sought to examine a system that retains the degeneracy of the π system and yet is amenable to handling in air for extended periods without decomposition. A structure such as that in Figure 2d was a good candidate. In the ground state, one ring of this π-system is aromatic, but the other ring is quinoidal. In the charge separated state, the aromatic-quinoidal nature of each ring is reversed, preserving the π-electron degeneracy.

Dimethylaminoindoaniline (DIA) is a commercially available dye with the electronic degeneracy outlined above. The steric interactions between the ortho hydrogens of the rings preclude the rings from being coplanar. This undoubtedly results in a decreased oscillator strength and polarizability compared to an equivalent coplanar system. Nevertheless, this is a logical starting point to test our hypothesis. EFISH measurements on this compound in chloroform at 1.907 μm yielded β of 190×10^{-30} esu. The low absorption edge leads to a dispersive contribution. Correcting for this using the two level model[5] gives the zero frequency value $\beta_0 = 95 \times 10^{-30}$ esu. In contrast, for dimethylaminonitrostilbene (DANS) $\beta = 75 \times 10^{-30}$ esu. and $\beta_0 = 52 \times 10^{-30}$ esu.[8] Thus, even though DIA is two atoms shorter than DANS, is bent (resulting in a substantially shorter charge separation axis compared to a more linear molecule like DANS), and the rings are noncoplanar, β is roughly a factor a three larger and β_0 is almost a factor of two larger. Another revealing comparison is to 5-(4-N,N-dimethylamino)phenylpentadienal that is structurally similar to DIA but lacks the double bond critical for the degeneracy of the π-system. β for this compound is 52×10^{-30} esu. These results are strongly suggestive that the degeneracy of the π system is, to a large extent, responsible for the enhanced nonlinearity of DIA. This can be viewed either as increasing the polarizability of the π system or increasing the acceptor strength of the carbonyl moiety.

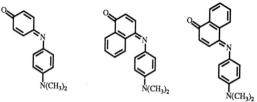

DIA   IPB 1   IPB 2

We hypothesized that structural changes that broke the degeneracy of the π system (but kept the overall length of the charge transfer constant) would in general lower the hyperpolarizability of the molecule. To test this hypothesis we examined indophenol blue (IPB). This dye is commercially sold as

a mixture of isomers that can be separated chromatographically.
The molecular structures are assigned on the basis of UV-visible
spectroscopy using steric arguments. The isomer with the aniline
ring situated over the fused ring (IPB 1) would have severe steric
interactions between the ortho hydrogen of the aniline ring and
the hydrogen ortho to the ring juncture carbon leading to a large
deviation from planarity; in contrast the other isomer has the
aniline ring oriented away from the fused ring (IPB 2) and can
therefore adopt a more planar configuration. We therefore
tentatively assign the former structure, IPB 1, to the material
that absorbs at higher energy (560nm) and the latter, IPB 2, to
the material that absorbs at lower energy (610nm). We predicted
that both IPB 1 and 2 would have lower β than DIA since one of the
double bonds of the "quinone" is already involved in aromatic
bonding. Thus the gain of aromaticity in the charge separated
forms of IPB 1 and 2 would be expected to be less than in the case
of DIA. The measured β values of 78×10^{-30} esu for IPB 1 and
90×10^{-30} esu for IPB 2 are consistent with the hypothesis.
Further, the lower β value of IPB 1 as compared to IPB 2 is
indicative of diminished coupling between the donor and the
acceptor, consistent with the assigned structures.

In conclusion we suggest a new methodology for enhancing β
that does not rely on the use of stronger donors or acceptors in
the normal sense of the word, but rather on the judicious tuning
of the π system.

ORGANOMETALLIC MATERIALS FOR NONLINEAR OPTICS

Organometallic compounds allow us to explore new variables for
the engineering of nonlinear optical (NLO) hyperpolarizabilities.
We can change the transition metal element, its oxidation state,
the number of d electrons, examine the differences between
diamagnetic and paramagnetic complexes and the effect of new
bonding geometries and coordination patterns. Initial efforts[11-17]
to evaluate the potential of organometallic compounds for
quadratic nonlinear optics have been restricted to the Kurtz
second-harmonic generation (SHG) powder test.[18] We found that (Z)-
{1-ferrocenyl-2-(4-nitrophenyl) ethylene, 1, has an SHG
efficiency 62 times urea[13] and the related salt (E)-{1-ferrocenyl-
2-(4-N-methyl pyridinium) ethylene} iodide, 2, has an SHG
efficiency roughly 220 times urea, the largest efficiency known
for an organometallic compound.[17] These results demonstrate that
organometallic compounds can exhibit large $\chi^{(2)}$. The magnitude of
the SHG signals obtained from Kurtz powder test is largely
determined by crystallographic, linear optical (*i.e.*
birefringence), and dispersive factors, therefore little insight
into molecular structure-property relationships can be inferred.[1-3]
EFISH[4,5] is therefore a more appropriate method for molecular
hyperpolarizability studies. Our desire to develop NLO
structure/property relationships for organometallic complexes due

to the general considerations outlined above, and specifically, the large observed powder efficiencies of several ferrocene complexes[13,16,17], motivated us to characterize β for a series of metallocene complexes. We emphasize that our goal was to probe the effect of systematic structural variation on β, *not to find optimized high β molecules.*

The low oxidation potential generally observed for ferrocene complexes[19] and the stability of α-ferrocenyl-substituted carbocations[20] lead us to speculate that ferrocene would be an effective charge-transfer donor for NLO systems. The donor strength of the metallocene will naturally depend on the oxidation potential of the metal center and additional substituents on both five-member rings. Variations including different metal centers, *cis* and *trans* isomers, extension of conjugation, variation of the acceptor group and symmetric electron donating substituents in the form of pentamethylcyclopentadienyl rings (Cp*) have been implemented; β values are summarized in Table I.

Table I: Summary of NLO optical data on metallocene derivatives. EFISH measurements were performed in dioxane, using 1.907μm input.

Cmpd #	M	X	n(isom.)	Y	λ_{CT}(nm)	$\mu(10^{-18}$esu)	$\beta(10^{-30}$ esu)
1	Fe	H	1(Z)	NO_2	325/480	3.4	14
3	Fe	H	1(E)	NO_2	356/496	3.6	34
4	Fe	Me	1(E)	NO_2	366/533	4.4	40
5	Fe	H	1(E)	CN	356/496	3.6	11
6	Fe	H	1(E)	CHO	356/496	3.6	34
7	Ru	H	1(E)	NO_2	350/390	4.5	16
8	Ru	Me	1(E)	NO_2	370/424	5.3	24
9	Fe	H	2(E)	NO_2	382/500	4.1	66

Substituted ferrocenes have two low energy bands in the UV-visible spectrum. Extended Hückel molecular orbital calculations on **1** provide insight about the nature of these optical bands and the chemical bonding in substituted metallocenes.[21] The HOMO is almost completely d_{zz} in nature with the next lower energy occupied orbitals being $d_{x^2-y^2}$ and d_{xy}. These orbitals are essentially nonbonding. Immediately below the d orbitals lies an orbital which has substantial metal and π ligand character.[21] The LUMO is largely localized on the nitro group and the next highest unfilled orbital has coefficients distributed throughout the π ligands. We therefore assign the lowest energy transition in these system as a metal (HOMO) to ligand (the orbital immediately

above the LUMO) *CT* band and the higher energy transition as being
effectively a ligand π (orbital immediately below the filled *d*
orbitals) to π* (LUMO) transition *with some metal character*.
Electron density is substantially redistributed in both
transitions and therefore *both* transitions will likely contribute
to β. Solvatochromic behavior has been observed for compounds **3**
and **4**. Upon changing the solvent from *p*-dioxane to acetonitrile
the lower energy bands shift bathochromically by about 8-10 nm and
the higher energy bands also shift but to a lesser extent. These
shifts are indicative of increased polarity in the Frank-Condon
excited states. The π–π* CT transition is analogous to the CT
transition in donor/acceptor substituted benzenes where electron
densities move from a filled, bonding π-orbital of benzene
perturbed by the donor, (here the iron atom) to an empty low-lying
orbital of the substituent. The lowest energy metal to ligand
(ML) CT band is fundamentally different because an electron almost
completely localized in one orbital is transferred upon
excitation. The donated electron density involved in both CT
bands depends strongly on the metal center and it is not
appropriate to consider the metal center as a counterion merely
providing a full electron to form a 5-member aromatic Cp anion.
We expect the higher energy band to be more sensitive to
variations in the extended π system and the lower energy band to
changes at the metal center. Using compound **3** as a reference,
pentamethyl substitution of one ring leads to 36 and 10 nm
bathochromic shifts to the lower and the higher energy bands
respectively. Replacement of iron by ruthenium, lowers the energy
of the nonbonding *d* orbitals thus increasing the metals redox
potential and lowering its donating strength. As expected, the
lower energy band is hypsochromically shifted by 106 nm but the
higher energy band is only shifted by 6 nm. In contrast the
higher energy band of E,E [1,-ferrocenyl-4-(4-nitrophenyl)-
butadiene], **9**, is shifted bathochromically (26nm) and
hyperchromically relative to **3** and the lower energy band shifts
slightly (4nm). Variation of the acceptor strength affects the π–
π*CT and MLCT bands as expected.

The compounds in Table 1 show nonlinearities of the same order
of magnitude as nitrostilbene (β = 9.1×10^{-30}esu), 4-4'-
methoxynitrostilbene (β = 29×10^{-30}esu), and DANS (β = 75×10^{-30}esu).[8] Since these organometallic compounds have long
wavelength absorption bands, the measured nonlinearity has a small
dispersive enhancement. The *cis* compound **1** is found to be less
nonlinear than the *trans* compound **3**. It is expected that the *cis*
geometry compound would exhibit lower β for two reasons: 1) the
steric interactions between the ortho Cp and ortho benzene
hydrogens preclude the two rings being coplanar (this was seen in
the crystal structure of **1**[13]) resulting in a diminution of
coupling between the donor and the acceptor, 2) the through-space
distance between the donor and the acceptor is less in the *cis*
compound, therefore the change in dipole moment per unit charge

separated will be less. Permethylation of the opposite ring significantly increases both the dipole moment and the nonlinearity, resulting from the destabilization of the high-lying occupied orbitals as evidenced by the large spectral red shift and lowered oxidation potentials.[19] The ruthenium compounds are less nonlinear than their iron counterparts, consistent with the higher oxidation potential[19] of ruthenocene (vs. ferrocene). In agreement with structural trends observed in stilbene derivatives, the effect of increasing the number of conjugated double bonds between the donor and the acceptor is dramatic, with compound **9** exhibiting significantly higher β than **3**. The changes in β as a function of acceptor strength parallel to those observed for donor-acceptor substituted stilbenes, with stronger acceptors leading to higher β. For a given metallocene fragment, β appears to correlate with the oxidation potential of the complex associated with the increasing strength of the acceptor. However, when the acceptor (p-nitrophenyl) is kept constant and the metallocene fragment is varied, β scales inversely with redox potential of the complex. This suggests that the energy of the filled metal orbitals correlates with β.

In summary: (1) structural variations that effect the π-π*CT and the MLCT transitions lead to changes in β. Since both transitions make substantial contributions to the observed nonlinearity it is not appropriate to use a two-state model when considering metallocenes such as those discussed here. The energy and extinction coefficients of both bands are sensitive to changes about the metal center (the MLCT band being more sensitive). The results therefore demonstrate that *the metal center plays an important role in the nonlinear polarizabilities* of these molecules. (2) Based on binding energies and redox potential alone the molecular hyperpolarizabilities might be expected to be larger than the observed values. *Poor coupling between the metal center and the substituent because of the π geometry most likely lowers the effectiveness of the metal center as a donor.* Future studies of organometallic systems for NLO applications should focus on improving the coupling between the metal center and the organic fragment.

Acknowledgements:

The authors thank H. Jones and T. Hunt for expert technical assistance. We also thank Dr. Joseph Perry (JPL) for helpful discussions. The research described in this paper was performed, in part, by the Jet Propulsion Laboratory, California Institute of Technology as part of its Center for Space Microelectronics Technology which is supported by the Strategic Defense Initiative Organization, Innovative Science and Technology Office through an agreement with the National Aeronautics and Space Administration (NASA). Additional support was provided by the Department of

Energy's Energy Conversion and Utilization Technologies Division
(ECUT).

References and Footnotes:

1. D. J. Williams Angew. Chem. Int. Ed. Engl. 1984, 23, 690.
2. Nonlinear Optical Properties of Organic and Polymeric
   Materials; D. J. Williams, Ed.; ACS Symposium Series No. 233;
   American Chemical Society: Washington, DC, 1983.
3. Nonlinear Optical Properties of Organic Molecules and
   Crystals; D. S. Chemla; J. Zyss, Eds.; Academic: Orlando,
   1987; Vols. 1 and 2.
4. (a) J. Ward, J. Rev. Mod. Phys., 1965, 37, 1. (b) B. J. Orr;
   J. F. Ward Mol. Phys., 1971, 20, 513.
5. (a) J. L. Oudar; D. S. Chemla J. Chem. Phys., 1977, 66, 2664.
   (b) B. F. Levine; C. G. Bethea J. Chem. Phys., 1977, 66, 1070.
   (c) S. J. Lalama; A. F. Garito Phys. Rev. A, 1979, 20, 1179.
6. (a) J. L. Oudar; H. Le Person Opt. Commun., 1975, 15, 258.
   (b) B. F. Levine; C. G. Bethea Appl. Phys. Lett, 1974, 24,
   445. (c) K. D. Singer; A. F. Garito J. Chem. Phys., 1981, 75,
   3572.
7 (a) See for example: J. F. Nicoud and R. J. Twieg, ref 3, p255
   and ref therein. More recently: (b) R. A. Huijts and G. L. J.
   Hesselink, Chem. Phys. Lett., 1989, 156, 209; (c) M.
   Barzoukas, M. Blanchard-Desce, D. Josse, J.-M. Lehn, and J.
   Zyss J. Chem. Phys., 1989, 133, 323.
8 (a) L.Cheng; Tam, W.; Meredith, G. R.; Rikken, G. L.; Meijer,
   E. W. Proc. SPIE. 1989, 1147, 61.
9. (a) D. J. Williams in Electronic and Photonic Applications of
   Polymers; M. J. Bowden and S. R. Turner Eds.; Adv. Chem. Ser.,
   218,; American Chemical Society: Washington, DC, 1988. (b) D.
   N. Beratan in Materials for Nonlinear Optics: Chemical
   Perspectives, S. R. Marder, J. E. Sohn, G. D. Stucky, Eds.;
   ACS Sypm. Ser.; American Chemical Society: Washington, DC,
   1991, in press.
10. S. R. Marder; D. N. Beratan; L.-T. Cheng, in preparation.
11. C. C. Frazier; M. A. Harvey; M. P. Cockerham; H. M. Hand; E.
    A. Chauchard; C. H. Lee J. Phys. Chem. 1986, 90, 5703.
12. D. F. Eaton; A. G. Anderson; W. Tam; Y.Wang J. Am. Chem. Soc.
    1987, 109, 1886.
13. M. H. L. Green; S. R. Marder; M. E. Thompson; J. A. Bandy; D.
    Bloor; P. V. Kolinsky; R. J. Jones Nature 1987, 330, 360.
14. J. C. Calabrese; W. Tam Chem Phys. Lett. 1987, 133, 244.
15. A. G. Anderson; J. C. Calabrese; W. Tam; I. D. Williams Chem.
    Phys. Lett. 1987, 134, 392.
16. J. A. Bandy; H. E. Bunting; M. L. H. Green; S. R. Marder; M.
    E. Thompson; D. Bloor; P. V. Kolinsky; R. J. Jones In Organic
    Materials for Non-linear Optics; R. A. Hann; D. Bloor Eds.;
    Royal Society of Chemistry Special Publication No. 69; Royal
    Society of Chemistry: London, 1989.

17. S. R. Marder; J. W. Perry; W. P. Schaefer; B. G. Tiemann; P. C. Groves; K. J. Perry *Proc. SPIE*. 1989, **1147**, 108.

18. S. K. Kurtz; T. T. Perry *J. Appl. Phys*. 1968, **39**, 3798.

19. T. Kuwana; D. E. Bublitz; G. L. K. Hoh *J. Am. Chem. Soc*. 1960, **82**, 5811

20. T. D. Turbitt; W. E. Watts *J. Chem. Soc., Perkin 2* 1971, 177.

21. J. C. Green, unpublished results.

The Mediation of Intra- and Inter-ligand Charge Transfer by D_{4h} Co-ordination to Heavy Group II and Group IV Ions and its Influence Upon Non-linear Optical Properties

Colin L. Honeybourne

MOLECULAR ELECTRONICS AND SURFACE SCIENCE GROUP, BRISTOL
POLYTECHNIC, FRENCHAY, BRISTOL BS16 1QY, UK

INTRODUCTION.

In the simple working models selected for the work described below, the molecules are flat and in the xy plane.
The two-fold symmetry axis, the x-axis, is of prime interest in the context of charge transfer(CT). The ligands
selected are the oxalate dianion, and symmetrically di-
substituted glyoxals to represent inter-ligand CT and
a dicyano-dihydroxy-14-annulene and 4-cyano-4'-dimethyl
amino-2,2-bipyrimidyl to represent intraligand CT.
In order to model the effect of removing mobile-electron
density, we have utilised an empty p_z orbital in heavy
alkaline-earth ions, invoking an $sp_x p_y d$ hybridisation of
empty orbitals for the accommodation of the donated
ligand electrons in a square-planar regime. In order to
model the effect of enhancing the mobile-electron density,
we have utilised an occupied p_z orbital in, say Pb(II) or
Sn(II) with the same hybridisation model as given above.
The parameters for the metal p_z-orbital, H_{AA}^O, G_{AA}^O and β_A^O
(see(1) for defn.) were varied, in eV, between -12 to -25,
15 to 30 and -2 to -6 respectively. We here report the
results for -16, 25 and -4. The parameters for other
atoms are as given elsewhere(2,3).

THEORY.

Pugh and Morley(4) have given the expressions for calculating the elements of the second-order non-linear
optical tensor,ß, over sum-over-all-state-pairs for SHG
(eqn.47(4)), signal mixing(eqn.46(4)) and the Pockels
effect(eqn.48(4)): they have also given the simplified
expression for sum-over-same-state-only interactions
(the diagonal terms) for SHG, as shown below.

$$\text{(DIAG)}\ \beta_{ijk}(-2w,w,w) = -\frac{e^3}{2\hbar^4}\sum_n \{2(<ikj>+<ijk>)[f(3,1)+f(\bar{3},\bar{1})]\ +4<ijk>[f(\bar{1},1)]\}$$

We now present the analogous diagonal-term expressions
for the signal-mixing and Pockels-effect cases:

$$\text{(DIAG)}\ \beta_{ijk}(-w_3,w_1,w_2) = -\frac{e^3}{\hbar^2}\sum_n \{<ikj>[f(3,1)+f(\bar{3},\bar{1})]\ + <ijk>[f(3,2)+f(\bar{3},\bar{2})]\ + <kij>[f(\bar{2},1)+f(2,\bar{1})]\}$$

$$\text{(DIAG)}\ \beta_{ijk}(-w,w,0) = -\frac{e^3}{\hbar^2}\sum_n \{<ikj>[f(1,1)+f(\bar{1},\bar{1})]\ + <ijk>[f(1,0)+f(\bar{1},0)]\}$$

If one state dominates the contributions to the ß-tensors,
and if this state is a low-energy first-excited state
associated with extensive CT, then the above expressions
can further be simplified in which r^x_{no} is the transition
moment along the x-axis and Δr^x_{no} is the associated change
in molecular dipole moment(see Table 1). The meaning of
all other symbols in the foregoing expressions may be
found elsewhere(2-4) together with a description of the
computation of energies, eigenfunctions and the matrix
elements therefrom required for the estimation of ß(2).
In our models, we need only consider the vectorial parts
ß-x and ß-y of ß due to having C2v symmetry(xy plane).

POCKELS EFFECT	$\zeta \cdot \dfrac{2(3\omega_{no}^2 - \omega^2)}{(\omega_{no}^2 - \omega^2)^2}$
SIGNAL MIXING (SUM) (DIFF)	$-\zeta \cdot \dfrac{3\,\omega_{no}^4 - \omega_{no}^2\left(\omega_1^2 + \omega_2^2 \pm \omega_1\omega_2\right)}{\left(\omega_{no}^2 - \omega_2^2\right)\left(\omega_{no}^2 - \omega_1^2\right)\left(\omega_{no}^2 - \omega_1^2 - \omega_2^2 \mp 2\omega_1\omega_2\right)}$
S HG	$-\zeta \cdot \dfrac{3\,\omega_{no}^2}{2\left(\omega_{no}^2 - 4\omega^2\right)\left(\omega_{no}^2 - \omega^2\right)}$
ζ	$e^3 \left(r_{no}^x\right)^2 \left(\Delta r_{no}^x\right)\big/\hbar^2$

Table 1. Expressions for various β_{xxx}^{CT} .

Therefore, in this work, we are able to set the average values, ß-vec, to $((\beta\text{-x})^2 + (\beta\text{-y})^2)^{\frac{1}{2}}$ in which we have ß-x = $\beta_{xxx} + (\beta_{xyy} + \beta_{yxy} + \beta_{yyx})/3$ and ß-y = $\beta_{yyy} + (\beta_{yxx} + \beta_{xyx} + \beta_{xxy})/3$.

RESULTS AND DISCUSSION.

Interligand charge transfer.

In this model, we have disparate ligands generating a square-planar ligand field of four oxygens, two of which are from an oxalate dianion. Two other ligand oxygens are from glyoxal, dicyanoglyoxal or diaminoglyoxal(oxamide). In these cases, there is sufficient CT between the two ligands to generate a ß-vec, using nine excited states, of the order of $40 \times 10^{-30} \text{cm}^5 \text{esu}^{-1}$, with the major contribution coming from the third excited state and with significant contributions from states 2 and 5. In the dicyanoglyoxal case, the third excited state does indeed involve a pronounced movement of charge to the ligand with the cyano-groups, although the transition moment is not large. Thus the central metal provides a mechanism for interligand charge transfer. However, this CT does

not arise from the first excited state, and does not
dominate the contributions to the calculated values of
ß(SHG),i.e.ß(-2w,w,w), cited above as the average, ß-vec.
Thus, the expressions in Table 1 cannot be used.

<u>Disruption of axial CT in a pseudolinear ligand.</u>
The free ligand, 4-CN-4'-N(Me)$_2$-2,2'-bipyrimidyl(CDAP),
proves to be a very good example of an axial, pseudolinear
donor-acceptor dye with a ß-vec(SHG) dominated by a
first excited state exhibiting pronounced charge transfer.
The results given in Table 2, consist of ß-vec(SHG)
calculated using nine excited states (and the values of
ß$^{CT}_{xxx}$, obtained using the expressions in Table 1). It is
clear that the latter are applicable to this ligand.
When this neutral bidentate ligand is disposed around a
central metal in such a way that non-centrosymmetry is
maintained, we find that numerous excited states now
contribute to ß-vec, with state 1 only contributing 13%
of the total value. It is therefore not correct to use
ß$^{CT}_{xxx}$ once the heavy metal is introduced.

<u>The effect of complexation on CT in macrocyclic ligands.</u>
The asymmetrically donor-acceptor-substituted macrocyclic
ligand is 2,3-dicyano-9,10-dihydroxy-1,8-dihydro-(1,4,8,
11)tetraaza-(14)-annulene. The fully delocalised ligand
has equivalent N-ligand atoms with the x-axis in the zx-
mirror plane through the 2,3 and 9,10 bonds.
For the free ligand, we find that the two principal
contributions to ß-vec(Table 3) come from states 3 and 5.
Using the designation "H" for the highest-filled MO and
"L" for the lowest-empty MO, we find that the H-L
transition does not feature in state 3 or 5 which are
mainly comprised of H-(L+2) and (H-2)-L respectively.
When the ligand interacts with a metal with a vacant

p_z orbital, the values of ß-vec are very similar(Table 3).
Although state 1 now makes a contribution, the xxx-term
is zero, and it is not a CT state. States 2 and 4, which
do make xxx-contributions, are mainly comprised of the
(H-1)-L and H-(L+2) transitions respectively. Thus, the
expressions in Table 1 are not applicable.

2,3-DICYANO-9,10-DIHYDROXY-1,8-DIHYDRO-[1,4,8,11]TETRAAZA-[14]-ANNULENE.

INDIVIDUAL STATE CONTRIBUTIONS TO BETADIAG TERMS.

ST	EGY	XXXW	XXYW	XYXW	YXXW	XYYW	YXYW	YYXW	YYYW
1	0.51	0.00	0.00	0.00	-0.00	-0.48	0.33	0.33	0.00
2	2.62	0.00	0.00	0.00	0.00	5.43	9.81	9.81	0.00
3	3.06	9.31	0.00	0.00	0.00	-0.00	0.00	0.00	-0.00
4	3.40	0.00	-0.00	-0.00	-0.00	19.66	27.39	27.39	-0.00
5	3.74	28.40	0.00	0.00	0.00	0.00	0.00	0.00	0.00
6	4.53	0.00	-0.00	-0.00	-0.00	0.79	0.94	0.94	0.00
7	4.77	0.14	0.00	0.00	0.00	0.00	0.00	0.00	-0.00
8	5.21	0.10	0.00	0.00	0.00	0.00	0.00	0.00	0.00
9	5.49	0.00	0.00	0.00	0.00	0.03	0.03	0.03	-0.00

BETADIAG ELEMENTS IN 10*-30 cm*5 esu*-1
LASER WAVELENGTH IN nm = 1600

XXXW	XXYW	XYXW	YXXW	XYYW	YXYW	YYXW	YYYW
37.95	-0.00	-0.00	-0.00	25.43	38.51	38.51	0.00

AVERAGE VALUES OF VECTOR PART OF BETA-DIAG
AT LASER WAVELENGTH AND IN 10*-30 cm*5 esu*-1

BETA-X	BETA-Y	BETA-VEC
72.10	-0.00	72.10

DICYANO-DIHYDROXY-[14]-ANNULENE WITH Ca(II) or Mg(II).

INDIVIDUAL STATE CONTRIBUTIONS TO BETADIAG TERMS.

ST	EGY	XXXW	XXYW	XYXW	YXXW	XYYW	YXYW	YYXW	YYYW
1	2.71	0.00	-0.00	-0.00	-0.00	22.49	38.88	38.88	-0.00
2	3.26	13.66	0.00	0.00	0.00	0.00	0.00	0.00	0.00
3	3.83	0.00	0.00	0.00	0.00	0.52	0.67	0.67	-0.00
4	4.33	27.25	-0.00	-0.00	-0.00	-0.00	-0.00	-0.00	0.00
5	4.35	0.00	0.00	0.00	0.00	1.99	2.42	2.42	-0.00
6	4.89	0.57	0.00	0.00	0.00	0.00	0.00	0.00	0.00
7	5.22	0.00	-0.00	-0.00	-0.00	0.12	0.13	0.13	0.00
8	5.45	0.00	0.00	0.00	0.00	3.31	3.74	3.74	0.00
9	5.78	0.08	-0.00	-0.00	-0.00	0.00	0.00	0.00	-0.00

BETADIAG ELEMENTS IN 10*-30 cm*5 esu*-1
LASER WAVELENGTH IN nm = 1600

XXXW	XXYW	XYXW	YXXW	XYYW	YXYW	YYXW	YYYW
41.56	0.00	0.00	0.00	28.42	45.84	45.84	-0.00

AVERAGE VALUES OF VECTOR PART OF BETA-DIAG
AT LASER WAVELENGTH AND IN 10*-30 cm*5 esu*-1

BETA-X	BETA-Y	BETA-VEC
81.60	-0.00	81.60

Table 3. The calculation of ß-vec(SHG) using
nine excited states with laser wavelength = 1600nm.

DICYANO-DIHYDROXY-14-ANNULENE WITH Sn(II) or Pb(II).

INDIVIDUAL STATE CONTRIBUTIONS TO BETADIAG TERMS.

ST	EGY	XXXW	XXYW	XYXW	YYXW	XYYW	YXYW	YYXW	YYYW
1	2.28	166.66	0.00	0.00	-0.00	-0.00	-0.00	-0.00	0.00
2	2.49	-0.00	-0.00	-0.00	-0.00	-8.10	-15.81	-15.81	-0.00
3	3.09	-0.00	0.00	0.00	0.00	-0.61	-0.92	-0.92	0.00
4	4.07	19.27	0.00	0.00	0.00	0.00	0.00	0.00	0.00
5	4.47	0.00	-0.00	-0.00	-0.00	3.99	4.81	4.81	-0.00
6	4.61	0.00	-0.00	-0.00	-0.00	0.07	0.08	0.08	-0.00
7	4.83	-13.44	-0.00	-0.00	-0.00	0.00	0.00	0.00	0.00
8	5.61	8.87	0.00	0.00	0.00	0.00	0.00	0.00	0.00
9	6.48	-2.01	-0.00	-0.00	-0.00	-0.00	-0.00	-0.00	0.00

ST	EGY	XXX0	XXY0	XYX0	YXX0	XYY0	YXY0	YYX0	YYY0
1	2.28	79.00	-0.00	-0.00	-0.00	-0.00	-0.00	-0.00	0.00
2	2.49	-0.00	-0.00	-0.00	-0.00	-7.32	-7.32	-7.32	-0.00
3	3.09	-0.00	0.00	0.00	0.00	-0.57	-0.57	-0.57	0.00
4	4.07	15.87	0.00	0.00	0.00	0.00	0.00	0.00	0.00
5	4.47	0.00	-0.00	-0.00	-0.00	3.87	3.87	3.87	-0.00
6	4.61	0.00	-0.00	-0.00	-0.00	0.07	0.07	0.07	-0.00
7	4.83	-11.74	-0.00	-0.00	-0.00	0.00	0.00	0.00	0.00
8	5.61	8.04	0.00	0.00	0.00	0.00	0.00	0.00	0.00
9	6.48	-1.86	-0.00	-0.00	-0.00	-0.00	-0.00	-0.00	0.00

BETADIAG ELEMENTS IN 10*-30 cm*5 esu*-1
LASER WAVELENGTH IN nm - 1600

XXXW	XXYW	XYXW	YYXW	XYYW	YXYW	YYXW	YYYW
179.35	0.00	0.00	-0.00	-4.66	-11.84	-11.84	-0.00

XXX0	XXY0	XYX0	YXX0	XYY0	YXY0	YYX0	YYY0
89.30	-0.00	-0.00	-0.00	-3.95	-3.95	-3.95	-0.00

AVERAGE VALUES OF VECTOR PART OF BETA-DIAG
IN STATIC APPROXIMATION IN 10*-30 cm*5 esu*-1

BETA-X	BETA-Y	BETA-VEC
85.35	-0.00	85.35

AVERAGE VALUES OF VECTOR PART OF BETA-DIAG
AT LASER WAVELENGTH AND IN 10*-30 cm*5 esu*-1

BETA-X	BETA-Y	BETA-VEC
169.91	-0.00	169.91

FIRST EXCITED STATE BETA-XXX VALUES.
HOMO-LUMO(H-L) CONTRIBUTION = 0.0%
CT STATE WITH 95.8% H-(L+1).
SHG AND POCKELS,LASER.   MIX: LASER WITH 1900nm

SHG	POCKELS	MIXSUM	MIXDIFF
166.7	-388.7	288.7	188.7

PSEUDOLINEAR AND CT APPROXIMATIONS HOLD
BETA(xxx)CT = 166.66 = 98.09% BETA-VEC

Table 4. The calculation of β-vec(SHG) using nine excited states in both the static, "laser off" mode and at laser wavelength of 1600nm. β$_{xxx}^{CT}$ for SHG, signal mixing and the Pockels effect at 1600nm are also presented.

A dramatic change occurs upon the introduction of a metal ion with a parameterisation modelled for the presence of two electrons in a p_z atomic orbital, which can, at least on symmetry grounds, take part in metal-to-ligand CT. The values of ß-x and ß-vec are now substantially enhanced above their previous values(Table 4), and are dominated by the $ß_{xxx}$ contribution from the first excited state. However, the first excited state does not contain a contribution from the H-L transition: in fact, it is almost a pure,single H-(L+1) transition(see Table 4). The latter is indeed a CT transition with substantial increases in charge, after excitation, at the CN carbon atoms and the ligand N-atoms adjacent thereto. The concurrent diminution of charge occurs at the C-OH carbon atoms and the ligand N-atoms adjacent thereto. The $ß^{CT}$ expressions(Table 1) can be applied in this case.

SUMMARY.

The influence of non-transition metals upon ligand CT is pronounced. When associated with metal-to-ligand CT, useful increases in calculated values of ß are noted. The topic is worthy of experimental and further theoretical investigations.

REFERENCES.

1.J.A.Pople and D.L.Beveridge, in Approximate Molecular Orbital Theory, McGraw-Hill, New York, 1970.

2.C.L.Honeybourne, J.Phys.D:Appl.Phys., 1990, 23, 245.

3.C.L.Honeybourne, J.Mater.Sci., 1990, 25, 3843.

4.D.Pugh and J.O.Morley, in Non-linear Optical Properties of Organic Molecules and Crystals,Volume I, edited by D.S.Chemla and J.Zyss,Academic Press, Orlando,1987, pp.193 (cf. equations 45-53).

Second Harmonic Generation Properties of Some Co-ordination Compounds Based on Pentadionato- and Polyene-Ligands

Curt Lamberth, Don M. Murphy, and D. Michael P. Mingos

INORGANIC CHEMISTRY LABORATORY, UNIVERSITY OF OXFORD, SOUTH PARKS ROAD, OXFORD OX1 3QR, UK

INTRODUCTION

In recent years considerable success has been achieved in designing organic materials which display nonlinear optical properties.[1] The molecular requirements for maximizing the nonlinear properties of organic molecules are well defined and the introduction of appropriate donor and acceptor functional groups can be achieved using conventional synthetic procedures. There are additional crystallographic requirements. Firstly the molecules must pack in a non-centrosymmetric space group and, secondly, they must satisfy phase matching criteria by packing close to a particular inclination to the principle axis. These requirements are difficult to achieve in practice.[2] Recently, some attempts have been made to transfer these molecular design concepts to coordination and organometallic compounds.[3,4] The results which have been achieved with substituted ferrocenes have been particularly promising in this regard.[5]

In this paper we describe the synthesis and characterization of substituted metal pentadionato-complexes which represented reasonable target systems for nonlinear optical systems. These complexes were chosen for the following reasons: (a) thermally stable pentadionato- complexes have been reported for the majority of metals; (b) the pseudo-aromatic nature of the metal pentadionato- ring can provide a range of derivatives via electrophilic substitution reactions and,[6,7] (c) a wide range of conjugated pentadionato-

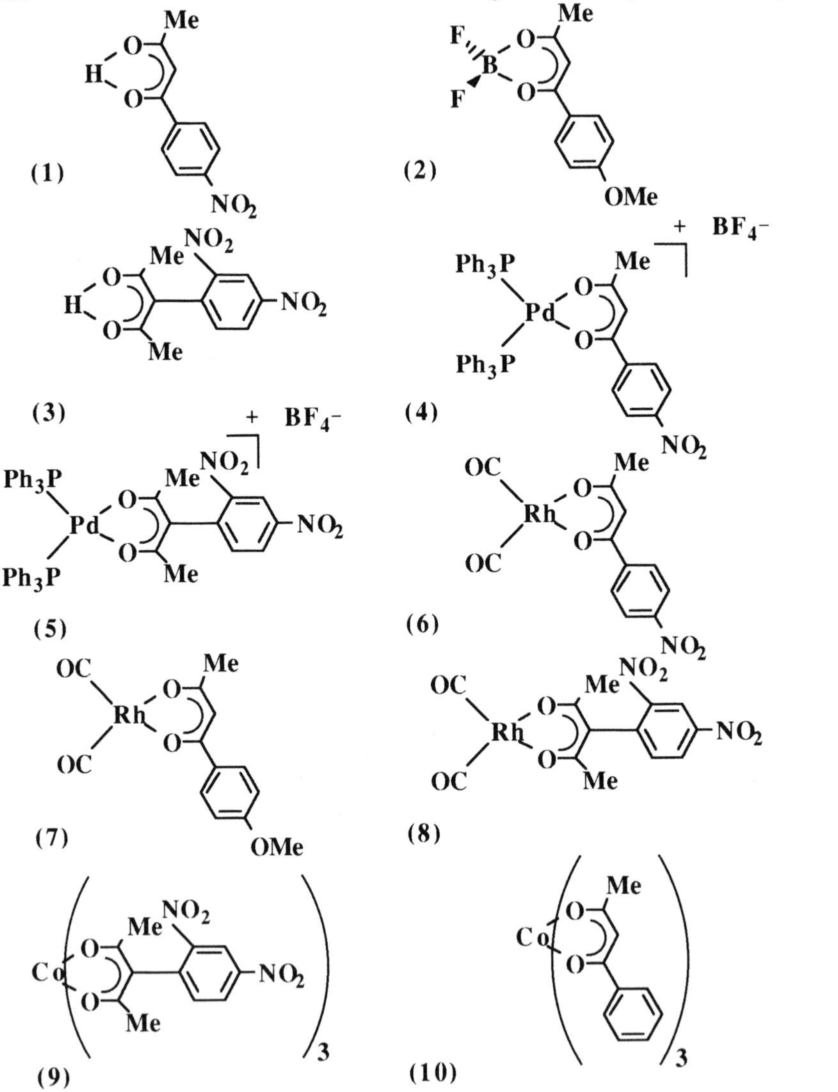

Me

H

(1)

NO$_2$

Me

F ,,, B
F
OMe

(2)

NO$_2$
Me
H
Me
NO$_2$

(3)

Me NO$_2$
Ph$_3$P
Pd
Ph$_3$P
Me
+ BF$_4^-$

(5)

Me
Ph$_3$P
Pd
Ph$_3$P
NO$_2$
+ BF$_4^-$

(4)

OC
Rh
OC
Me NO$_2$

(6)

OC
Rh
OC
Me
OMe

(7)

OC
Rh
OC
Me NO$_2$
NO$_2$

(8)

Co
Me NO$_2$
Me
NO$_2$

/ 3

(9)

Co
Me

/ 3

(10)

<u>Scheme 1</u>

ligands can be synthesized. Such complexes have been prepared and characterized by conventional analytical, spectroscopic and, in some cases, X-ray crystallographic techniques. Their nonlinear optical properties have been probed directly on powdered samples using a modification of the Kurtz powder second harmonic generation experiment and, less directly in solution, by solvatochromism measurements.[8-10]

Results and Discussion

Pentadionato- complexes with substituents at the 3 and 4 positions were synthesized using conventional techniques and some illustrative examples are shown in Scheme 1 as compounds (1) to (3).[11,12] The coordination compounds (4) to (8) derived from these ligands, were synthesized by halide abstraction and ligand replacement reactions. The cobalt(III) pentadionato- complexes (9) and (10) were made directly from $CoCO_3.H_2O$, the ligand and hydrogen peroxide in aqueous solution.[13] Compounds (4) to (8) have d^8 metals in square-planar environments and have been characterized by ^{31}P and ^1H n.m.r. spectroscopic studies. In addition the structures of (4) and (5) have been determined by single crystal X-ray structural determinations. These square planar complexes are yellow crystalline solids.[14] The octahedral cobalt(III) complexes (9) and (10) are dark green crystalline solids and have been characterized by ^1H n.m.r. spectroscopy. Pentadionato- complexes undergo electrophilic substitution reactions at the 3-position and this route has been used to synthesize the substituted octahedral complexes of cobalt and rhodium(III) shown in Scheme 2.[13] Compounds (12) to (14) have Cl, NO_2 and CHO substituents in the 3-position. All of the cobalt compounds are green, while the rhodium compound is yellow. These compounds have also been characterized by spectroscopic techniques, and have the physical properties stated in the original publications.

The range of compounds illustrated in Schemes 1 and 2 demonstrate the ease with which a range of donor and acceptor groups may be introduced into pentadionato-coordination compounds. This is important if a number of such compounds are to be screened for SHG. All the compounds are air stable, and are not light sensitive.

Second Harmonic Generation (SHG) Properties

The SHG properties of the ligands and coordination compounds illustrated in Schemes 1 and 2 were measured using a modification of the Kurtz powder method and the results are presented in Table 1. Samples were ground, but not sieved, and exposed to incident light of 1064 nm.

Table 1   Measured second harmonic signals at 1064 nm
fundamental laser wavelength; all relative to urea,
unless stated otherwise.

Compound	SHG	Colour
(1)	0.60	Yellow
(2)	0.70	Dark green
(3)	0.01	Pale yellow
(4)	0.22	Yellow
(5)	0.73	Yellow
(6)	0.00	Yellow
(7)	0.00	Orange
(8)	0.00	Orange-brown
(9)	0.00	Dark green
(10)	0.00	Dark green
(11)	detected	Dark green
(12)	detected > (11)	Green
(13)	detected > (12)	Dark green
(14)	0.00	Light green
(15)	0.00	Yellow
(16)	0.12	Yellow
(17)	not isolated	Yellow solution
(18)	0.01	Tan brown
(19)	detected	Pale yellow
(20+21)	0.02	Yellow
(22+23)	0.03	Off white

The source was a low power Nd/YAG laser (less than
0.5mW), attenuated by a series of calibrated neutral
optical density filters, which irradiated a silica cell,
containing the sample. The cell faces were at
right angles to the incident beam, and were located at
the centre of a concave mirror, which collected all
reflected light. The reflected light was then passed
through an interference filter centred at 532nm (B.W. 2
nm). The intensity of the second harmonic frequency was
detected by a photomultiplier tube. The amplified anode
signal was measured with a summing oscilloscope. Each
measurement was the average of 64 pulses, and this
ensured that the pulse to pulse variation was removed.
The sample signals were compared to those from graded
(100 micron) reference samples. To avoid any ambiguity
the test was repeated with a different incident laser
power, and a new measurement taken. If the relationship
between the two measurements was second order, with
respect to the incident power, then the signal was
presumed to be truly a second harmonic.

In part the small SHG effects may result from the
crystallization of the compounds in centric space groups

therefore not fulfilling the symmetry requirements for SHG. The single crystal structures of (4) and (5) have demonstrated that they crystallize in the centric space group P$\bar{1}$, although the ligands are quite asymmetric, either because the substituent is in the 4-position of the pentadionato- ligand or because of the asymmetric nitro- groups. The crystal structures have revealed that the phenyl group in (IV) is coplanar with the pentadionato- ring and therefore is fully conjugated with it. In contrast, the dinitrophenyl ring in (5) is perpendicular to the pentadionato- ring. Although the compounds crystallize in centric space groups they show significant, and reproducible residual SHG effects (see Table 1) of approximately 0.7 x urea. Such effects have been observed previously and attributed to co-crystallization of centric and noncentric modifications.[16,17]

All the octahedral cobalt(III) and rhodium(III) compounds show essentially zero SHG effects despite the presence of favourable groups within the molecule. The solvatochromism of the pentadionato- ligands and their complexes (1) to (8) was also measured. All the compounds showed significant shifts of electronic transition state bands when the polarity of the solvent was changed. Typically the following solvents were used; hexane, dichloromethane and ethanol. The bands were found to shift by 15 to 70 nm. These results suggest that the excitations associated with these bands involve a reasonable change in dipole moment. The results support the above suggestion that the poor SHG properties observed for these compounds have their origin in the mode of crystallization rather than in the molecular properties.

Current research is centred on the synthesis of coordination compounds derived from conjugated polyenes. In particular β-ionylidene cyanoacetic acid has been used to synthesize the coordination compounds illustrated in Scheme 3. Recent work on chiral amine salts of alkylidenecyanoacetic acids has resulted in compounds which show a SHG signal of 1.2 x urea.[18] The free acid was synthesized by a modification of a literature route,[19] and reacted directly with base to form the sodium salt. The sodium salt reacted in ethanol/water with silver nitrate to form the silver salt (18), while (16) reacted with $ZnCO_3$, to give the zinc salt (19), and with *cis*-Pt(PPh$_3$)$_2$CO$_3$ in dichloromethane to give an inseparable mixture of *cis*- and *trans*-

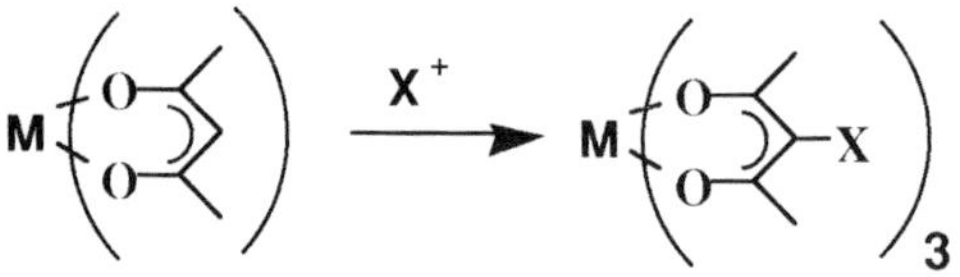

Compound Number:		Group type:
	M=	X=
(11)	Co	H
(12)	Co	Cl
(13)	Co	NO_2
(14)	Co	2 x H, 1 x CHO
(15)	Rh	H

Scheme 2

platinum(II) carboxylates, (20) and (21). Compounds (22) and (23) were synthesized by halide abstraction reactions from cis-Pt(PPh$_3$)$_2$Cl$_2$ and silver β-ionylidenecyanoacetate. Compounds (17) to (23) were characterized mainly by [1]H n.m.r. spectra, while the platinum(II) complex mixtures, (20-21) and (22-23) were analysed, in addition, by [31]P n.m.r. spectra. The SHG signals of the pure and mixed compounds (16) to (23) were

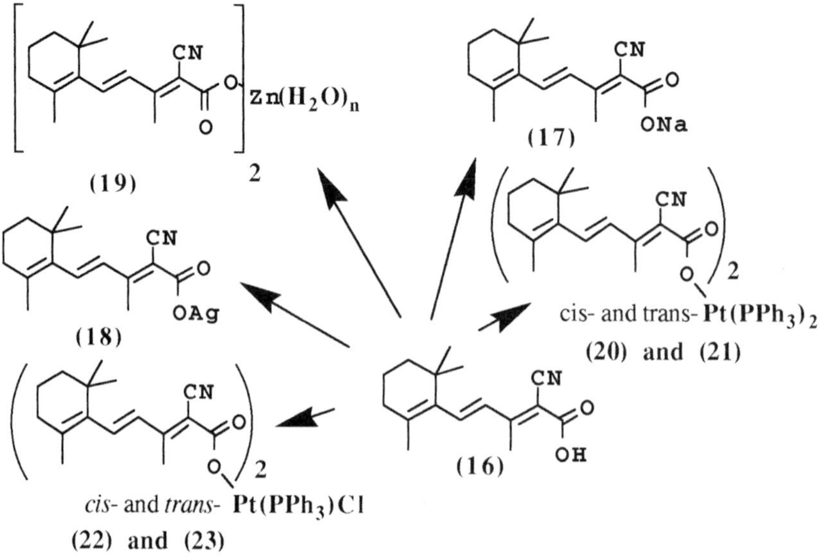

Scheme 3

measured at 1064 nm, and the results are given in Table 1. The SHG signals for these compounds were low, and so were possibly due to the compounds crystallizing in centric space groups, resulting in zero SHG signals.[20]

Acknowledgements

We wish to thank ICI and S.E.R.C. and for financial support and generous cooperation (to C.L.), and also to the A.F.O.S.R. for their financial support (to D.M.M.).

REFERENCES

1. D.S. Chemla and J. Zyss, "Nonlinear Optical Properties of Organic Molecules and Crystals", Academic Press, Orlando, FL, 1987, Vol. 1 and 2 and references therein.
2. W. C. Egbert, "Advances in Nonlinear Polymers and Inorganic Crystals", Liquid Crystals and Laser Media", Proc. SPIE - Int. Soc. Opt. Eng., 1987, 824, 107.
3. W. Tam and J.C. Calabrese, Chem. Phys. Lett., 1988, 144(1), 79.
4. D.F. Eaton, A.G. Anderson, W. Tam and Ying Wang, J. Am. Chem. Soc., 1987, 109, 1886.
5. M.L.H. Green, S.R. Marder, M.E. Thompson, J.A. Bandy, D. Bloor, P.V. Kolinsky and R.J. Jones, Nature, 1987, 330, 360.
6. J.P. Collman, R.L. Marshall and W.L. Young(III), Chem. Ind. (London), 1962, 1380.
7. J.P. Collman, R.A. Moss, S.D. Goldby and W.S. Trahanovsky, Chem. Ind. (London), 1960, 1213.
8. S.K. Kurtz and T.T. Perry, J. Appl. Phys., 1968, 39, 3798.
9. B. Koutek, Coll. Czech. Chem. Commun., 1984, 49, 1680.
10. D. Broussoux, E. Chastaing, S. Esselin, P. Le Barny, P. Robin, Y. Bourbin, J.P. Pocholle and J. Raffy, Rev. Tech. Thomson-CSF, 1989, 20-21(1), 151.
11. C. Mao and C.R. Hauser, "Organic Syntheses", Wiley, New York, 1962, 51, 90.
12. Y. Nakano and S. Sato, Inorg. Chem., 1980, 19, 3391.
13. J.P. Collman, R.P. Blair, R.L. Marshall and L. Slade, Inorg. Chem., 1963, 2, 576.
14. C. Lamberth, D.M. Murphy and D.M.P. Mingos, to be published.
15. B.E. Bryant and W.C. Fernelius, Inorg. Syn., 1957, 5, 188.
16. J.A. Bandy, H.E. Bunting, M.H.Garcia, M.L.H. Green, S.R. Marder, M.E. Thompson, D. Bloor, P.V. Kolinsky and R.J.Jones, Spec. Publ.- R. Soc. Chem., 1989, 69 (Org. Mater. Non-linear Opt.), 225.
17. T. Watanabe and Seizo Miyata, Proc. SPIE - Int. Soc.Opt. Eng., 1989 (Pub. 1990), 1147 (Non-linear Opt. Prop. Org. Mater. 2), 101.
18. Y. Taketani, H. Matsuzawa and K. Iwata, Eur. Pat. Appl. EP 335641, 1989.
19. H.O. Huisman, A. Smit, S. Vromen and L.G.M. Fisscher, Rev. Trav. Chim., 1952, 71, 899.
20. C. Lamberth and D.M.P. Mingos, unpublished results.

Three Co-ordinate Boron as a π-Acceptor in Organic Materials for Non-linear Optics

Z. Yuan, N.J. Taylor, and T.B. Marder*

DEPARTMENT OF CHEMISTRY, UNIVERSITY OF WATERLOO, WATERLOO,
ONTARIO, CANADA N2L 3G1

I.D. Williams and S.K. Kurtz

MATERIALS RESEARCH LABORATORY, THE PENNSYLVANIA STATE
UNIVERSITY, UNIVERSITY PARK, PASADENA 16802, USA

L.-T. Cheng

CENTRAL RESEARCH AND DEVELOPMENT DEPARTMENT, E.I. DU PONT DE
NEMOURS AND CO., INC., EXPERIMENTAL STATION, P.O. BOX 80356,
WILMINGTON, DELAWARE 19880-0356, USA

1. INTRODUCTION

Conjugated π-donor-acceptor substituted organic molecules exhibit low-lying charge transfer states which can give rise to large second-order optical nonlinearities.[1,2] Most studies have concentrated on a relatively small class of π-donors such as MeS, MeO, H_2N, and Me_2N, and π-acceptors such as NO_2, CN, $C_2(CN)_3$, $[C_5H_5NMe]^+$, etc. Recent studies[2d] by Tam et al. have shown that weaker π-donors such as Br or I can provide large $\chi^{(2)}$ values in push-pull stilbene and tolan compounds. It is clear that other donors and acceptors warrant examination.

Three co-ordinate boron possesses an empty p-orbital which should be a powerful π-acceptor. Previous workers[3,4] have demonstrated the potential acceptor properties of the $B(mes)_2$ (mes = 2,4,6-$Me_3C_6H_2$) moiety in 4-X-C_6H_4-$B(mes)_2$ (X = π-donor)[3] based on UV-VIS spectroscopic studies, and in the $[CH_2C_6H_2(3,5$-$Me_2)\{4$-$B(mes)_2\}]$ anion,[4] by X-ray structure analysis.

As boron has a slightly lower electronegativity than carbon, BR_2 groups may even be weak donors in an inductive sense. Molecular hyperpolarisabilities, β, can arise from both inductive as well as charge transfer effects[1a] (Equation 1).

$$\beta_{tot} = \beta_{add} + \beta_{CT} \qquad \text{Equation 1}$$

Using the sterically hindered $B(mes)_2$ group, in which the mesityl π-systems are rotated out of conjugation with the empty p-orbital on boron, would provide a series of compounds in which inductive and charge transfer components of β can be easily separated. Thus, molecular hyperpolarisabilities in the B(mes)$_2$ compounds must arise almost exclusively from β_{CT}, and ground-state dipole moments should be relatively small compared with other conventional acceptor groups. In addition, *via* modifications of the mesityl group with strong acceptor substituents at appropriate locations, it should be possible to influence μ_g and β_{add} without significantly affecting β_{CT} or the positions of the charge transfer absorptions.

With these factors in mind, we initiated a study of the spectroscopic, structural, and nonlinear optical properties of a series of three co-ordinate organoboranes of the general form D-Y-B(mes)$_2$, where D represents a π-donor and Y, a conjugated organic π-system. It is easy to prepare such species containing aromatic, alkenyl, and alkynyl units, or combinations thereof, as the conjugated linker Y.

2. SYNTHESIS

The aryl boranes, 1a-d, were prepared by lithiation of the appropriate 4-D-C$_6$H$_4$-Br reagent, followed by reaction with commercially available FB(mes)$_2$. This route had been employed previously[3a] for the preparation of 1a,d and several other derivatives (Table 1; Scheme 1).

Table 1. Substituents and Yields for Conjugated D-Y-B(mes)$_2$ Compounds.

Cpd #	D	Yield (%)	Cpd #	X	Yield (%)
1a	Me$_2$N	80	4a	Me$_2$N	82
1b	MeS	59	4b	MeS	76
1c	MeO	58	4c	MeO	68
1d	Br	65	4d	H	75
2a	Me$_2$N	50	4e	H$_2$N	84
2b	MeS	65	4f	NC	60
2c	MeO	66	4g	O$_2$N	61
2d	H	60	5	-	93
3	-	75	6	-	57

Similarly, lithiation of the donor-substituted alkynes D-C≡C-H (D = 4-X-C$_6$H$_4$, X = Me$_2$N, MeS, MeO, H; D = Ph$_2$P) followed by reaction with FB(mes)$_2$, yielded the dimesityl alkylnylboranes 4-X-C$_6$H$_4$-C≡C-B(mes)$_2$ (X = Me$_2$N, MeS, MeO, H) 2a-d, and Ph$_2$P-C≡C-B(mes)$_2$ 3 in good yields. Compound 2d was previously prepared by this procedure,[5] whereas the remaining ethynylboranes are new. The *trans*-alkenyl boranes, 4a-g, 5, and 6, were prepared by hydroboration of terminal alkynes Y-C≡C-H (Y = 4-X-C$_6$H$_4$, X = Me$_2$N, H$_2$N, MeO, MeS, H, NC, O$_2$N; Y = Ph$_2$P, [(η^5-C$_5$H$_5$)Fe(η^5-C$_5$H$_4$-)]) with dimesitylborane, 7, in THF at room

Scheme 1

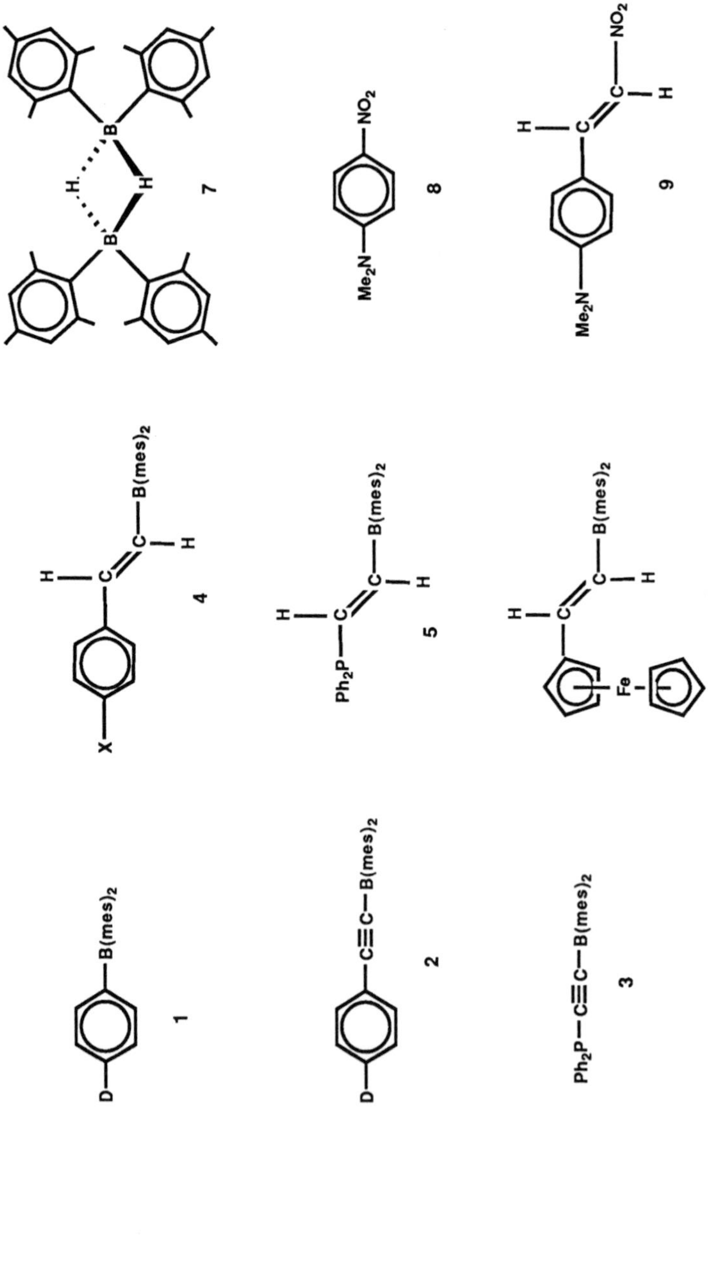

temperature. Again, compound **4d** had been reported previously,[6] whereas the remaining derivatives had not been described. The new boron compounds were characterised by conventional spectroscopic techniques, including IR, UV-VIS, and multinuclear FTNMR, and, in several cases, by single-crystal X-ray diffraction experiments. It is important to note that most of the organoboranes described herein are stable to the atmosphere; we have thus far observed hydrolysis only for solid samples of **2b,c**. Studies are currently in progress to determine the stability of related species with less sterically demanding boron substituents.

3. SOLID STATE STRUCTURES AND POWDER SHG RESULTS [7]

The crystal and molecular structures of **1a**, **3**, **4c**, **4d**, and **5** have been determined by single-crystal X-ray diffraction analysis. Only **5** crystallises in a non-centro-symmetric space group, namely $P2_12_12_1$, and exhibits powder SHG of 1.064 μm laser light by the Kurtz Powder technique.[8] We measured an efficiency of *ca.* 1.0 x an optimized, index-matched quartz sample (62 μm particle size). For comparison, an unoptimized urea sample gave a signal of 1.5 x quartz under these conditions. Selected structural data for D-Y-B(mes)$_2$ compounds are given in Table 2.

Table 2. Selected Bond Lengths and Torsion Angles in D-Y-B(mes)$_2$ Compounds.

Cpd #	Bond Length (Å)		Torsion Angles (°) [c]	
	B-C(mes) [a]	B-C [b]	Vinyl	Phenyl
1a	1.589(3)	1.557(3)	-	19.6
3	1.572(4)	1.526(4)	-	-
4c	1.586(6)	1.546(6)	21.4	23.7
4d	1.580(3)	1.554(3)	27.6	37.8
5	1.581(5)	1.561(5)	16.8	-
d	1.609(12)	1.522(10)	-	25.8
B(mes)$_3$ [e]	1.579(2)	-	-	49.9

[a] Average B-C(mes) distance.
[b] B-C(Y) distance.
[c] With respect to B-C$_3$ plane.
[d] [H$_2$CC$_6$H$_2$(3,5-Me$_2$){4-B(mes)$_2$}]$^-$, from reference 4.
[e] M.M. Olmstead and P.P. Power, *J. Am. Chem. Soc.*, 1986, **108**, 4235.

The B-C(mes) and B-C(Y) distances are fairly similar for all compounds with the exception of **3** for which the B-(C≡C) bond to the sp hybridized carbon is shorter, as expected. Of interest are the torsion angles between the vinyl and phenyl π-systems and the B-C$_3$ plane. Maximum overlap of the π-systems with the empty boron p orbital would be achieved at an angle of 0°; however, steric interaction apparently prevent these angles from falling much below *ca.* 17°. A comparison of the values for 4c and 4d indicate that attachment of the MeO donor group in 4c results in a significant drop in the torsion angles due to improved conjugation.

4. UV-VIS DATA AND MOLECULAR HYPERPOLARISABILITIES

The important charge transfer absorptions in these compounds, in $CHCl_3$ solvent, are listed in Table 3. Interestingly, λ_{max} for 1a in $CHCl_3$ occurs at 358 nm, whereas for 4-Me_2N-C_6H_4-NO_2, 8, λ_{max} = 392 nm. Likewise, for 4a, λ_{max} (in $CHCl_3$) = 401 nm *vs.* 438 nm for E-4-Me_2N-C_6H_4-CH=CH-NO_2, 9. Thus the $B(mes)_2$ compounds are considerably blue shifted from their NO_2 analogues. In addition, a bathochromic shift of *ca.* 10-20 nm was observed for compounds 1a, 2a, and 4a in DMF *vs.* cyclohexane. Based on the assumption of a two state model, other workers have recently suggested[9] a correlation between β and the magnitude of such solvatochromic shifts in UV-VIS absorption spectra.

Table 3. UV-VIS, Ground-State Dipole Moments, and Results of EFISH and
THG Measurements. [a]

Cpd #	$\lambda_{CT\,(nm)}$	$\mu\ 10^{-18}$ (esu) [b]	$\alpha\ 10^{-23}$ (esu)	$\beta\ 10^{-30}$ (esu) [b]	$\gamma\ 10^{-36}$ (esu)[c]
1a	309, 359	3.0	4.6	11	24
1b	249, 335	1.6	4.4	2.5 ± 0.5	32
1c	274, 317	1.6	4.2	3.3	23
1d	269, 314	1.5	4.3	1.6	20
2a	398	3.6	4.8	25	25 ± 30%
2b	368	1.3 ± 20%	4.6	7.1 ± 20%	33
2c	356	1.1 ± 20%	4.5	9.3	13 ± 30%
2d	336	1.5	4.2	3.4	27 ± 30%
3	338	3.6	5.6	3.3	15
4a	401	3.4	5.0	33	93
4b	358	1.5	4.9	8.6 ± 0.5	81
4c	347	1.6	4.8	9.3	54
4d	290, 330	1.0	4.4	5.1	37
4e	368	2.6	4.7	18	41
5	244, 342	3.4	5.6	2.6 ± 1	8
8 [10]	392	6.4	2.2	12	28
9 [11]	438	6.5	3.2	50	-

[a] In $CHCl_3$.
[b] ±10% unless otherwise indicated.
[c] ± 20% unless otherwise indicated.

Using a facility which has been described previously,[10] we undertook a detailed investigation of the first hyperpolarisabilities *via* the Electric Field Induced Second Harmonic Generation (EFISH) technique. Values for γ were obtained simultaneously *via* Third-Harmonic Generation (THG) measurements. All optical measurements were performed at 1.908 μm. Ground-state dipole moments, μ_g, were obtained using a capacitance bridge. These results are summarised in Table 3.

As expected, the ground-state dipole moments are relatively small compared with those of the NO_2 analogues. Thus, $\mu_g = 3.0 \times 10^{-18}$ esu for 1a and 3.4 x 10^{-18} esu for 4a, whereas corresponding values for 8 and 9 are 6.4 and 6.5 x 10^{-18} esu. Values of β are comparable for the $B(mes)_2$ and NO_2 compounds 1a and 8. However, β is *ca.* 50% larger for the extended NO_2 analogue 9 compared with 4a.

It is clear that the $B(mes)_2$ group is a very effective π-acceptor in push-pull organic compounds, and that large nonlinearities can be obtained with reasonable optical transparencies. Further studies will examine the effects of longer π-systems and alternative BR_2 groups on the optical properties and stabilities of this class of materials, and on methods to achieve bulk dipole alignment.

ACKNOWLEDGEMENTS

We thank the Natural Sciences and Engineering Research Council of Canada, the Ontario Centre for Materials Research, and the University of Waterloo for funding, Johnson Matthey for a loan of Pd salts, Drs. S.R. Marder and A.E. Stiegman (JPL-Caltech), and W. Tam (DuPont) for helpful discussions and preprints of several manuscripts, and Drs. A. Meyer and A.J. Carty (Waterloo) for a sample of Ph_2-C≡C-H.

REFERENCES

1.   a) D.J. Williams, Angew. Chem. Int. Ed. Engl., 1984, 23, 690. b) 'Nonlinear Optical Properties of Organic and Polymeric Materials', Williams, D.J., Ed., ACS Symposium Series 233, American Chemical Society, Washington, D.C., 1983. c) 'Nonlinear Optical Properties of Organic Molecules and Crystals', Vol. 1 and 2, D.S. Chemla and J. Zyss, Eds., Academic Press, Orlando, FL, 1987. d) 'Nonlinear Optical Properties of Organic Materials', Proc. SPIE, 971, The International Society for Optical Engineering, Washington, DC, 1988. e) 'Organic Materials for Non-linear Optics', R.A. Hann and D. Bloor, Eds., Spec. Publ. No. 69, The Royal Society of Chemistry, London, England, 1989.
2.   a) H. Tabei, T. Kurihara, and T. Kaino, Appl. Phys. Lett., 1987, 50, 1855. b) C. Fouquey, J.-M. Lehn, and J. Malthête, J. Chem. Soc., Chem. Commun., 1987, 1424. c) J.W. Perry, A.E. Stiegman, S.R. Marder, and D.R.Coulter, in ref. 1e, p. 189. d) W. Tam, Y. Wang, J.C. Calabrese, and

R.A. Clement, in ref. 1d, p. 107.   e) T. Kurihara, H. Tabei, and T. Kaino, J. Chem. Soc., Chem. Commun., 1987, 959.

3.   a) J.C. Doty, B. Babb, P.J. Grisdale, M. Glogowski, and J.L.R. Williams, J. Organomet. Chem., 1972, 38, 229. b) A. Schulz and W. Kaim, Chem. Ber., 1989, 122, 1863.

4.   R.A. Bartlett and P.P. Power, Organometallics, 1986, 5 , 1916.

5.   N.M.D. Brown, F. Davidson, and J.W. Wilson, J. Organomet. Chem., 1981, 209, 1.

6.   A. Pelter, N.N. Singaram, and H. Brown, Tet. Lett., 1983, 24, 1433.

7.   Z. Yuan, N.J. Taylor, T.B. Marder, I.D. Williams, S.K. Kurtz, and L.-T. Cheng, J. Chem. Soc., Chem. Commun., in press.

8.   a) S.K. Kurtz and T.T. Perry, J. Appl. Phys., 1968, 39, 3798. b) J.P. Dougherty, and S.K. Kurtz, J. Appl. Cryst., 1976, 9, 145.

9.   M.S. Paley, J.M. Harris, H. Looser, J.C. Baumert, G.C. Bjorklund, D. Jundt, and R.J. Twieg, J. Org. Chem., 1989, 54, 3774.

10.   L.-T. Cheng, W. Tam, G.R. Meredith, G.L.J.A. Rikken, and E.W. Meijer, Proc. SPIE, 1989, 1147, 61.

11.   L.-T. Cheng, W. Tam, S.H. Stevenson, G.R. Meredith, G. Rikken, and S.R. Marder, in preparation.

Second-order Optical Non-linearities in Linear Donor–Acceptor Substituted Transition Metal Acetylides

G. Lesley, Z. Yuan, G. Stringer, I.R. Jobe, N.J. Taylor, L. Koch, K. Scott, and T.B. Marder*

DEPARTMENT OF CHEMISTRY, UNIVERSITY OF WATERLOO, WATERLOO, ONTARIO, CANADA N2L 3G1

I.D. Williams and S.K. Kurtz

MATERIALS RESEARCH LABORATORY, THE PENNSYLVANIA STATE UNIVERSITY, UNIVERSITY PARK, PASADENA 16802, USA

1. INTRODUCTION

Donor-acceptor substituted conjugated organic molecules possess low-lying charge transfer states which give rise to large changes in molecular dipole moments upon interaction with light. Associated with these properties are large first hyperpolarisabilities, β. An additional requirement for the bulk susceptibilities $\chi^{(2)}$ to be non-zero is the non-centrosymmetric alignment of the molecular dipoles, either in a crystal lattice or a thin-film polymeric medium. There is a general tendency for achiral molecules to crystallise in centrosymmetric space groups due to intermolecular dipole-dipole interactions.

Although organic molecules for $\chi^{(2)}$ have been examined in detail in recent years, transition metal containing systems have been relatively neglected. Interestingly, however, several organometallic complexes have recently been shown to exhibit large second-order optical nonlinearities,[1] and powder SHG intensities of up to ca. 220 x urea have been observed for ferrocene derivatives. Also relevant are reports that certain symmetric square-planar platinum and palladium acetylide complexes and their polyyne polymers exhibit[2] large third-order susceptibilities $\chi^{(3)}$ and can be incorporated into electro-optic devices.

Encouraged by these studies, and observations of promising second-order nonlinear optical behaviour[3] in donor-acceptor substituted diphenyl acetylenes and their diyne and triyne homologues, we undertook the synthesis and characterisation of a series of linear donor-acceptor substituted transition metal bis(acetylide) complexes.

2. SYNTHESIS AND CHARACTERISATION

Initial efforts in our laboratory were directed at the synthesis of octahedral rhodium complexes of the form *mer-trans*-[Rh(PMe$_3$)$_3$(H)(C≡CR$_1$)(C≡CR$_2$)], $\underline{1}$. Our previous work[4] had shown that symmetric analogues R$_1$ = R$_2$ could be prepared easily and in high yields, by reaction between two equivalents of a terminal alkyne RC≡CH, $\underline{2}$, [Rh(PMe$_3$)$_4$(CH$_3$)], $\underline{3}$, with loss of methane, and one equivalent of the volatile PMe$_3$ ligand. If the reaction was carried out using only one equivalent of the alkyne, the electron-rich Rh^I acetylides [Rh(PMe$_3$)$_4$(C≡CR)], $\underline{4}$, were formed,[4] also with loss of methane. Reaction of $\underline{4}$ with $\underline{2}$ cleanly yielded $\underline{1}$ (R$_1$ = R$_2$); however, when the second alkyne was dissimilar to the first (R$_2$ ≠ R$_1$), statistical scrambling resulted in immediate formation of a 1:2:1 mixture of $\underline{1}$ (R$_1$)$_2$:(R$_1$,R$_2$):(R$_2$)$_2$. We thus turned our attention to the synthesis of complexes of type $\underline{4}$ in which R is a strong π-acceptor moiety and the electron-rich Rh centre is the π-donor, and the preparation of the unsymmetrical bis(acetylides) of platinum of the form *trans*-[Pt(PMe$_2$Ph)$_2$(C≡C-D)(C≡C-A)], $\underline{5,6}$. Preliminary results on type $\underline{4}$ systems are promising and will be reported in due course.

Type $\underline{5}$ species can be prepared readily and in the absence of significant scrambling of acetylide ligands. Thus, reaction of *cis*-[Pt(PMe$_2$Ph)$_2$Cl$_2$] with H-C≡C-C$_6$H$_4$-4-A (A = NO$_2$, CN) at reflux in CHCl$_3$ in the presence of Et$_2$NH for 3 days yields the mono(acetylide) complexes *trans*-[Pt(PMe$_2$Ph)$_2$(Cl)(C≡C-C$_6$H$_4$-4-A)] (A = NO$_2$, $\underline{7a}$; CN, $\underline{7b}$) in excellent yield.

$$\underline{7}$$

These can be further reacted with a π-donor substituted terminal alkyne H-C≡C-D in CHCl$_3$ in the presence of small amounts of Et$_2$NH and CuI under carefully controlled conditions to provide $\underline{5,6}$. Only small quantities of symmetric bis(acetylides) were found to be present in the reaction mixtures. If the reactions were allowed to proceed for longer than *ca.* 1 hr, scrambling of the acetylide ligands resulted in the formation of symmetric products. In fact, when the two pure symmetric compounds were mixed in the presence of CuI and Et$_2$NH, scrambling of acetylide ligands was observed, and the unsymmetric donor-acceptor substituted compounds were formed. Thus, it is critical to workup the reaction mixtures after short time periods prior to ligand scrambling. Column chromatography gives rise to the desired unsymmetric bis(acetylides) in pure form.

$$PMe_2Ph$$

$$D—C{\equiv}C—Pt—C{\equiv}C—\langle C_6H_4 \rangle—A$$

$$PMe_2Ph$$

5 or **6**

(A = NO$_2$)   (A = CN)

D

a 4-MeO-C$_6$H$_4$
b 4-MeS-C$_6$H$_4$
c 4-H$_2$N-C$_6$H$_4$
d 4-Me$_2$N-C$_6$H$_4$
e 4-C$_5$H$_4$N
f ferrocenyl
g C$_6$H$_5$

This synthetic procedure was also used to prepare two extended chain complexes of the form *trans*-[Pt(PMe$_2$Ph)$_2$(C≡C-C≡C-R)(C≡C-C$_6$H$_4$-4-A)] (R = Ph, ferrocenyl; A = NO$_2$).

All of the complexes were characterised by IR, and ^{1}H and ^{31}P{^{1}H} nmr spectroscopy, and their UV-VIS spectra were recorded. With the exception of the ferrocenyl substituted compounds, which have absorption tails into the region around 532 nm, the remaining complexes are essentially transparent in this region. The lowest energy absorptions in the UV-VIS spectra occur in the range of *ca.* 325-390 nm. There is a larger dependence on the acceptor group than on the donor moiety; the nitro substituted complexes **5** are significantly red shifted from their cyano analogues, **6**. In addition, the replacement of the *trans*-Cl ligand in **7a,b** with the donor acetylide ligands in **5,6**, also results in a red shift of the charge transfer band. Thus, in CH$_3$CN, λ_{max} for **7a** occurs at 368 nm, whereas, for the *p*-MeO-C$_6$H$_4$-C≡C- complex **5a**, λ_{max} = 386 nm; for **7b**, λ_{max} = 326 nm, whereas, for **6a**, λ_{max} = 350 nm. Interestingly, however, the lowest energy absorption band for **5d** occurs at 378 nm which is blue shifted from that in **5a**. Clearly, detailed studies will be necessary to fully assign the absorption spectra of these complexes.

3. SINGLE CRYSTAL X-RAY STRUCTURE DETERMINATIONS

In order to gain additional insight into both molecular geometries and crystal packing, we have carried out several single crystal X-ray diffraction studies on the platinum acetylide complexes. Complex **7b** crystallises in the centrosymmetric space group P1̄ and displays a slightly distorted square-planar geometry about the platinum centre. Of interest was the observation of a significant intermolecular interaction between the CN moiety on one molecule with the *ipso* carbon of the cyanophenylacetylide ligand on an adjacent molecule. The interplanar separation between the acetylide phenyl groups is 3.46 Å, which is consistent with π-π pairing.

We have examined the structures of the two symmetrically substituted *trans*-bis(acetylide) complexes *trans*-[Pt(PMe$_2$Ph)$_2$(C≡C-C$_6$H$_4$-4-X)$_2$] (X = MeO, NO$_2$). Both crystallise in space group P$\overline{1}$ with Z = 1 and the Pt atom sitting on the inversion centre. This leads to two important features, namely, that the acetylide phenyl rings are coplanar within the molecule, and that all acetylide ligands are perfectly colinear within the entire crystal.

We then carried out a structure determination on the unsymmetric complex *trans*-[Pt(PMe$_2$Ph)$_2$(C≡C-C$_6$H$_4$-4-OMe)(C≡C-C$_6$H$_4$-4-NO$_2$)], 5a. Complex 5a also crystallises in the triclinic system with Z = 1. However, the lack of a molecular centre of inversion precludes the platinum atom from being located at a crystallographic inversion centre. Having collected a complete diffraction data set, including Friedel pairs, we were able to refine the structure of 5a successfully in space group P1 despite the severe pseudosymmetry. Thus, with the exception of the MeO and NO$_2$ end groups, the structure is essentially centrosymmetric. However, upon inclusion of the end groups, difference Fourier analysis allowed resolution of all hydrogen atoms including those on the OMe group. In the final difference map, there was no evidence of the false symmetry image derived from O(2) near O(3), confirming the solution in P1. An ORTEP diagram of the molecular structure of 5a and a packing diagram showing the contents of 8 unit cells are presented in Figure 1. Selected crystal data collection and refinement parameters are as follows: triclinic, a = 5.985(1), b = 9.065(2), c = 14.105(3) Å, α = 84.27(2), β = 88.12(2), γ = 78.35(2)°, V = 745.7 Å^3, P1, Z = 1, Dc = 1.617 g cm^{-3}, μ(MoKα) = 48.9   cm^{-1}, R = 0.0286, Rw = 0.0292 for 12,978 observed reflections (6,489 Friedel pairs) with F > 6σ(F), 2θ_{max} = 70.0°, T = 170 K.

Thus, we have achieved perfect parallel alignment of all molecular dipoles in a single crystal. Interestingly, the Me$_2$N/NO$_2$ analogue, 5d, crystallises in the monoclinic system. Space groups Cc or C2/c were consistent with the systematic absences. The similarity of the scattering from the end groups, combined with the possibility of some end group disorder, precluded successful refinement in the acentric space group Cc. A powder SHG signal of similar magnitude to that for 5a was observed for 5d, indicating the lack of a true centre of symmetry. Successful refinement in C2/c nonetheless indicates nearly perfect parallel packing of all acetylide moieties, albeit that the dipole arrangement could not be determined.

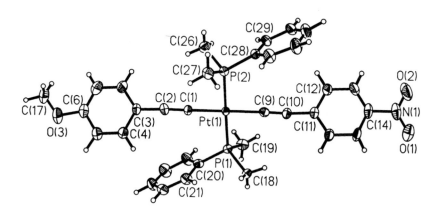

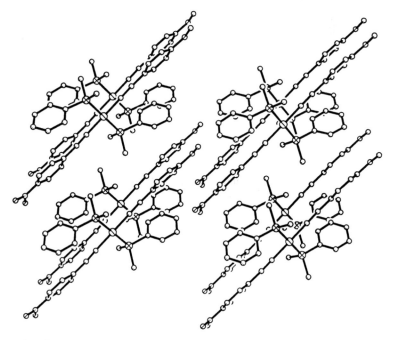

<u>Figure 1</u>. ORTEP diagram of the molecular structure of <u>5a</u> (top) and packing diagram showing the contents of 8 unit cells (bottom).

4. POWDER SHG MEASUREMENTS

All samples were evaluated on a modified version of the Dougherty and Kurtz second harmonic analyser using the Kurtz powder technique.[5] SHG intensities (1.064 μm → 0.532 μm) were determined relative to a graded 62 μm quartz powder which was immersed in an index-matching liquid (n = 1.544). An unoptimised urea sample gave a signal of *ca.* 2-3 x the quartz reference under the conditions used. Sample particle sizes in the *ca.* 50-100 μm range were employed. The index of refraction for 5a was estimated to be *ca.* 1.50 *via* the Becke line method, and the n = 1.544 index matching fluid was used throughout. Samples of 7a,b gave signals of < 0.01 x quartz; 7b is known to be centrosymmetric. In contrast, samples of 5a,d,f,g and 6b,g were analysed: all gave SHG intensities in the range of *ca.* 0.25-1.0 x the quartz reference.

5. CONCLUSION

An important conclusion is that all samples of the unsymmetrically substituted *trans*-bis(acetylides) examined thus far crystallise in non-centrosymmetric space groups. It should also be noted that perfect dipole alignment in P1 is not optimum for phase-matched SHG.[6] However, this is an advantageous arrangement for second-order electro-optic (EO) effects.

Subsequent studies will focus on analysis of additional samples of this class of materials, crystal growth, and measurements of molecular hyperpolarisabilities and bulk EO properties.

ACKNOWLEDGEMENTS

We thank the Natural Sciences and Engineering Research Council of Canada, the Ontario Centre for Materials Research, and the University of Waterloo for funding, Johnson Matthey for a loan of Pd and Pt salts, and the DuPont Company for a gift of materials and supplies.

REFERENCES

1. C.C. Frazier, M.A. Harvey, M.P. Cockerham, H.M. Hand, E.A. Chauchard, and C.H. Lee, J. Phys. Chem., 1986, 90, 5703; M.L.H. Green, S.R. Marder, M.E. Thompson, J.A. Bandy, D. Bloor, P.V. Kolinsky, and R.J. Jones, Nature, 1987, 330, 360; D.F. Eaton, A.G. Anderson, W. Tam, and Y. Wang, J. Am. Chem. Soc., 1987, 109, 1886; J.C. Calabrese and W. Tam, Chem. Phys. Lett., 1987, 133, 244; A.G. Anderson, J.C. Calabrese, W. Tam, and

I.D. Williams, ibid., 1987, 134, 392; W. Tam and J.C. Calabrese, ibid., 1988, 144, 79; B.J. Coe, C.J. Jones, J.A. McCleverty, D. Bloor, P.V. Kolinsky, and R.J. Jones, J. Chem. Soc., Chem. Commun., 1989, 1485; J.A. Bandy, H.E. Bunting, M.L.H. Green, S.R. Marder, M.E. Thompson, D. Bloor, P.V. Kolinsky, and R.J. Jones, in 'Organic Materials for Non-linear Optics', R.A. Hahn and D. Bloor, eds., Spec. Publ. No. 69, The Royal Society of Chemistry, London, 1989, p. 219; J.A. Bandy, H.E. Bunting, M.H. Garcia, M.L.H. Green, S.R. Marder, M.E. Thompson, D. Bloor, P.V. Kolinsky, and R.J. Jones, ibid., p. 225.

2.   C.C. Frazier, S. Guha, W.P. Chen, M.P. Cockerham, P.L. Porter, E.A Chauchard, and C.H. Lee, Polymer, 1987, 28, 553; C.C. Frazier, E.A. Chauchard, M.P. Cockerham, and P.L. Porter, Mat. Res. Soc. Symp. Proc., 1988, 109, 323; S. Guha, C.C. Frazier, K. Kang, and S.E. Finberg, Optics Lett., 1989, 14, 952; C.C. Frazier, S. Guha, and W. Chen, P.C.T. Int. Appl. WO 89 01,182, Feb. 1989; U.S. Appl. 81,785, Aug. 1987, Chem. Abstr. 1989, 111, 105446p.

3.   H. Tabei, T. Kurihara, and T. Kaino, Appl. Phys. Lett., 1987, 50, 1855; T. Kurihara, H. Tabei, and T. Kaino, J. Chem. Soc., Chem. Commun., 1987, 959; C. Fouquey, J.-M. Lehn, and J. Malthête, ibid., 1987, 1424; J.W. Perry, A.E. Stiegman, S.R. Marder, D.R. Coulter, D.N. Beratan, D.E. Brinza, F.L. Klavetter, and R.H. Grubbs, Proc. SPIE, 1988, 971, 17; W. Tam, Y. Wang, J.C. Calabrese, and R.A. Clement, ibid., 1988, 971, 107.

4.   P. Chow, D. Zargarian, N.J. Taylor, and T.B. Marder, J. Chem. Soc., Chem. Commun., 1989, 1545.

5.   S.K. Kurtz and T.T. Perry, J. Appl. Phys., 1968, 39, 3798; J.P. Dougherty, and S.K. Kurtz, J. Appl. Cryst., 1976, 9, 145.

6.   J. Zyss and J.L. Oudar, Phys. Rev. A, 1982, 26, 2028.

Studies Directed Towards Improving Conjugation in Organometallic Rigid-rod Polymers for $\chi^{(3)}$

Helen B. Fyfe, Michael Mlekuz, David Zargarian, and T.B. Marder*

DEPARTMENT OF CHEMISTRY, UNIVERSITY OF WATERLOO, WATERLOO, ONTARIO, CANADA N2L 3G1

1. INTRODUCTION

Transition metal acetylides represent an interesting class of linear organometallic molecules which are known to exhibit metal-to-ligand charge transfer transitions. Transition metal fragments with 6-8 d-electrons form very strong σ-bonds to acetylide ligands resulting in complexes with significant thermodynamic stability. In addition, the incorporation of two acetylide ligands in a *trans*-disposition about the metal centre, in complexes of the form *trans*-[R-(C≡C)$_n$-ML$_n$-(C≡C)$_n$-R], extends both the conjugation and the linearity of the molecules due to the 180° bond angles present. Recent studies in our laboratory[1] have demonstrated second-order optical nonlinearities in a class of donor-acceptor molecules of the general form *trans*-[D-C≡C-Pt(Me$_2$Ph)$_2$-C≡C-A]. A related class of materials are the soluble rigid-rod polyyne polymers *trans*-[Pt(P^{n}Bu$_3$)$_2$-C≡C-X-C≡C-]$_n$ originally prepared[2] by the Hagihara group. These novel polymers have been shown to exhibit both liquid crystalline behaviour[3] and large third-order optical nonlinearities.[4]

In order to examine the role of the transition metal and its ancillary ligands, as well as that of the acetylide linker groups on the physical and nonlinear optical properties of polyyne polymers, new synthetic routes to this class of materials must be found. The procedures should: (1) be general, such that a variety of transition metal centres could be incorporated into the polymer backbone; (2) proceed in high yields; and (3) produce polymeric products which are soluble in common organic solvents and are easily separable from reaction byproducts. With these goals in mind, we decided to turn our attention to the preparation of a series of rhodium acetylide polymers of the general form *mer-trans*-[Rh(PR$_3$)$_3$(H)(-C≡C-X-C≡C-)]$_n$. It was also of interest to synthesise small model species containing one metal centre linking two acetylide units and two metal centres linked by one acetylide moiety.

We believe that larger third-order optical nonlinearities will result from improved conjugation along the polymer chain. As the conjugation is a result of π-backbonding from filled metal d-orbitals into empty acetylide π^*-orbitals, raising the energies of the metal d-levels or lowering the energies of the acetylide π^*-levels will enhance these interactions. The use of transition metals to the left of platinum, and increased numbers of donor ligands (eg. 3 or 4, rather than 2) will both serve to raise metal d-energies. In this paper, we outline our preliminary synthetic findings.

2. SYNTHESIS OF MODEL MONO- AND DI-RHODIUM COMPOUNDS

Previous work in our laboratory demonstrated[5] that terminal alkynes $RC\equiv CH$ react cleanly with $[Rh(PMe_3)_4]Cl$, $\underline{1}$, *via* C-H oxidative addition, yielding cationic *cis*-hydrido acetylide complexes *cis*-$[Rh(PMe_3)_4(H)(C\equiv CR)]Cl$, $\underline{2}$ (Equation 1).

Equation 1

$$[Rh(PMe_3)_4]Cl \quad + \quad RC\equiv CH \quad \longrightarrow \quad \left[\begin{array}{c} H \\ | \\ Me_3P\!-\!\!-\!Rh\!-\!\!-\!C\equiv CR \\ | \\ PMe_3 \end{array} \right]^{+} Cl$$

$\underline{1}$

$\underline{2}$

Deprotonation of *cis*-hydrido acetylides with aqueous KOH leads[6] to neutral Rh^I complexes $[Rh(PMe_3)_4(C\equiv CR)]$, $\underline{3}$, which are capable[7] of a second oxidative addition of terminal alkynes (Equation 2).

Equation 2

$$[Rh(PMe_3)_4(C\equiv CR)] \quad + \quad HC\equiv CR \quad \longrightarrow \quad RC\equiv C\!-\!Rh\!-\!C\equiv CR$$

$\underline{3}$

$\underline{4}$

We have structurally characterised several of the resulting *mer-trans*-$[Rh(PMe_3)_3(H)(C\equiv CR)_2]$, $\underline{4}$, complexes. Application of this route using dialkynes such as butadiyne $[H\text{-}C\equiv C\text{-}C\equiv C\text{-}H]$, $\underline{5a}$, or diethynyl arenes $[H\text{-}C\equiv C\text{-}X\text{-}C\equiv C\text{-}H]$ (X = p-C_6H_4, $\underline{5b}$; p-C_6H_4-C_6H_4, $\underline{5c}$; p-C_6F_4, $\underline{5d}$) would yield polymeric rigid-rod species. This is, however, a multistep synthesis and a more convenient and general preparation could be envisaged.

Thus, a more useful precursor is the neutral Rh^I complex [Rh(PMe$_3$)$_4$(CH$_3$)], **6**. Complex **6** is known to react with one equivalent of a terminal alkyne yielding **3** with methane as the only byproduct. Reaction of two equivalents of **6** with **5b,c** yields the dirhodium complexes **7b,c** (Scheme 1), again with loss of CH$_4$. These complexes have been characterised by IR, and ^{1}H and ^{31}P{^{1}H} nmr spectroscopy. In addition, a crystal structure determination on **7b** was carried out. The bond lengths and angles for **7b** are similar to those in [Rh(PMe$_3$)$_4$(C≡CPh)] and are presented schematically in Figure 1.

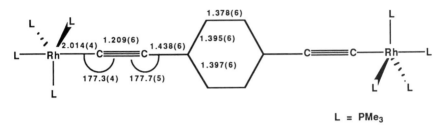

L = PMe$_3$

Figure 1   Solid-state structure of **7b**. Bond lengths are in Å and angles are in degrees.

The distribution of bond lengths in the central benzene ring suggest the possibility of a contribution from a quinoidal resonance form. Unfortunately, the accuracy of the structure determination does not permit this feature to be discussed with confidence. We might suggest, however, that this distortion would be expected as a result of strong π-backbonding from filled d-orbitals on the electron-rich [Rh(PMe$_3$)$_4$] fragments into the empty π*-system of the diacetylide linker. A previous structure determination[8] on the related species *trans*-[{Pt(PEt$_3$)$_2$(NCS)}$_2$(μ-C≡C-*p*-C$_6$H$_4$-C≡C)] was significantly less accurate than that for **7b**.

Using the reverse stoichiometry, i.e. two equivalents of **5b,c** to one equivalent of **6**, yields the neutral hydrido-bis(acetylide) complexes *mer-trans*-[Rh(PMe$_3$)$_3$(H)(C≡C-X-C≡CH)$_2$] (X = *p*-C$_6$H$_4$, **8b**; *p*-C$_6$H$_4$-C$_6$H$_4$, **8c**) (Scheme 1). Thus, it can be seen that both the dialkynes and complex **6** can serve effectively as bi-functional monomers. It is also worth noting that the synthesis of *trans*-[Fe(Me$_2$PCH$_2$CH$_2$PMe$_2$)$_2$(C≡C-*p*-C$_6$H$_4$-C≡CH)$_2$] has also been reported recently.[9]

3. SYNTHESIS OF RIGID-ROD POLYMERS

When complex **6** and the dialkynes **5b,c** are mixed in equimolar amounts, the expected polymeric products *mer-trans*-[Rh(PMe$_3$)$_3$(H)(C≡C-X-C≡C-)]$_n$ (X = *p*-C$_6$H$_4$, **9b**; *p*-C$_6$H$_4$-C$_6$H$_4$, **9c**) are formed in quantitative yields. Unfortunately, however, the new complexes were found to be insoluble in common organic

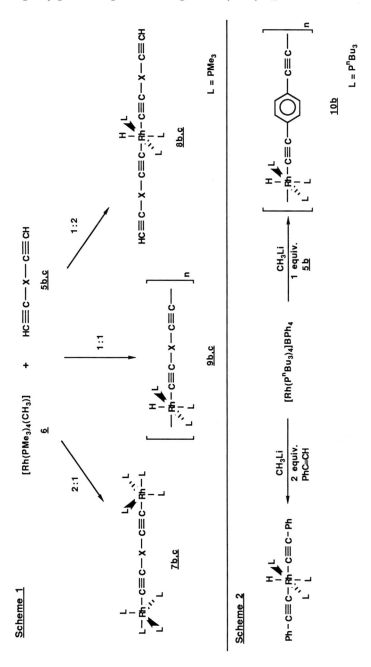

Scheme 1

Scheme 2

solvents and they were thus characterised by IR spectroscopy as Nujol mulls, and by solid-state ^{31}P and ^{13}C nmr spectroscopy. The data is consistent with our formulation of these species as the desired rigid-rod polyyne polymers. The appearance of very weak peaks at 3296 cm^{-1} (9b) and 3274 cm^{-1} (9c) in the IR spectra are consistent with the presence of [C≡C-X-C≡CH] end groups in the polymers. The C≡C stretches at 2084 cm^{-1} (9b) and 2085 cm^{-1} (9c), and Rh-H stretches at 1941 cm^{-1} (9b) and 1942 cm^{-1} (9c) are identical to those found in complexes 8b and 8c, respectively.

Clearly, the lack of solubility of the polymers makes them of little utility, although their successful preparation demonstrates that the general synthetic methodology is sound. We therefore decided to carry out related syntheses using the longer chain phosphine ligand tri-n-butyl phosphine (P^{n}Bu$_3$). This ligand was responsible for the solubility of the platinum polymers discussed in the Introduction.

4. SYNTHESIS OF SOLUBLE RIGID-ROD POLYMERS

Treatment of the known[10] complex [Rh(P^{n}Bu$_3$)$_4$]BPh$_4$ with methyl lithium generates the rhodium methyl complex analogous to 6. This new complex reacts with equimolar quantities of diethynylbenzene, 5b (Scheme 2), yielding the soluble polyyne mer-trans-[Rh(P^{n}Bu$_3$)$_4$(H)(C≡C-p-C$_6$H$_4$-C≡C-)]$_n$, 10b, analogous to 9b. Polymer 10b is soluble in tetrahydrofuran but not in benzene. The new material forms free-standing films upon evaporation of solvent. Complex 10b has been characterized by solution ^{1}H, ^{31}P{^{1}H}, and ^{13}C{^{1}H} nmr spectroscopy, and by an IR spectrum of a neat film. Solutions of this material are deep red; however, the colour is likely to be due largely to impurities carried over from the synthesis of the [Rh(P^{n}Bu$_3$)$_4$]BPh$_4$ starting material. We are currently repeating this reaction to obtain a polymer free from impurities.

5. CONCLUSIONS

We have demonstrated a new synthetic route to soluble conjugated organometallic rigid-rod polymers which are expected to possess large third-order optical nonlinearities. The route should be easily applied to a wide variety of electron-rich transition metal methyl complexes and dialkynes. The advantage of our new procedure is the fact that the only byproducts of the polymerisation reactions are methane and one equivalent of the phosphine ligands PR$_3$. Extension of this methodology to complexes such as [Rh(P^{n}Bu$_3$)$_3$(CH$_3$)] or trans-[M(R$_2$PCH$_2$CH$_2$PR$_2$)$_2$(CH$_3$)$_2$] (M = Fe, Ru) would eliminate the phosphine byproduct entirely. These synthetic studies, as well as an examination of the physical and optical properties of the new soluble polymers, will be reported in due course.

ACKNOWLEDGEMENTS

We thank the Natural Sciences and Engineering Research Council of Canada, the Ontario Centre for Materials Research, and the University of Waterloo for funding, Johnson Matthey for a loan of precious metal salts, the DuPont Company for a gift of materials and supplies.

REFERENCES

1. T.B. Marder, G. Lesley, Z. Yuan, H.B. Fyfe, P. Chow, G. Stringer, I.R. Jobe, N.J. Taylor, I.D. Williams, and S.K. Kurtz, in 'New Materials for Nonlinear Optics', G.D. Stucky, S.R. Marder, and J. Sohn, eds., 'ACS Symp. Ser.', XX, Americal Chemical Society, Washington, D.C., 1990, in press; G. Lesley, Z. Yuan, G. Stringer, I.R. Jobe, N.J. Taylor, L. Koch, K. Scott, T.B. Marder, I.D. Williams, and S.K. Kurtz, in these proceedings.

2. See, for example: N. Hagihara, K. Sonogashira, and S. Takahashi, <u>Adv. Polym. Sci.</u>, 1981, <u>41</u>, 149; S. Takahashi, H. Morimoto, E. Murata, S. Kataoka, K. Sonogashira, and N. Hagihara, <u>J. Polym. Sci., Polym. Chem. Ed.</u>, 1982, <u>20</u>, 565.

3. S. Takahashi, Y. Takai, H. Morimoto, and K. Sonogashira, <u>J. Chem. Soc., Chem. Commun.</u>, 1984, 3, and references therein.

4. C.C. Frazier, S. Guha, W.P. Chen, M.P. Cockerham, P.L. Porter, E.A Chauchard, and C.H. Lee, <u>Polymer,</u> 1987, <u>28</u>, 553; C.C. Frazier, E.A. Chauchard, M.P. Cockerham, and P.L. Porter, <u>Mat. Res. Soc. Symp. Proc.</u>, 1988, <u>109</u>, 323; S. Guha, C.C. Frazier, K. Kang, and S.E. Finberg, <u>Optics Lett.</u>, 1989, <u>14</u>, 952; C.C. Frazier, S. Guha, and W. Chen, P.C.T. Int. Appl. WO 89 01,182, Feb. 1989; U.S. Appl. 81,785, Aug. 1987, <u>Chem. Abstr.</u> 1989, <u>111</u>, 105446p.

5. T.B. Marder, D. Zargarian, J.C. Calabrese, T. Herskovitz, and D. Milstein, <u>J. Chem. Soc., Chem. Commun.</u>, 1987, 1484.

6. D. Zargarian, P. Chow, N.J. Taylor, and T.B. Marder, <u>J. Chem. Soc., Chem. Commun.</u>, 1989, 540.

7. P. Chow, D. Zargarian, N.J. Taylor, and T.B. Marder, <u>J. Chem. Soc., Chem. Commun.</u>, 1989, 1545.

8. U. Behrens, K. Hoffmann, J. Kopf, and J. Moritz, <u>J. Organomet. Chem.</u>, 1976, <u>117</u>, 91.

9. L.D. Field, A.V. George, T.W. Hambley, E.Y. Malouf, and D.J. Young, <u>J. Chem. Soc., Chem. Commun.</u>, 1990, 931.

10. L. Haines, <u>Inorg. Chem.</u>, 1970, <u>9</u>, 1517.

Synthesis and Non-linear Optical Properties of Inorganic Co-ordination Polymers

William Chiang, Mark E. Thompson,* and Donna Van Engen

DEPARTMENT OF CHEMISTRY, PRINCETON UNIVERSITY, PRINCETON, NEW JERSEY 08544, USA

It has become evident that organic compounds can have significant advantages over ionic inorganic solids for use in nonlinear optical (NLO) applications.[1] Organic materials can have much larger NLO coefficients, due partly to the high polarizability of their π-electron networks. Organic compounds are easily substituted with acceptor and donor groups to give them dipolar character in either the ground or excited state, leading to large second order hyperpolarizibilities. Unfortunately, it is very difficult to independently vary the electronic and solid-state structural properties of both organic and ionic inorganic materials. To circumvent this problem measurements have been carried out in solution or in polymer matrices, but these techniques usually yield species with lower temporal stabilities than pure crystalline compounds.

Several transition metal complexes have recently been reported to have second and third order NLO properties comparable to organic compounds.[2] In these complexes the metal atom typically interacts with the π-system of the ligands, acting as either an acceptor or donor. These metal-ligand interactions in complexes and polymers lead to extended π conjugated systems and moderate third order susceptibilities.[2d,2e] Metal compounds are easily prepared as dipolar complexes, similar to organic molecules, but have a much greater diversity of electronic properties due to the variation in electronic structure of the transition metals. A given metal atom can act as either an acceptor or a donor, depending on its oxidation state. By keeping the oxidation state constant and varying the metal atom one can change the acceptor or donor strength systematically in a series of compounds. Interestingly, with transition metal complexes it may be possible to independently vary the electronic and structural properties, since two different metal complexes prepared with identical ligands are often isomorphous.[3]

In this paper we will discuss the synthesis and characterization of a class of dipolar inorganic coordination polymers. These materials were prepared for Cr, Mn

* author to whom correspondence should be addressed.

and Fe. The Cr and Mn complexes show weak second harmonic generation (SHG) when irradiated with 1907 nm light. The crystal structure of the Mn complex demonstrates that coordination polymers are being formed in the solid-state.

Dipolar Coordination Polymers:

Octahedral coordination is very common in transition metal complexes. Five coordinate complexes often become octahedral in solution by coordinating a solvent molecule to fill out their coordination sphere. In the solid-state it is not uncommon for the sixth coordination site to be filled by a Lewis basic ligand of an adjacent molecule, giving a coordination polymer. An example of this type of polymer is (SALEN)CoCH$_2$C≡N, *1*. SALEN is a planar dianionic tetradentate ligand, making *1* a five coordinate complex as written. *1* crystallizes as a coordination polymer, in the space group P2$_1$2$_1$2$_1$.[4] Each Co atom is octahedrally coordinated by the SALEN, the CH$_2$C≡N ligand and the nitrile nitrogen of an adjacent molecule, leading to the polar polymeric chains shown in figure 1. There are two crystallographically independent chains in the unit cell, running parallel to each other.

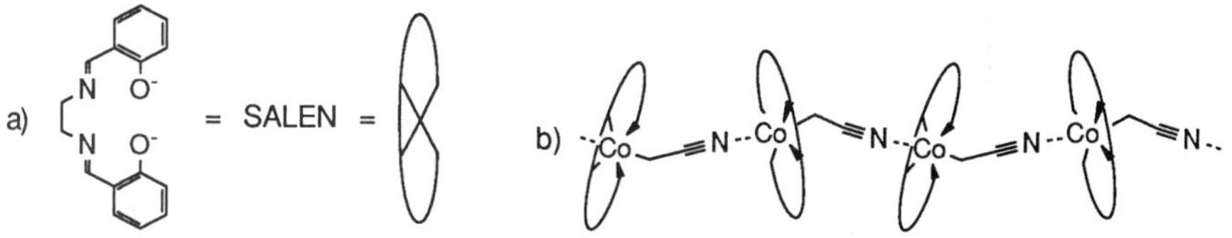

Figure 1: Structure of a) SALEN and b) (SALEN)CoCH$_2$CN.

The structure of *1* suggests that (SALEN)MX compounds may be ideal for preparing polar coordination polymers. Parallel chains of head-to-tail oriented molecules are formed in the crystal. A donor/acceptor system can be envisaged with a Co^{III} acceptor, however, the identity of the donor in this complex is not as clear. The CH$_2$C≡N group will certainly not act as a donor. The phenoxy groups of the SALEN ligand are the most likely donors, but will probably not lead to significant second order nonlinearity, since adjacent SALEN ligands along the chain are oriented opposite each other. Due to this ligand arrangement we expect any nonlinearity due to the ligand itself to be minimal as well. To confirm these suspicions we are preparing a sample of *1* for NLO testing. It would be desirable to prepare complexes similar to *1* with a simple organic donor (*e.g.* amine, alkoxide, *etc.*) built into the ligand connecting adjacent metal atoms along the polymeric chain. The coordination polymerization only guarantees polar alignment along the polymer axis; the molecules are free to rotate about the bond formed between monomer units. With these considerations in mind, the compounds shown in figure 2 were prepared. The metals in these compounds are trivalent, making them good acceptors. The metal atom acceptor is electronically connected to the amine donor *via* the coordinated pyridine,

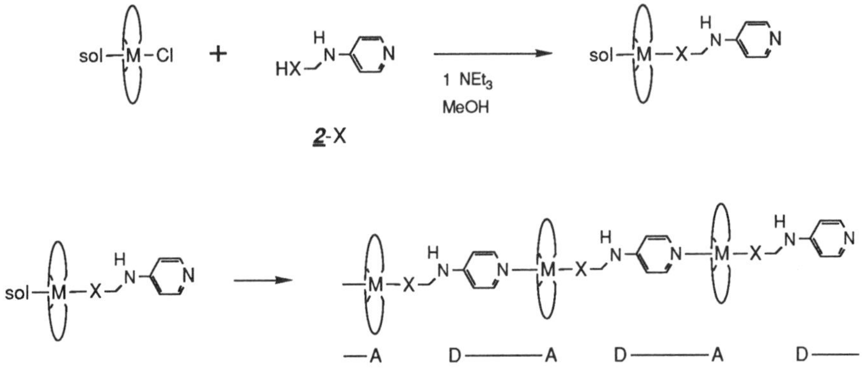

Figure 2: Synthesis of monomers (top) and dipolar coordination polymers **3**-Cr, **3**-Mn and **3**-Fe (bottom). Ellipse = SALEN; sol = solvent molecule; M = Cr, Mn, Fe; X = O_2C, O_3S.

and insulated from the donor in the opposite direction by the methylene group between X and the amine. This arrangement should lead to the dipolar alignment shown in figure 2 (bottom). Alteration of the metal atom should change the electronic structure of the complex and thus the NLO properties, but not the solid state structure.

The sulfonate bridging pyridyl ligand (X = O_3S, **2**-O_3S) is prepared by treating 4-aminopyridine with $HOCH_2SO_3Na$ in water. The carboxylic acid derivative, **2**-O_2C, is prepared by exchanging the sulfonate for a cyano group followed by hydrolysis in 6M HCl. Refluxing the pyridyl ligand, **2**-O_3S or **2**-O_2C, one equivalent of Et_3N and the appropriate SALEN metal halide complex in MeOH gives the solvated square pyramidal complex, figure 2 (top). On crystallization the molecule of solvent that fills the sixth coordination site in solution is replaced by the pyridyl group of an adjacent complex, to give the desired coordination polymer, figure 2 (bottom).

Compounds have been prepared from both **2**-O_3S and **2**-O_2C, but the materials with a **2**-O_3S bridging group have been characterized the most thoroughly. IR spectroscopy indicates that the ligands are coordinated to the metal atoms as shown in figure 2. To confirm this an X-ray structural determination was carried out on crystals of (SALEN)$MnO_3SCH_2NHC_5H_4N$, **3**-Mn, grown from methanol. The monomer unit of **3**-Mn is shown in figure 3. As expected, **3**-Mn forms a coordination polymer in the solid-state. The pyridyl nitrogen of the adjacent manganese complex, labeled N30a in figure 3, coordinates to the Mn giving it octahedral coordination. The Mn-N30a bond length is 2.313(6) Å, which falls in the middle of the range observed for Mn^{III}-N(pyridine) bond lengths in neutral compounds (2.15 - 2.44 Å).[5] The most closely analogous compound to **3**-Mn is (tetraphenylporphyrin)Mn(py)Cl, which has a Mn-N(py) bond length of 2.44 Å.[5c] The bond length in **3**-Mn is

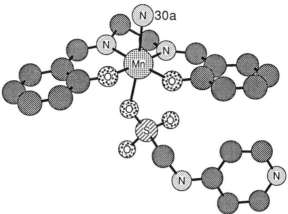

Figure 3: Structure of *3*-Mn monomer. Unlabeled atoms are C; hydrogen atoms are not shown. N30a is the pyridyl nitrogen of an adjacent monomer.

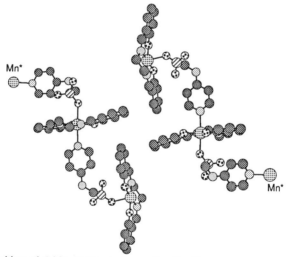

Figure 4: Packing of *3*-Mn chains in the unit cell. Mn atoms in adjacent cells also shown, labeled Mn*.

significantly shorter than this, which may be due to a donor/acceptor interaction between the pyridylamine and the Mn. Unfortunately, the accuracy of our structure was too low (R = 0.087, R_w = 0.081) to look for the expected bond alternation within the pyridylamino group, brought about by a donor/acceptor interaction. The bond distances and angles within the SALEN ligand and between the SALEN ligand and the Mn atom are similar to other Mn^{III}(SALEN) compounds.[6] The polar chains pack with an antiparallel arrangement in the solid state (space group $P2_1/c$), as shown in figure 4, making them ineffective for second order NLO. The crystal structure of **3**-Mn shows that the $^-O_3SCH_2NHC_5H_4N$ bridging ligand leads to a polymer in which adjacent dipolar axes along the polymeric chain are inclined at *ca* 90° to each other. Use of **2**-O_2C rather than **2**-O_3S may lead to polymers with more acute angles between adjacent dipoles, which would be very important for phase matching in these materials.[7]

Samples of **3**-Cr, **3**-Mn and **3**-Fe crystallized from methanol do not give SHG when irradiated with 1064 nm light. If the same compounds are recrystallized from a mixture of NEt_3 and methanol, **3**-Cr and **3**-Mn give weak SHG from 1064 nm light. The weakness of the SHG may be due to absorption, since both of these compounds have measurable absorption at 532 nm (**3**-Cr and **3**-Mn are orange and green, respectively). Kurtz powder tests were also carried out at 1907 nm, so that the sample would be transparent at both the fundamental and second harmonic frequencies. The SHG efficiencies at this wavelength were low, but measurable at 0.18 and 0.012 × urea for **3**-Cr and **3**-Mn, respectively. We have been unable to get X-ray quality crystals of these SHG active materials, so we can not yet say if the low SHG efficiency is due to a low molecular hyperpolarizability or an unfavorable alignment of the NLO active chromophores.

There are two likely possibilities for the identity of these SHG active materials. The first is that the added NEt_3 causes **3**-Mn and **3**-Cr to crystallize in a noncentrosymmetric space group. The other possibility is that the SHG active materials are monomeric NEt_3 adducts of the metal complexes, which crystallize noncentrosymmetrically. Both the SHG active and inactive forms of **3** give identical IR spectra, suggesting the former possibility is correct. The elemental analyses are not consistent with NEt_3 adducts either, since they are too low in both carbon and nitrogen for NEt_3 adducts.

There is a good chance that the Cr and Mn complexes will be isostructural, since the only difference between them is the identity of the metal atom. If the two complexes are isostructural, any difference in their NLO properties must be due solely to electronic differences, and a Kurtz powder measurement will be a valid way to compare the two materials. Even though it is tempting to conclude that Cr^{III} is a better acceptor for use in nonlinear optical coordination compounds than Mn^{III}, based on SHG efficiencies, at this point we do not have enough information on the NLO active forms of **3**-Cr and **3**-Mn to be sure that they are chemically and structurally equivalent. Moreover, considering magnitude of the SHG efficiencies we measured, they could be due to slight deviations from centrosymmetry. Similar SHG efficiencies have been reported for organometallic complexes which appeared from X-ray crystallography to be centrosymmetric.[8]

Coordination polymerization of metal complexes may be a useful way to prepare polar materials for nonlinear optics. While this approach has not yet led to transition metal complexes with large optical nonlinearities, we have demonstrated that coordination polymerization can be used to form polar chains. By modifying the bridging ligand, planar equatorial ligand (SALEN in this paper) or crystallization conditions we hope to be able to prepare materials with large optical nonlinearities, and use this class of compounds to examine the relative efficiencies of the transition metals to act as active components of NLO materials.

<u>Experimental:</u>

(SALEN)MCl (M = Cr, Mn, Fe)[9] were prepared by literature procedures. Solvents were dried with 3 Å molecular sieves, and all anaerobic manipulation were carried out using Schlenk techniques. Powder SHG measurements were made on unsized powders, which had been lightly ground and placed between glass cover slips. A two-channel optical system was used to provide intensity normalization of the SHG signals on every laser shot. Powder SHG efficiencies were determined relative to urea powder (50-100 μM particle size).

Synthesis of $HXCH_2NHC_5H_4N$: Ohta reported the synthesis of $HO_2CCH_2NHC_5H_4N$ from 4-aminopyridine.[10] One of the intermediates in his procedure was $HO_3SCH_2NHC_5H_4N$, which was not isolated, but used directly. Moderate yields of $HO_3SCH_2NHC_5H_4N$ are obtained by refluxing 4-aminopyridine with $HOCH_2SO_3Na$ in water for 1 hour, and crystallizing the product straight from the reaction mixture (35% yield). ^1H NMR (δ, D_2O): 4.42 (s, 2H, CH_2), 6.73 (d, ^3J = 6Hz, 2H, py), 8.05 (d, ^3J = 6Hz, 2H, py). Contrary to the reported synthesis, we found that the procedure for conversion of the sulfonate to the carboxylic acid (treatment with KCN followed by acidic hydrolysis) had to be repeated several times to convert all of the sulfonate to the carboxylate.

Synthesis of **3**-Cr, **3**-Mn *and* **3**-Fe: **3**-Mn was prepared by treating (SALEN)MnBr (0.80 g, 2 mmol) with $HO_3SCH_2NHC_5H_4N$ (0.50 g, 2.7 mmol) and triethylamine (0.38 mL, 2.7 mmol) in methanol (5O mL). The resulting solution was refluxed under argon overnight. A dark solid formed upon cooling to room temperature and was collected by filtration. The solid was washed with methanol and dried in vacuo to give a greenish brown powder (yield: 85%). Melting point 260-262°C. Elemental analysis, found: C, 50.74; H, 4.32; N, 10.56; calc. for (SALEN)$MnO_3SCH_2NHC_5H_4N \cdot CH_3OH$ ($C_{23}H_{24}N_4MnO_6S$): C, 51.11; H, 4.66; N, 10.37. IR: 3432(s), 3334(s), 1625(s), 1599(s), 1543(m), 1529(m), 1446(s), 1333(w), 1294(s), 1242(w), 1220(m), 1207(m), 1199(m), 1168(m), 1148(m), 1125(w), 1043(s), 1000(s), 820(w), 797(w), 760(m), 595(w), 467(m) cm^{-1}. Mag. Susc.: 5.2±0.3 μ_B. **3**-Cr and **3**-Fe were prepared analogously from (SALEN)Cr($H_2O)_2$Cl and (SALEN)FeCl, respectively.

Crystal data for **3**-Mn: A 0.08×0.25×0.35 mm crystal of **3**-Mn was grown from methanol and had one methanol and one half of a water molecule per formula unit. $C_{23}H_{25}MnN_4O_{6.5}S$, M = 548.5, monoclinic, space group P2$_1$/c, a = 10.826(2), b = 18.305(5), c = 12.833(2), β = 107.89(2)°, V = 2420.0(9) Å^3, Z = 4,

$D_c = 1.505$ Mg/m³, $F(000) = 1136$ MoKα ($\lambda = 0.71073$Å). At convergence R = 0.087, $R_w = 0.081$ and the goodness-of-fit = 1.62 for 4267 reflections.

Acknowledgement: The authors would like to thank Joseph W. Perry and Kelly J. Perry (Jet Propulsion Laboratory, California Institute of Technology, Pasadena, CA) for carrying out the nonlinear optical measurements described in this paper. This work was supported the Airforce Office of Scientific Research (AFOSR-90-0122) and the Advanced Technology Center for Photonics and Optoelectronic Materials (Princeton University).

1 J. Zyss *J. Mol. Elec.* **1985**, *1*, 25-45. D.J. Williams *Angew. Chem. Int. Ed.* **1984**, *23*, 690-703. J. Simon; P. Bassoul; S. Norvez *New J. Chem.* **1989**, *13*, 13-31.

2 a) M.L.H. Green; S.R. Marder; M.E. Thompson; J.A. Bandy; D. Bloor; P.V. Kolinsky; R.J. Jones *Nature* **1987**, *330*, 360-362. b) Judith A. Bandy, Heather Bunting, Malcolm L.H. Green, Seth R. Marder, Mark E. Thompson, David Bloor, P.V. Kolinsky, R.J. Jones "Organic Materials for Nonlinear Optics", R.A. Hann, D. Bloor (Eds.), Royal Chemical Society, London, **1988**, p 219-224. c) W. Tam; J.C. Calabrese *Chem. Phys. Lett.* **1988**, *44*, 79-82. d) S. Ghosal; M. Samoc; P.N. Prasad; J.J. Tufariallo *J. Phys. Chem.* **1990**, *94*, 2847-2851. e) C.C. Frazier; S. Guha; W.P. Chen; M.P. Cockerham; P.L. Porter; E.A. Chauchard; C.H. Lee *Polymer* **1987**, *24*, 553-555.

3 For example see: Costellano, E.E.; Hodder, O.J.R.; Prout, C.K.; Sadler, P.J. *J. Chem. Soc. (A)***1971**, 2620-2627.

4 M. Cesari; C. Neri; G. Perego; E. Perrotti; A. Zazzetta *J. Chem. Soc., Chem. Comm.* **1970**, 276-277.

5 a) L.H. Vost; A. Zalkin; D.H. Templeton *Inorg. Chem.* **1967**, *6*, 1725-1730. b) J.B. Vincent; K. Folting; J.C. Huffman; G. Cristou *Inorg. Chem.* **1986**, *25*, 996-999. c) K.F. Kirner; W.R. Scheidt *Inorg. Chem.* **1975**, *14*, 2081.

6 V.L. Pecoraro; W.M. Butler *Acta. Cryst. (C)* **1986**, *42*, 1151. J.W. Ghodes; W.H. Armstrong *Inorg. Chem.* **1988**, *27*, 1841.

7 J. Zyss; J.L. Oudar *Phys. Rev. A* **1982**, *26*, 2028-2048.

8 J.A. Bandy; H.E. Bunting; M.H. Garcia; M.L.H. Green; S.R. Marder; M.E. Thompson; D. Bloor; P.V. Kolinsky; R.J. Jones "Organic Materials for Nonlinear Optics", R.A. Hann, D. Bloor (Eds.), Royal Chemical Society, London, **1988**, p 225-231.

9 A. van den Bergen; K.S. Murray; M.J. O'Connor; B.O. West *Austr. J. Chem.* **1969**, *22*, 39-48. P. Coggon; A.T. McPhail; F.E. Mabbs; A. Richards; A.S. Thornley *J. Chem. Soc.(A)* **1970**, 3296. J. Lewis; F.E. Mabbs; A. Richards *J. Chem. Soc. (A)* **1967**, 1014.

10 M. Ohta; M. Masaki *Bull. Chem. Soc. Jap.* **1960**, *33*, 1150.

Resonance Enhanced Third-order Non-linearities in Metal Dithiolenes

C.A.S. Hill and A.E. Underhill

DEPARTMENT OF CHEMISTRY, UNIVERSITY OF WALES, BANGOR, GWYNEDD LL57 2UW, UK

C.S. Winter, S.N. Oliver, and J.D. Rush

BRITISH TELECOM RESEARCH LABORATORIES, MARTLESHAM HEATH, IPSWICH IP5 7RF, UK

1 INTRODUCTION

Study of the third order nonlinear optical properties of organic, or organometallic, materials has been largely confined to conjugated polymer systems (1, 2). Generally, in these systems the optical nonlinearities are studied far from resonance with the nonlinear response arising from correlated electron movements within the conjugated system. The extent of the pi-electron delocalization is believed to determine the magnitude of the nonlinearity. However, the correlated electron effects can only extend over a limited number of atoms, thus constraining the maximum possible size of the nonlinearities (3).

The size of the coefficients observed to date in these polymers limits their potential application to planar waveguide devices with interaction lengths of the order of mm-cm's, and places tight constraints on the acceptable ratio of linear and nonlinear loss to the nonlinear refractive index (4). A wide range of hybrid glass waveguide-nonlinear overlay devices exist that are compatible with existing device and system architectures and can be used in conjunction with fibre-laser amplifiers to reduce the linear loss requirements. These devices require larger coefficients than have been achieved with polymers due to the shorter interaction lengths involved.

A number of alternative approaches have been tried-organic metals (5), two-dimensional pi-electron systems (6,7) and organic charge transfer complexes (8). Of these, only the first was shown to possess a large

coefficient ($>10^{-7}$ cm^2/kW), however, the linear
absorption was too high to be useful. Theoretical
analysis of phthalocyanine and naphthalocyanine type
systems showed that the correlated electron effects in
these materials would be smaller than in linear
conjugated systems (7).

A further approach is to examine materials where the
laser source is tuned close to or into a single photon
resonance in the material. This was shown to lead to
large nonlinearities in solutions of infra-red absorbing
dyes (9, 10). This paper reports an extension of this
approach were trade-offs between enhanced nonlinearities
and increased optical loss are examined to determine
whether such resonance enhancement can be exploited. The
family of compounds chosen for study are the metal
dithiolenes of which only the saturable absorber BDN has
been previously studied (9, 10, 11).

The metal bis-dithiolenes (fig 1) are known to
exhibit a very intense ($\epsilon{\sim}10^4{-}10^5$) absorption band in the
near infra-red region of the spectrum. By use of
suitable substituents it is possible to 'tune' this band
between 650 nm - 1140 nm (12, 13). Furthermore, they are
known to exhibit great stability when exposed to intense
laser light in the near infra-red region, exemplified by
the use of the material BDN as a saturable absorber for
the Nd/YAG laser. Due to the foregoing properties, the
metal bis-dithiolenes were particularly attractive for a
study of the effect of resonance enhancement on the
third-order nonlinear susceptibility. We have previously
reported (14, 15) how suitable nickel dithiolenes can
possess a large optical nonlinearity and acceptable
absorption coefficients. We report here on an extension
of this work to include platinum and palladium complexes.

2   EXPERIMENTAL

Metal bis-dithiolenes were prepared from the appropriate benzoins or acyloins by refluxing with P_2S_5 in xylene or 1,4-dioxane, according to the method of Schrauzer and Mayweg (16). Purification was effected by column chromatography (silica gel), and/or recrystallisation. All gave satisfactory analyses after purification. Solutions of the metal bis-dithiolenes were made up in dichloromethane. Linear absorption spectra were measured using a Perkin-Elmer Lambda 9 spectrophotometer.

Third-order nonlinear optical coefficients were measured by two techniques, de-generate four wave mixing (DFWM) and intensity dependent transmission. The measurements were performed using a Nd:YAG operating at 1064 nm. The intensity dependent transmission studies were carried out using Q-switched TEM_{00} mode, 10 nm laser pulses. The laser beam was telescoped to give a 300 μm diameter collimated beam at the sample. By measuring the transmission as a function of input power, the two photon absorption coefficient can be calculated (17). The pulse width was measured using a 200 ps rise time diode and the energy of the pulse measured with a pyroelectric energy meter designed to measure single pulses. The detector system could resolve differences in transmission with an accuracy of about 2%. DFWM studied were carried out using a mode-locked Q-switched pulse train from which a single 100 ps mode-locked pulse was selected. A retro-reflection DFWM set up was used (18,19), this being a preferable arrangement for absorbing samples (18). The full details of the experimental arrangement are given in reference 14. The nonlinear response of the material was measured for both parallel and crossed polarisations of the pump and probe beams ($\chi_{1111}^{(3)}$, $\chi_{1221}^{(3)}$). This allows certain possible contributions to the nonlinearity to be distinguished. By altering the position of the rear mirror, it is possible to measure the decay time response of the induced gratings and thus distinguish fast from slow contributions to the nonlinearity. To correct for any spatial or temporal irregularities in the beam, all measurements were compared to a CS_2 standard for which a value of $\chi^{(3)} = 2.5 \times 10^{-20} m^2 V^{-2}$ was used.

3   RESULTS AND DISCUSSION

The absorption maxima (λmax), linear absorption coefficients (α), third order nonlinear susceptibilities ($\chi_{1111}^{(3)}$, $\chi_{1221}^{(3)}$) and Stegeman figure of merit (W) for the Pt and Pd bis dithiolenes studied are listed in table 1. All values are normalised to a concentration of 10^{18} molecules per cm^3. It can be seen that the values of the nonlinear susceptibility are in general related to the position of the absorption maximum for the compound; that is the nearer λmax is to 1064 nm the larger is the value of $\chi^{(3)}$. This is to be expected if a near resonance enhancement effect is taking place.

TABLE 1

M	n	R	λmax (nm)	$\chi_{1111}^{(3)}$	$\chi_{1221}^{(3)}$	W	α
				(x $10^{-20}m^2/V^2$)			(cm^{-1})
I	Pt	0	CH_3	780	0.34	0.206	1.75	0.04
II	Pt	0	C_6H_5	780	0.78	0.099	1.95	0.07
III	Pd	0	C_6H_5	885	2.25	0.340	0.84	0.49
IV	Pt	-1	CN	865	7.43	1.123	0.50	2.72
V	Ni	0	CH_3	770	1.11	0.36	5.0	0.04

The central metal has important effects on the properties of these compounds. Whereas the platinum and nickel complexes with the methyl substituted ligand have λmax values very close to one another the $\chi_{1111}^{(3)}$ of the nickel complex is three times that of the platinum complex. However, the palladium complex with the phenyl substituted ligand has a very different absorption maximum compared with the similarly substituted platinum complex. The shift of λmax to the red is responsible for the much higher value of $\chi_{1111}^{(3)}$ compared with the platinum complex. The monoanion platinum complex (compound IV) has an absorption at longer wavelengths compared with the neutral platinum compounds studied. This compound has the highest value of both $\chi_{1111}^{(3)}$ and $\chi_{1221}^{(3)}$ of all compounds studied.

Time delay studies have shown that $\chi_{1221}^{(3)}$ signal decays as the autocorrelation function of the laser pulses. This suggests that the signal arises from electronic phenomena rather than from molecular rotation, for which the calculated relaxation time is of the order 0.5 - 5.0 ns; far longer than the pulse duration.

Although a large nonlinear susceptibility is desirable, it is not the only criterion upon which material selection must be made. Precise requirements depend upon device format, a low linear absorption coefficient is essential. For this reason, the Stegeman figure of merit (W) is often quoted in the literature. This is defined as:-

$$W = \frac{\Delta n \ sat}{\alpha \lambda}$$

Where Δn sat is the maximum observed change in refractive index, here taken as the product of n_2 and the damage intensity for sublimed films of metal dithiolenes (0.8 - 1.0 GW/cm^2). α is the linear absorption coefficients at the operations wavelength (λ).

For device use (e.g. Mach-Zehnder) a W>2.5 is required, corresponding to a π phase change. To date the largest value of W obtained from an organic material has been of the order of 1.5 for polydiacetylene. Both the methyl and the phenyl substituted derivatives platinum (I and II) exceed this value. That of the tetramethyl nickel derivative exceeds this figure by a large margin.

A further requirement for efficient device operation is a low two photon absorption coefficient (β). A figure of merit based upon this parameter has been defined as:-

$$B = \frac{n_2}{2\beta\lambda}$$

Where n_2 is the nonlinear refractive index, and β the two-photon absorption coefficient.

For two photon absorption not to be a limiting factor, B must exceed unity or the device will be inefficient. The nonlinear transmission coefficient of each sample was measured, from which was calculated the two photon absorption coefficient. Although precise measurements are not yet available, preliminary studies indicate that B exceeds unity for most materials studied and is of the order of ten for a λmax of 770 nm.

3 CONCLUSIONS AND FUTURE WORK

The observed figures of merit and $\chi^{(3)}$ values are most encouraging and we are now starting to make solid-state measurements. Studies on sublimed thin films of metal bis(dithiolenes) have proved unsatisfactory, due to high scatter, and we have turned our attentions to liquid-crystalline dithiolene and to guest-host polymer systems.

REFERENCES

1. J. Zyss; <u>J. Mol. Electron</u>, 1985, <u>1</u>(1), 25.
2. G.T. Boyd; <u>J. Opt. Soc. Am. B</u>., 1985, <u>6</u>(4), 685.
3. A.F. Garito; J.R. Heflin; K.Y. Wong;
    I. Zamani-Khamiri; Nonlinear Optical Properties of
    Polymers: <u>Proc. Mat. Res. Soc. Symp. Proc</u>., 1988,
    Vol. <u>109</u>, 91.
4. G.I. Stegeman; R. Zanoni; C.T. Seaton; Nonlinear
    Optical Properties of Polymers, <u>Proc. Mat. Res. Soc.
    Symp. Proc</u>., 1988, <u>109</u>, 53.
5. P.G. Huggard; W. Blau; D. Schweitzer; <u>Appl. Phys.
    Lett</u>., 1987, <u>51</u> (26), 2183-5.
6. Z.Z. Ho; C.Y. Zu, W.M. Hethernington; <u>J. Appl. Phys.</u>
    1987, <u>62</u>(2), 716-8.
7. J.W. Wu; J.R. Heflin; R.A. Norwood; K.Y. Wong;
    O. Zamani-Khamiri; A.F. Garito; P. Kalyanaraman;
    J. Sounik; <u>J. Opt. Soc. Am. B</u>, 1989, <u>6</u>(4), 707-720.
8. T. Gotoh; T. Kondoh; K. Kubodera; <u>J. Opt. Soc. Am.
    B</u>, 1989, 703-706.
9. C. Maloney; H. Burne; W.M. Dennis, W. Blau;
    J.M. Kelly; <u>Chem. Phys</u>., 1988, <u>121</u>, 21-39.
10. C. Maloney; W. Blau; <u>J. Opt. Soc. Am. B</u>., 1987,
    <u>4</u>(6), 1035-39.
11. G. Yili; Y, Minyan; P. Shengmin; Tongbao Kexue,
    1984, <u>29</u>(6), 731-4.
12. J.A. McCleverty, <u>Prog. Inorg. Chem</u>., 1968, <u>10</u>,
    49-221.
13. U.T. Mueller-Westerhoff; B. Vance; <u>Comprehensive
    Coordination Chemistry</u>, Volume 2; G. Wilkinson;
    R.D. Gillard, J.A. McCleverty; 1986, (Eds),
    pp 595-631, <u>Pergamon</u>.
14. C.S. Winter; S.N. Oliver; J.D. Rush; R.J. Manning;
    C.A.S. Hill; A.E. Underhill; ACS Symposia on
    Novel Nonlinear Optical Materials (1990) in print.
15. S.N. Oliver; C.S. Winter; J.D. Rush; C.A.S. Hill;
    A.E. Underhill; SPIE, Vole. 1337, (1990), in print.
16. G.N. Schrauzer, V.P. Mayweg; W. Heinrich; <u>Inorg.
    Chem</u>., 1965, <u>4</u>, 1615.
17. W.L. Smith, Handbook of Laser Science and Technology
    WEBER, M.J. CRC Press, Boca Raton, Vol III, Part 1,
    1986, 229.
18. D.M. Pepper, A. Yariv, Optical Phase Conjugation,
    R.A. Fisher (Ed). Academic Press, London, 1983,
    p. 23.
19. C. Maloney, W. Blau, <u>J. Opt. Soc. Am. B.</u> 1987, <u>4</u>,
    1935.

Polymers

Recent Progress in Our Studies on Organic Materials for Non-linear Optics

H. Nakanishi

RESEARCH INSTITUTE FOR POLYMERS AND TEXTILES, 1-1-4 HIGASHI,
TSUKUBA, IBARAKI 305, JAPAN

1 INTRODUCTION

During the past decade there has been growing interest
in the field of organic and polymeric nonlinear optical
materials, and great progress has been seen not only
in the materials themselves but also in their device
applications.

In Japan, more than fifty groups from universities,
governmental institutes, and industries have already
made any contributions at the academic meetings of this
field. And the number is still increasing. Following
is a report from the recent Japanese activities.

Great efforts have been made in the survey of
second-order nonlinear optical crystals, aiming at the
highly efficient frequency conversion and various
second-order active crystals have been found, or
characterized. For example crystals with large d_{ij} for
waveguide configuration are 4'-nitrobenzylidene-3-
acetoamino-4-methoxyaniline (MNBA) (d_{11}=454pm/V, λ_{co}=
505nm) by Toray Industries Inc.,[1] 1-(4-nitrophenyl)-
pyrole (NPRO) (d_{33}=657×10^{-9}esu, λ_{co}=410nm) by Fuji Photo
Film Co.,[2] 2-methoxy-6-nitrophenol (MNP) (d_{33}=d_{11}(MNA),
λ_{co} =460nm) by Konika Corp.,[3] cyclobutenedione
derivatives (DAD) by Fuji Xerox Co.,[4] 2-(cyano-
(ethoxycarbonyl)methylene)-4,5-dimethyl-1,3-dithiole by
Nogami et al.[5] and so on. Recently, Umegaki and Fuji
Photo Film Co. have succeeded in the generation of a blue
laser by the frequency doubling of a semiconductor
laser, using the cored fiber of 3,5-dimethyl-1-(4-
nitrophenyl)pyrazole (DMNP) (d_{32}=90pm/V, λ_{co}=450nm).[6]
Fabrication of a channel waveguide of MNBA crystals has
been demonstrated by using a lithographic technique.[1]
Studies on the orientation-controlled crystal growth for

waveguides   are   also   in progress   by   Nishihara[7])   and
others.   Crystals   for bulk use also   have   been   found,
e.g., N-methoxymethyl-4-nitroaniline (MMNA)[8]) and   N,N'-
bis(4-nitrophenyl)-methanediamine (NMDA)[9]) by Miyata   et
al., 2-(2,2-dicyanovinyl)anisole   (DIVA) by   Sasabe   et
al.,[10]) chalcone derivatives by Nippon Oil & Fats Co.[11])
and so on. Sum-frequency generation using DIVA   crystals
and   intracavity   SHG   using   a   chalcone   have   been
demonstrated.   Several   groups   investigated   Langmuir-
Blodgett   films   and met difficulty   in   attaining   good
optical   quality.[12])   Very   recently   Tsutsui   et   al.
succeeded in the deposition of more than a hundred layers
of hetero Y-type film of 5-(4-dodecanoxyphenyl)pyrazine-
2-carboxylic acid with good optical quality and the d_{33}
of the film exceeded d_{33} of LiNbO$_3$.[13]) So far only a few
about   the polymers for   second-order effects   have   been
reported,   e.g.,   pNA-polymer   complexes   by   Miyata   et
al.[14]) and several poled polymers by NTT.[15]) The   number
of   relevant papers are going to increase.

    As   for third-order materials, studies are   in   the
very   academic   stage and less efforts   have   been   made
around   dye-stuffs,   chromophore   containing   polymers,
conjugated   polymers   and so on.   For   examples,   charge
transfer   crystals   were   found   to   have $\chi^{(3)}$ very
comparable to those of conventional polydiacetylenes.[16])
Processable conjugated polymers of   polyarylenevinylenes
have   been   well characterized.[17]) Takeda, Koda   et   al.
succeeded   in   the   preparation   of   polycrystalline
almost perfectly orientation-controlled   polydiacetylene
thin   films   with the area of   ca.   10×10cm^2.[18]) Device
functions such as bistability, optical coupling,[19]) Kerr
shutter,[20]) phase conjugation, pulse compression and   so
on have also been studied.

    It   should   be   noted   that   as   for   the   device
applications   of   organic materials, a   lot   of   leading
contributions   so   far   have been done   by   Umegaki   and
Sasaki,   and   currently the numbers of   researchers   are
increasing.   In addition, remarkable contributions   have
been   made   also   from the   analytical   and   theoretical
sides, i.e., the evaluation method of nonlinear   optical
properties   and nonlinear optical dynamics by   Kobayashi
et   al. and the proposal of quantum confined   structures
to be tackled by Hanamura and Gohnokami-Kuwata et al.[21])

    Recent   progress in our preparative studies are   as
follows.

2 ORGANIC DYES FOR NONLINEAR OPTICS

2.1 Cyanines-p-toluenesulfonic acid complexes for
    second-order nonlinear optics.

Mero- and hemi-cyanine molecules which have extraordinarily large hyperpolarizabilities and large dipoles usually crystallize in centrosymmetric space groups and therefore the bulk materials exhibit no second-harmonic activities. Only exception was reported for ion complexes of 4-dimethylamino-N-methyl-4-stilbazolium.[22] We have done a series of studies focused on such an ion complex system,[23-25] and found that p-toluenesulfonate anion gives noncentrosymmetric crystals of complexes at high probability. Among the complexes, an interesting one was the merocyanine-p-toluenesulfonate (MC-pTS) which crystallizes in the space group P1.

Recently, the d_{11} of the crystal has been evaluated to be ca. 500pm/V.[26] In the synthetic expansion of the pTS complexes,[27] we have found another P1 crystal of methoxy derivative (2 in Table 1 and Figure 1). For it, larger d_{11} are expected than that of MC-pTS. In the case of cyanostilbazolium complex (8) which has λ_{co} around 400nm, ca. four times larger d_{33} than POM has been clarified.

Table 1   Stilbazolium p-toluenesulfonate complexes and their properties

No.	X-	mp($^{\circ}$C)	λ_{co}(nm)	λ_{max}(nm)	SHG(×Urea)
1	HO-	247	480	392	8.9
2	CH$_3$O-	238	470	383	29.5
3	C$_2$H$_5$O-	241	455	379	–
4	CH$_3$-	258	425	361	–
5	H-	199	410	347	–
6	Cl-	283	410	349	–
7	Br-	288	410	348	0.5
8	NC-	279	390	335	3.7

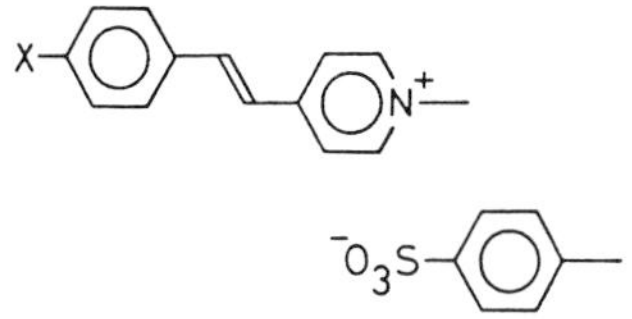

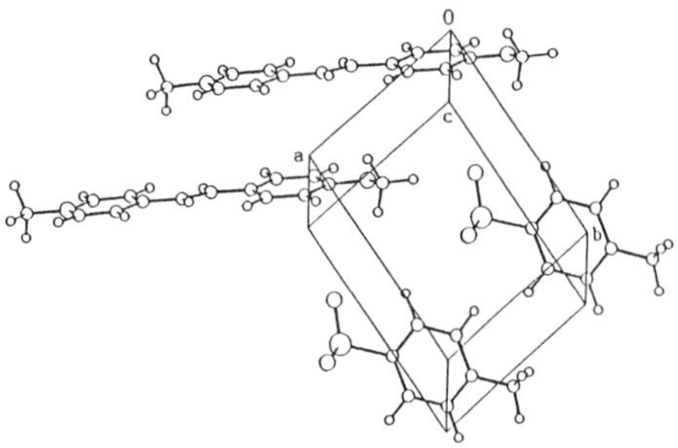

Figure 1   Crystal structure of 2 projected onto (001) plane

228   *Organic Materials for Non-linear Optics II*

2.2 Ionic dye-polymer ion complexes for third-order
nonlinear optics.
From the chemist's view, there are several possible
ways to prepare polymer-chromophore systems. In the case
of ionic dye-polymer ion complexes, following advantages
would be expected because of electrostatic interaction:
(1) The complex can be easily prepared by ion exchange
reaction. (2) Dyes can be dispersed homogeneously up to
high concentration with time stability. (3) Various
combinations of ionic polymers and ionic dyes are
possible. Actually, it was demonstrated that stable
dispersion at high concentration is possible.[28]
   In our recent study,[29] an interesting result has
been obtained. As seen from Figure 2, OC-1 and HC-1 have
almost the same wavelength of λ_{max} (i.e., band gap) and
similar molecular volume, and the absorption (i.e., the
transition moment) of OC-1 is three times stronger and
sharper than that of OC-1, but $\chi^{(3)}$ of OC-1 is ten times
smaller than that of HC-1 (Table 2). The result can be
explained only by assuming that HC-1 of unsymmetrical
structure has far larger $\Delta\mu_{eg}$ than OC-1 of symmetrical
structure. This may be an useful information for
molecular design.

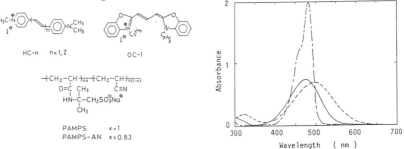

Figure 2   Absorption spectra of cyanines in methanol;
—— for HC-1 (1.7×10^{-5}mol/l), ----- for HC-2
(1.7×10^{-5}mol/l) and —·— for OC-1 (1.4×10^{-5}
mol/l)

Table 2 $\chi^{(3)}$s of HC-1, HC-2 and OC-1 polymer complexes

Dye	Polymer	Dye content (mmol/cm³)	$\chi^{(3)} \times 10^{12}$(esu) Pumping wavelength(μm)		
			1.5	1.9	2.1
HC-1	PAMPS-AN	3.47	18.6	4.57	2.34
HC-2	PAMPS	2.74	24.6	13.8	5.62
OC-1	PAMPS	4.30	1.29	1.70	-

2.3 Phthalocyanines for third-order nonlinear optics

Comparatively large $\chi^{(3)}$ values of the order of 10^{-11} esu (by THG in the resonant region) were found for chlorogallium, fluoroaluminum and vanadium-oxo phthalocyanines.[30,31] Molecular structural characteristics of those, compared with simple methallophthalocyanines showing smaller $\chi^{(3)}$, seem to show the existence of a permanent dipole based on the axial ligand.

Recently, we have done a comprehensive survey of phthalocyanines (Pcs) and found that not only do the above Pcs with axial ligand but also soluble Pcs with tetraalkylthio substituents give comparatively large $\chi^{(3)}$ (Table 3).[32] Closer examination of absorption spectra and X-ray diffraction leads to the unexpected view that substituents or axial ligands force the molecules not to form cofacial tight stack but to form any loose split-packing, and give large $\chi^{(3)}$.

<u>Table 3</u> $\chi^{(3)}$ values of phthalocyanine derivatives

Compound	Film thickness (μm)	$\chi^{(3)} \times 10^{12}$ (esu) Pumping wavelength(μm)		
		1.5	1.9	2.1
CuPc	0.53	1.3	1.5	1.1
O=VPc	0.28	8.6	30	40
CuPc(C_4S)$_4$	0.34	2.6	3.7	6.2
CuPc(C_6S)$_4$	0.23	2.5	20	20
CuPc(C_7S)$_4$	0.10	4.0	10	12
CuPc(C_8S)$_4$	0.26	3.4	23	50
CuPc(C_{10}S)$_4$	0.12	4.4	26	14
CuPc(C_{12}S)$_4$	0.58	2.0	8.7	13
O=VPc(C_6S)$_4$	0.37	3.3	9.8	14
O=VPc(C_8S)$_4$	0.20	4.1	18	31

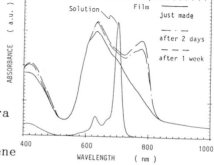

<u>Figure 3</u>   Absorption spectra of CuPc(C_8S)$_4$ in both films and 1-chloronaphthalene solution

3 POLYDIACETYLENES FOR THIRD-ORDER NONLINEAR OPTICS

Among many other conjugated polymers, polydiacetylenes
(PDAs), which can be obtained often as large single
crystals by the solid-state polymerization, have[3,33] caught
much attention owing to their large $\chi^{(3)}$, fast
response[34] and device applicability.[35] Aiming at the
enlarged $\chi^{(3)}$, we have been synthesizing a lot of new
PDAs.

3.1 Polydiacetylenes with directly bound aromatic rings

To achieve larger $\chi^{(3)}$ at molecular level, PDAs
with aromatic substituents directly bound to the main
chain (APDAs) and therefore increased numbers of π-
electrons per repeating unit must be a better
candidate.[36] We synthesized a lot of APDAs on the basis
of crystal engineering technique (Figure 4) to have
aromatic-substituted diacetylene monomers crystallized
in a solid-state polymerization stack, and demonstrated
that $\chi^{(3)}$ can be enlarged depending on the dihedral
angle between chain and aromatic substituents.[37-39]

Recently we have synthesized an unsymmetrically
substituted APDA (poly-MADF) with smaller dihedral angle
than before (Figure 5) and found the almost one order of
magnitude larger $\chi^{(3)}$ than the well-known PTS (Table
4).[40]

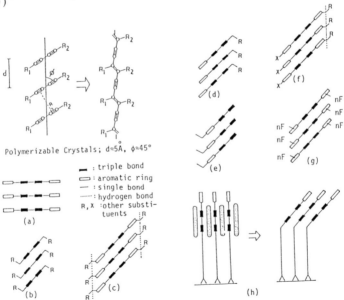

Polymerizable Crystals; d≈5Å, φ≈45°

━ : triple bond
▭ : aromatic ring
— : single bond
⋯ : hydrogen bond
R,X :other substi-
tuents

Figure 4 Crystal engineering for APDAs

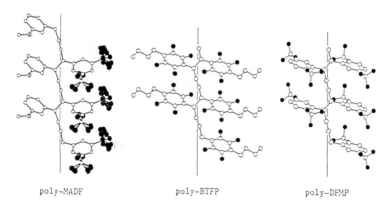

poly-MADF   poly-BTFP   poly-DFMP

<u>Figure 5</u> Molecular structure of some APDAs

<u>Table 4</u> $\chi^{(3)}$ values of PDA thin films

Poly-DA	$\chi^{(3)} \times 10^{11}$(esu) Pumping wavelength(μm)		
	1.96	2.10	2.16
Poly-DFMP		4.5	
Poly-BTFP	25	18	11
Poly-MADF	30	23	23
Poly-PTS		4.0	

3.2 Polydiacetylenes with directly bound acetylene moieties

Because of van der Waals interaction, aromatic rings of APDAs can not take coplanar conformation with respect to the main chain (Figure 6). The smallest dihedral angle calculated using plausible geometry is $44°$, and has been already attained by poly-MADF. However, if the substituent is an acetylene (AcePDA), we can expect always perfect π-conjugation between the main chain and the acetylene moiety.

Thus we have synthesized tri- and tetra-yne monomers with long alkyl chains and found their solid-state polymerizabilities.[41] Expectedly the longest absorption maximum of poly-triyne shifts ca. 40nm to the longer wavelength than that of poly-diyne. It is exactly the case between poly-triyne and poly-tetrayne. Structures of all these polymers have been identified to be an AcePDA by NMR spectroscopy.[42] Further investigations including NLO properties of these polymers are in progress.

232

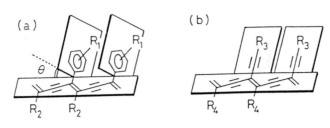

<u>Figure 6</u> Comparison of the π-conjugation between main chain and substituents of APDA and AcePDA.

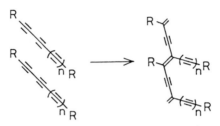

<u>Figure 7</u> Real structures of the polymers from di-, tri- and tetra-ynes.

REFERENCES

1) T. Gotoh, T. Tsunekawa, T. Kondoh, S. Fukuda, H. Mataki, M. Iwamoto, and Y. Maeda, Preprints of International Workshop on Crystal Growth of Organic Materials, 1989, p.234.
2) T. Katoh, A. Harada, M. Ishihara, M. Okazaki, K. Kamiyama, M. Tonogaki, S. Ohba, and Y. Saito, Extended Abstracts (The 37th Spring Meeting, 1990); The Japan Society of Applied Physics and Related Societies, 1990, p.1027.
3) K. Asano, M. Morita, Y. Nagasawa, and H. Ninomiya, ibid., 1990, p.1028.
4) L. S. Pu, I. Ando, K. Ageishi, Extended Abstracts (The 50th Autumn Meeting, 1989); The Japan Society of Applied Physics, 1989, p.984.
5) Y. Shimizu, T. Uemiya, N. Yoshie, H. Yoshioka, K. Nakatsu, T. Nogami, and Y. Shirota, 'Nonlinear Optics of Organics and Semiconductors', Springer-Verlag, Berlin, 1989, p.210.
6) A. Harada, Y. Okazaki, K. Kamiyama, and S. Umegaki, CLEO'90, 1990, CFE3.
7) T. Suhara, N. H. Hwang, and H. Nishihara, Extended Abstracts (The 37th Spring Meeting, 1990); The Japan

Society of Applied Physics and Related Societies, 1990, p.1035.

8) H. Hosomi, T. Suzuki, H. Yamamoto, T. Watanabe, H. Sato, and S. Miyata, Proc. SPIE, 1990, 1147, 124.

9) H. Yamamoto, T. Hosomi, T. Watanabe, and S. Miyata, Nippon Kagaku Kaishi, 1990, 789.

10) T. Wada, C. H. Grossman, S. Yamada, A. Yamada, A. F. Garito, H. Sasabe, Proc. SPIE, 1337, to be published.

11) G. Z. Zhang, T. Kinoshita, K. Sasaki, Y. Goto, and M. Nakayama, Proc. SPIE, 1990, 1147, 116.

12) S. Okada, H. Nakanishi, H. Matsuda, M. Kato, T. Abe, and H. Ito, Thin Solid Films, 1989, 178, 313.

13) M. Era, K. Nakamura, T. Tsutsui, S. Saito, H. Niino, K. Takehara, K. Isomura, and H. Tanigushi, Extended Abstracts (The 37th Spring Meeting, 1990); The Japan Society of Applied Physics and Related Societies, 1990, p.1030; private communication.

14) H. Nakanishi, M. Kagami, N. Hamazaki, T. Watanabe, Hisaya, H. Sato and S. Miyata, Proc. SPIE, 1990, 1147, 84.

15) M. Amano and T. Kaino, Electron. Lett., to be published.

16) T. Gotoh, T. Kondoh, K. Egawa, and K. Kubodera, J. Opt. Soc. Am., 1989, B6, 703.

17) T. Tsutsui, H. Murata, T. Momii, S. Saito, T. Kaino, Polym. Preprints, Jpn., 1990, 39, 484.

18) K. Takeda, T. Hasegawa, T. Kanetake, K. Ishikawa, Y. Tokura, T. Koda, Proc. SPIE, 1337, to be published.

19) K. Sasaki et al., Extended Abstracts (The 37th Spring Meeting, 1990); The Japan Society of Applied Physics and Related Societies, 1990, p.1046.

20) H. Kanbara, H. Kobayashi, K. Kubodera, T. Kurihara, and T. Kaino, Extended Abstracts (The 37th Spring Meeting, 1990); The Japan Society of Applied Physics and Related Societies, 1990, p.1044.

21) Kotaibutsuri, 1989, 24, 815-1008.

22) G. R. Meredith, 'Nonlinear Optical Properties of Organic and Polymeric Materials', ACS Symposium Series 233, 1983, p.27.

23) S. Okada, H. Matsuda, H. Nakanishi, S. Takaragi, R. Muramatsu, and M. Kato, Paper presented at the 2nd SPSJ International Polymer Conference (IPC86), 1986, Presentation No. 2C04; Japan Patent Application No. Toku-Gan-Sho 61-192404.

24) H. Nakanishi, H. Matsuda, S. Okada, and M. Kato, Proc. of the MRS Int. Mtg. on Adv. Mat., 1989, 2, 97.

25) S. Okada, A. Masaki, H. Matsuda, H. Nakanishi, M. Kato, R. Muramatsu, and M. Otsuka, Jpn. J. Appl.

Phys., 1990, 29, 1112.
26) S. Okada, A. Masaki, H. Matsuda, H. Nakanishi, T. Koike, T. Ohmi, N. Yoshikawa, and S. Umegaki, Proc. SPIE, 1337, to be published.
27) K. Sakai, N. Yoshikawa, T. Ohmi, T. Koike, S. Umegaki, S. Okada, A. Masaki, H. Matsuda, and H. Nakanishi, Proc. SPIE, 1337, to be published.
28) H. Matsuda, S. Okada, T. Nishiyama, H. Nakanishi, and K. Kato, 'Nonlinear Optics of Organics and Semiconductors', Springer-Verlag, Berlin, 1989, p.188.
29) H. Tomiyama, S. Okada, H. Matsuda, and H. Nakanishi, Proc. SPIE, 1337, to be published.
30) Z. Z. Ho, C. Y. Ju, and W. M. Hetherington III, J. Appl. Phys., 1987, 62, 716.
31) Y. Matsuoka, T. Wada, K. Shigehara, Y. Yamada, H. Sasabe, and A. F. Garito, Proc. of IUPAC 32nd Int. Symp. on macromolecules, 1988, p.581.
32) H. Matsuda, S. Okada, A. Masaki, H. Nakanishi, Y. Suda, K. Shigehara, and A. Yamada, Proc. SPIE, 1337, to be published.
33) C. Sauteret, J. P. Hermann, R. Frey, F. Pradere, J. Ducuing, R. H. Baughman and R. R. Chance, Phys. Rev. Lett., 1976, 36, 956.
34) T. Hattori and T. Kobayashi, Chem. Phys. Lett., 1987, 133, 230.
35) M. Thakur, B. Verbeek, G. C. Chi, and K. J. O'Brien, Mat. Res. Soc. Symp. Proc., 1988, 109, 41.
36) S. B. Clough, S. Kumar, X. F. Sun, S. Trypathy, H. Matsuda, H. Nakanishi, S. Okada, and M. Kato, 'Nonlinear Optics of Organics and Semiconductors', Springer-Verlag, Berlin, 1989, p.149.
37) H. Matsuda, H. Nakanishi, T. Hosomi, and M. Kato, Macromolecules, 1988, 21, 1238.
38) H. Nakanishi, H. Matsuda, S. Okada, and M. Kato, 'Nonlinear Optics of Organics and Semiconductors', Springer-Verlag, Berlin, 1989, p.155.
39) H. Nakanishi, H. Matsuda, S. Okada, and M. Kato, 'Frontiers of Macromolecular Science', Blackwell Scientific Publications, Oxford, 1989, p.469.
40) S. Okada, M. Ohsugi, A. Masaki, H. Matsuda, S. Takaragi, and H. Nakanishi, Mol. Cryst. Liq. Cryst., 1990, 183, 81.
41) S. Okada, H. Matsuda, H. Nakanishi, and M. Kato, Mol. Cryst. Liq. Cryst., in press.
42) S. Okada, H. Matsuda, A. Masaki, H. Nakanishi, and K. Hayamizu, to be submitted to Bull. Chem. Soc. Jpn.

Electro-optical Resins

S. Allen, D.J. Bone, N. Carter, T.G. Ryan, and R.B. Sampson

ICI WILTON MATERIALS RESEARCH CENTRE, PO BOX 90, WILTON,
MIDDLESBROUGH, CLEVELAND TS6 8JE, UK

D.P. Devonald and M.G. Hutchings

ICI COLOURS AND FINE CHEMICALS RESEARCH CENTRE, HEXAGON
HOUSE, BLACKLEY, MANCHESTER M9 3DA, UK

1. INTRODUCTION

Rapid advances have been made in recent years in the
development of amorphous polymeric materials which, on
poling with d.c. electrical fields, have high second-order
nonlinear optical properties. Most work has concentrated on
the use of side chain polymer systems, and materials of
this type have been reported as having electro-optic
coefficients in excess of those of lithium niobate[1]. A
significant problem, however, with these side-chain
polymers is the tendency for slow relaxation of the
electrically induced orientation, particularly at elevated
temperatures. Recently there have been reports[2,3] of more
stable nonlinear optical compounds, produced by the
incorporation of nonlinear optical chromophores within a
rigid cross-linked polymeric network. Stability of the
nonlinear optical effects for long periods of time at
temperatures of 80°C have been observed in such compounds[3].
In this paper we report the synthesis and characterisation
of two functionalised nonlinear optical chromophores, which
can be cross-linked to form rigid resin networks. The
temperature stability of nonlinear optical materials so
produced is superior to any previously reported in the
literature.

2. MOLECULAR DESIGN AND SYNTHESIS

The design of functionalised monomers for inclusion in the
cross-linked resin was influenced by the following factors:
ease of synthesis, high molecular nonlinearity β, chemical
and physical stability, transparency at the relevant
operating wavelength, dipole moment high enough to allow

orientation in the poling field and functionalisation so as
to permit ready cross-linking, and in such positions in the
molecule to prevent relaxation of orientation once the
poling field is removed. Donor/acceptor substituted
azobenzene derivatives satisfy many of these criteria and
incorporation of acrylate and methacrylate esters leads to
readily cross-linkable functionality. Initial experiments
on monomethacrylated molecules suggested that the ordering
of the chromogens in the cross-linked film could only be
retained after poling if the linking groups were included
at both ends of the molecular dipole. Most of our
development work has been carried out with films based on
the azobenzene derivatives (1) and (2). The former contains
one methacrylate group at each end of the molecule, while
(2) contains two such groups at each end. A conventional
dialkylamino donor group is used, while a combination of
nitro and sulphonamide act as the acceptor groups. The
latter functionality is also a convenient group through
which the methacrylate group(s) can be attached. The
syntheses of (1) and (2) are straightforward and are based
on readily available azo dyestuff intermediates.

3. FILM PREPARATION

A typical formulation used in the preparation of an
electro-optic film consists of 87.5% by weight of either
(1) or (2), 5% of a thermoplastic polymer, which is used to
ensure the coherency of the film at the onset of curing and

to improve its mechanical properties, and free radical thermal initiators to ensure efficient network formation. Films are deposited by spin coating from a solution containing 25% solids by weight in toluene. For use in the planar waveguide devices described in section 5, film thicknesses of about 1.5 microns are used, but thicker films may be deposited for refractive index and optical loss characterisation. Before electrical poling is commenced the film is pre-cured for 10 minutes at 80°C to increase its molecular weight and electrical resistivity. Electrical poling is achieved by use of a positive corona discharge[4] under a nitrogen atmosphere at 150°C for a period of about 45 minutes. The nitrogen atmosphere is required to prevent oxygen inhibition of the cross-linking.

In many cases the substrate onto which the film is deposited (either directly or with a thin buffer layer between the NLO layer and the substrate) will have an electrode coated onto its top surface. Corona poling is then effected by connecting this electrode to earth via a 10 MOhm current-limiting resistor so that the field is applied across the NLO film and buffer layer if present. Corona pin voltages of about 12 kV are used, with a pin-to-film distance of about 25-30 mm. This method of poling can be used with a wide variety of substrates.

At other times it may be preferable to utilise a substrate which has no electrode coating. In this case poling may take place through film and substrate, with the electrical earth being connected to the metallic hotplate on which the substrate is positioned. The efficiency of poling will depend on the relative resistivities of the sample and substrate. Glass microscope slides have been found to have resistivities of about 10 MOhm at 150°C, thus serving the function of the current-limiting resistor used in the first method of poling. Measurements of second harmonic generation from samples poled by the two methods show that they have comparable levels of NLO activity.

4. LINEAR OPTICAL PROPERTIES

The optical absorption spectra of compounds (1) and (2) are essentially identical in solution (being derived largely from the common chromophore). As shown in Figure 1 the wavelength of peak absorption is at about 480 nm.

The refractive indices of the films have been measured at a number of wavelengths using both waveguide and critical

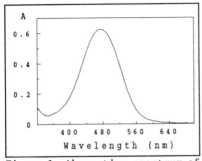

Figure 1. Absorption spectrum of a 0.01% solution of (1) in toluene.

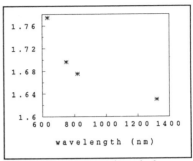

Figure 2. Refractive index n_{TE} of (1) as a function of wavelength.

angle techniques. In the first method planar waveguides are fabricated by spin-coating the material of interest directly onto glass substrates. The films are thick enough to support several waveguide modes. Laser light is coupled into and out of planar waveguides using high index strontium titanate prisms, and coupling angles measured. From these data the film thickness and refractive index can be determined, for both TE and TM waveguide modes. The second method involves measurement of the critical angle for total internal reflection at the interface between a high refractive index prism and the film of interest. This method can be used at wavelengths where the absorption is too strong to permit waveguide propagation. In unpoled films there is a slight anisotropy of the refractive indices due, presumably, to the alignment of the molecules in the spinning process. Typically n_{TE} is about 0.002 - 0.004 higher than n_{TM}. A plot of refractive index against wavelength is shown in Figure 2.

Optical loss measurements have also been carried out on monomode waveguide samples of the compositions containing molecules (1) and (2) using prism coupling techniques. At a wavelength of 1.32 microns losses around 1 dB/cm can be achieved routinely. At 820 nm the losses are much higher (about 12 dB/cm). These high losses are due to the fact that the weak tail of the absorption edge shown in Figure 1 extends out as far as this wavelength.

5. ELECTRO-OPTICAL CHARACTERISATION

Planar electro-optic waveguide devices have been constructed from these materials. A buffer layer, about 1.5

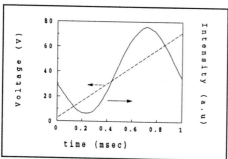

Figure 3. Applied voltage (dashed line) and corresponding photodiode ouput (solid line) from a planar waveguide modulator at 1.32 microns

microns thick, is coated onto a glass substrate which has a conducting electrode (indium tin oxide or gold) on its top surface. This buffer layer is an aromatic urethane acrylate resin, which is cross-linked *in situ* by exposure to high intensity uv light forming a solvent resistent, mechanically tough, optically clear film of refractive index about 1.51. The NLO layer is then coated, poled and cross-linked as described in section 3. A thickness of about 1.5 microns is used, to form a single mode waveguide. A top buffer layer of a thermoplastic polymer, again about 1.5 microns thick, is spin coated on top of the NLO layer. This top layer is removed from a portion of the film to allow coupling of the input and output prisms onto the NLO layer. Finally an aluminium top electrode 1 cm in length is evaporated onto the top buffer. The half wave voltage for this device is measured by the application of ramped voltages to the electrodes. The intensity of light monitored by a photodiode in the output beam has the form shown in Figure 3. Results for the two compositions, at wavelengths of 820 nm and 1.32 microns are given in Table 1.

Table 1: Summary of Electro-optic Properties

Component molecule	Wavelength (nm)	half-wave voltage	r_{33} (pm/V)
(1)	820	7.7	15.4
(1)	1320	23.4	8.7
(2)	820	16.8	7.8
(2)	1320	49.0	4.3

The lower activity of (2) is due largely to the greater degree of dilution of the molecular nonlinearity by the methacrylate functionalities.

6. THERMAL STABILITY

The thermal stability of the induced nonlinear optical

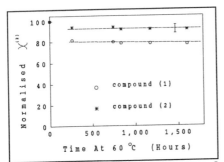

Figure 4. Long term stability of poled films.

activity in films of the above formulations has been studied using second harmonic generation. The long term stability has been monitored by storing samples at 60°C for long periods, and periodically measuring the level of SHG against a reference sample. The results are shown in Figure 4 for both compounds (1) and (2), where $X^{(2)}$ is normalised against its initial value. For both materials there is an initial drop in activity (between the starting point, before the application of heat, and the first measurement after a period in the oven), but the level of activity then remains constant within the limits of experimental error. In a second experiment samples of the two film formulations were heated on a hotplate, in air, whilst monitoring the level of SHG produced by a laser incident at 45° to the film normal. The results are shown in Figure 5. Both films are stable up to 75°C, but the composition containing (1) relaxes between this temperature and 150°C. That containing compound (2), however, retains a significant level of activity up to 190°C, where it is under threat of decomposition in air. The difference between the two sets of results can be related to the form of the cross-linked network formed by the two materials. Compound (2), with four functional groups will form a more highly cross-linked network than molecule (1), having only two. IR spectroscopy shows that after curing about 60% of the functional groups have reacted, in both cases. This is sufficient to give a tough, highly solvent-resistant film. However, modelling of the process of network formation suggests that in the case of compound (1), at this level of cure only about 10% of the molecules are linked into the network at both ends. The network is therefore in this case similar to a conventional side chain polymer, and the relaxation process shown in Figure 5 agrees with this picture. On the other hand the modelling suggests that for (2) about 60% of the molecules will be linked into the network at both ends, and so molecular reorientation is much more highly inhibited.

7. CONCLUSIONS

We have reported in this paper the production of cross-

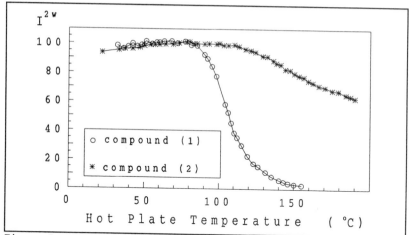

Figure 5. Temperature stability of poled films on heating

linked resin films having high levels of nonlinear optical
activity and showing excellent thermal stability. These
films have been fabricated into planar waveguide electro-
optic modulators having low operational voltages. The
importance of the cross-link density to the achievable
thermal stability has been clearly demonstrated. The
highest levels of thermal stability are achieved at the
expense of some nonlinear optical activity. The magnitude
of the electro-optic coefficient in these films is limited
by the degree of orientation that can be induced in the
molecules (1) and (2). Our continuing research will address
the problem of increasing the level of activity whilst
retaining the excellent thermal and mechanical properties
achieved to date.

Financial support for this work from the European Commission, through
the ESPRIT programme is acknowledged.

REFERENCES

1. D Haas, H Yoon, H Man, G Cross, S Mann and N Parsons, <u>Proc. SPIE</u>,
1989, <u>1147</u>, 222.
2. M Eich, B Reck, D Y Toon, C G Willson and G C Bjorklund, <u>J. Appl.
Phys.</u>, 1989, <u>66</u>, 3241.
3. D Jungbauer, B Reck, R Twieg, D Y Yoon, C G Willson and J D Swalen,
<u>Appl. Phys. Lett.</u>, 1990, <u>56</u>, 2610.
4. H L Hampsch, J L Torkelson, S J Bethke and S G Grubb, <u>J. Appl.
Phys.</u>, 1990, <u>67</u>, 1037.

Non-linear Optical Properties of Organic Polymers – Origins and Nature

H.J. Byrne and W. Blau

DEPARTMENT OF PURE AND APPLIED PHYSICS, TRINITY COLLEGE, DUBLIN 2, IRELAND

1. INTRODUCTION.

The π-electron backbone of organic conjugated polymeric systems is predicted to be highly polarisable, giving a large ultrafast third order optical nonlinearity [1,2]. Such a nonlinearity should be real in nature and so should be broadband in the near infrared. Despite intense research activity few guidelines for the design of nonlinear optical polymers have emerged. The influence of electron-lattice coupling, electron correlation and interchain effects on the molecular nonlinearity needs to be elucidated. In order to increase the polymeric nonlinearity towards values feasible for technological exploitation, a relationship between the influence of such effects and backbone structure must be achieved. In this study, a range of nonlinear optical studies in the near infrared of π-conjugated materials is reviewed in an attempt to identify such relationships.

2. EXPERIMENTAL METHOD.

The experimental method employed in these studies was that of forward degenerate four wave mixing [3]. The light source is a Nd^{3+} : YAG laser emitting linearly polarised pulses of 75 ± 25psec duration and of wavelength $\lambda = 1.064\mu m$. The experimental method.[4] is based on the formation of a transient grating in the material as a result of the reponse of the nonlinear refractive index to the interference of two spatially and temporally overlapped beams Under thin grating conditions [3] an expression relating the diffraction efficiency, η to the third order material nonlinearity, for transparent materials, may be derived;

$$|\chi^{(3)}| = \frac{4\varepsilon_0 c \, n^2 \, \lambda \, \sqrt{\eta}}{3 \, \pi \, d \, I_0} \qquad\qquad 1$$

where c is the speed of light, ε_0 is the permittivity of free space, n is the refractive index of the sample, d is the sample thickness (= 1mm) and I_0 is the input pulse intensity. Verification of the presence of a true third order nonlinear process may be performed by monitoring the intensity dependence of the diffraction efficiency. For a true third order process, η is proportional to I^2. Such a verification is important as fifth and seventh order processes, originating in two and three photon resonant enhancement of the material nonlinearities, have been observed in organic conjugated materials [4,5].

$|\chi^{(3)}|$ may have both real and imaginary components originating from the polymer as well as a contribution from the solvent, $\chi^{(3)}_{solv.}$, which is purely real and positive in the case of most organic solvents [6]. Thus,

$$|\chi^{(3)}| = \{ \, (\chi^{(3)}_{solv} + Re \, \chi^{(3)}_{pol})^2 + (Im \, \chi^{(3)}_{pol})^2 \}^{1/2} \qquad\qquad 2$$

where $Re\chi^{(3)}_{pol}$ and $Im\chi^{(3)}_{pol}$ are the real and imaginary components of the material nonlinearity. The concentration dependence of $|\chi^{(3)}|$ therefore yields the contribution $\chi^{(3)}_{solv}$ due to the solvent, the magnitude and sign of $Re\chi^{(3)}_{pol}$ and the magnitude of $Im\chi^{(3)}_{pol}$.

3. NONLINEAR OPTICAL STUDIES OF ENYNE OLIGOMERS.

For the purpose of investigation of the dependence of the molecular hyperpolarisability on chain length, a series of enyne oligomers, which serve as a model for the much studied polydiacetylene [7], were studied. Each compound has (4n+2) conjugated carbons and hence, for convenience, these compounds will be referred to as c[4n+2]. In this study compounds with n = 1,2,3 and 5 were used. The oligomers are readily soluble in n-hexane, with the exception of c[22] which is less soluble.

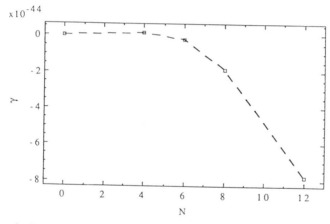

Figure 1 : Real component of γ as a function of N

Measurements of the diffraction efficiency as a function of pump intensity and oligomer concentration were performed. For all solutions the intensity dependences of the diffraction efficiency are found to be characteristic of a well behaved third order nonlinear process. Similar behaviour is observed for neat solvent. The real and imaginary components of the material nonlinearity may be determined by fitting equation 2 to the observed concentration dependences. In the case of c[22] the real component of the oligomer nonlinearity is clearly negative. A similar behaviour is observed for both c[10] and c[14] solutions. For c[6] solutions, however, the dependence fits well to a positive real component of the nonlinearity. Figure 1 shows the calculated values of γ_R, the real part of the molecular hyperpolarisability for the range of oligomers as a function of N.

244   *Organic Materials for Non-linear Optics II*

Within the formulation of the free electron model [8,9] the molecule is treated as a one dimensional box of length 2L, being the so-called conjugation length, containing 2N degenerate electrons. Such a simple free electron gas model appears adequate to describe the linear optical properties of conjugated systems [10]. Similarly, the nonlinear optical properties may be described by the free electron model. This model however, predicts a positive real nonlinearity for N>3 and makes no accommodation for an imaginary component. However, as is shown by figure 1, the value of γ_R is positive for N = 4, but changes sign to become negative for N≥6. Moreover, the presence of a non zero minimum in the concentration dependence of η/I^2 for c[22] is indicative of a strong imaginary contribution to the molecular nonlinearity. For each oligomer, best fits to these concentration dependences are obtained with a value of $\gamma_I \sim 2|\gamma_R|$. Clearly this may not be accounted for by the free electron model.

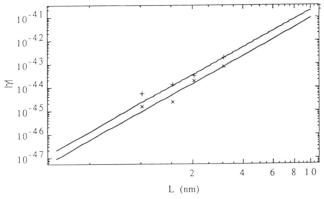

Figure 2 : Magnitude of the molecular hyperpolarisabilities as a function of oligomer chain length : (x)- |real γ|, (+)- |γ| , (-) - slope 4.

A more recent theoretical treatment of third order molecular hyperpolarisabilities of conjugated linear chains has been presented by Grossman et al [11]. In this treatment, the third order nonlinear susceptibilities of conjugated organic molecules are described in terms of virtual electronic excitations occuring within the π-electron states. The theory explicitly takes into account electron-electron Coulomb interactions and describes electron correlation. Application of this model to the case of short chain polyenes predicts a power law dependence of γ on the chain length L, with the dependence $\gamma \propto L^{4.6\pm0.2}$. In figure 2 the magnitude of the real component of γ is plotted as a function of the chain length. The curve fits well to a power dependence of order 4.0±0.5. Figure 2 also shows a similar curve for |γ|. Again a power of 4.0±0.5 dependence fits well. This shows reasonably good agreement with the model of Grossman et al, Also, a similar dependence has been observed in oligothiophenes [12]. More significantly, the above model can accommodate a contribution to the nonlinear susceptibility of imaginary contributions originating in virtual electronic excitations. In the case of one dimensional conjugated molecules a localised transition may be treated in the format of Su et al.[13], giving rise to lattice deformations in the form of solitons or bipolarons. A virtual transition, being similarly localised should also produce such lattice deformations, giving rise to an imaginary contribution to the molecular hyperpolarisability.

4. NONLINEAR OPTICAL STUDIES OF CONJUGATED POLYMERS.

The polymeric materials studied include polythiophene which has a hetero-cyclic backbone in which the sulphur stabilises the cis acetylenic like structure [14]. Polythiophene itself is insoluble and therefore the polymer used in this study was the substituted poly (3-butyl thiophene) which was studied in chloroform solution. A particular series of polydiacetylenes which has been extensively studied is the nBCMU series [15] which has the diacetylenic backbone structure with sidegroups

$$R=R'= -(CH_2)_n-OC-ONHCH_2COO(CH_2)_3CH_3$$

The material employed in this study has $n=4$. Poly 4BCMU is soluble in organic solvents such as chloroform and dimethyl formamide, forming strongly yellow coloured solutions. A third polymer investigated in this study is the "rigidised backbone" diacetylene type polymer APPE [16]. Similar to the case of polythiophene, the cyclic structures of the anthracene and phenol rings in the backbone should provide a rigidisation of the backbone, in so doing, increasing the π electron delocalisation. APPE is soluble in many solvents, including toluene.

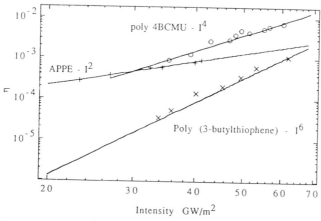

Figure 3: Intensity dependences of the diffraction efficiency.

In the case of neat solvent, the intensity dependence of the diffraction efficiency is found to vary as the square of the input intensity for both chloroform and chlorobenzene, as is expected for a third order nonlinear process. The intensity dependence for poly 4BCMU and poly (3-butyl thiophene) solutions is however of higher order. Figure 3 shows the intensity dependence of the diffraction efficiency for some of the polymer solutions as well as for chloroform and toluene. In the range of polythiophene solutions, a clear dependence of I^6 is observed at certain concentrations. Such behaviour is indicative of a seventh order nonlinear process and may be associated with a three photon resonant enhancement of the third order nonlinear process [4]. In polydiacetylene solutions, processes characteristic of fifth order nonlinearities have been observed at 1.064μm [5,16] and have been identified with the influence of a two photon resonance centred on 1.35μm [17]. This behaviour may be seen as an I^4 dependence of the diffraction efficiency on intensity in figure 3. Also shown in figure 3 is the intensity dependence of the diffraction efficiency of a solution of APPE in toluene. All solutions of APPE exhibit an intensity dependence which is characteristic of a true third order process.

The linear absorption spectra of solutions of each polymer are singly peaked in the visible region and this broad absorption may be associated with a one dimensional semiconductor interband transition [13]. The problem therefore arises of understanding the great differences in the nonlinear optical properties at 1.064μm of three polymers which show virtually identical linear optical properties. It has been shown [18,19], that in order to correctly account for the ordering of the 1^1B_u π-electron state and the lower lying, two photon 2^1A_g state in polyenes [20], a proper treatment of the electron-electron interactions must be introduced, and that independent particle theories such as Huckel theory are inadequate. The importance of these electron-electron interactions is a direct result of the dimensionality of these systems which leads to strongly correlated electron motion. The relative contributions of these interactions must be considered also a function of the electron density within a specific monomer unit. In the case of polythiophene, the monomer unit contains only double bonds and so the electron density is comparatively low compared with other polymers. This results in a positioning of the three photon level at an energy sufficiently high to influence the four wave mixing process at 1.064μm. Polydiacetylene has a very much higher electron density per monomer unit and this leads to a two photon level which is positioned close to 1.064μm. In the case of the APPE solutions, the absence of a multiphoton resonant contribution to the nonlinearity is consistent with the large electron density of the monomer unit resulting from its complex aromatic structure. This gives rise to a strong electron correlation which shifts the influence of multiphoton resonances far from the laser wavelength.

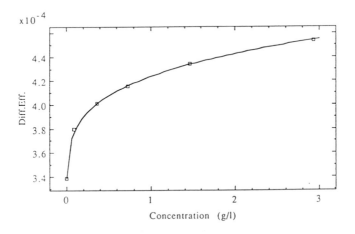

Figure 4: Concentration dependence of the diffraction efficiency of APPE solutions at a constant intensity of 3×10^{13} Wm^{-2}

It has been pointed out in the previous section that the diffraction process in some polymer solutions is strongly influenced by the proximity of multiphoton resonances. The degree of this influence is, however seen to be strongly concentration dependent. In poly (3-butyl thiophene) solutions, for example, a three photon resonant enhancement is observed at low and intermediate concentrations but this enhancement decreases at higher concentrations [4]. Indeed, for all polymers examined, the concentration dependence of the diffraction efficiency is seen to deviate considerably from the parabolic dependence predicted by equation 2. This deviation is most pronounced in polymers containing bulky aromatic groups in the backbone. Figure 4 shows the concentration dependence of the diffraction efficiency for the range of APPE solutions in toluene. Although the diffraction efficiency increases rapidly from that observed for pure solvent, indicating a large nonlinearity, this increase rapidly dies off, and even at moderate concentrations the molecular hyperpolarisability has been considerably reduced.

For all polymers examined, the concentration dependence of the diffraction process is seen to obey a sublinear power law dependence of the form

$$\chi^{(3)} = \Gamma_{eff} C^{(1-\xi)}$$

3

where Γ_{eff} is a concentration independent nonlinear parameter. ξ may be identified with the degree of deviation of the dependence from linearity and is seen to be 0.9 for APPE solutions, 0.5 for polythiophene solutions and 0.3 for the poly 4BCMU solutions.

Although equation 3 does not extrapolate towards molecular parameters the parameter ξ may be taken as a measure of the degree of molecular interaction and the extent of the influence on the molecular hyperpolarisability of the material. As can be seen in figure 4 this influence can be considerable, reducing the bulk nonlinearity to a small fraction of that of the sum of individual molecular nonlinearities. Thus is highlighted the importance of consideration of intermolecular as well as intramolecular effects in the design of nonlinear optical molecular systems.

Although, the concentration dependence of the material nonlinearity is complicated by the influence of interchain effects, a best fit of a well behaved third order nonlinearity to the low concentration regime of the concentration dependence yields a value of the magnitude of the molecular hyperpolarisability calculated per monomer unit. For poly 4BCMU solutions, this is calculated to be $|\gamma| = 1.5 \pm 0.5 \times 10^{-42}$ m^5/V^2. Comparison of this to the power law dependence observed for oligomeric enyne series predicts a value of the chain length of order 70Å. Since the polymer is of considerably longer chain length, a deviation from this power law dependence may be inferred. This deviation is not, however, surprising due to the differing nature of the nonlinearities. In both cases, the nonlinearity is dominated by imaginary components, in the case of the polymer being influenced by the proximity of a two photon resonance. As has been illustrated by Su et al.[13], the electronic transitions in one dimensional conjugated polymers are localised due to lattice coupling and give rise to lattice deformations in the form of solitons or bipolarons. The spatial extent of such lattice deformations is predicted to be of the order of 15 repeat units [13], which may be estimated to be ≈70Å. If, therefore, polarisation of the π-electron system involves a localisation of the charge distribution resulting in solitonic-type lattice deformations, then this polarisation itself limits the π-electron delocalisation. Notably, the chain length of the polymer estimated from a fit of the nonlinearity to the power law dependence of the oligomers compares rather favourably with the estimated deformation extent. Deviation from the power law dependence is then interpreted by a limitation of the effective chain length, or conjugation length, by the nature of the polymer response to the polarisation.

Previous studies therefore suggest that an improvement on the electronic molecular hyperpolarisability over those of polydiacetylenes may not simply be achieved by increasing the charge density per monomer unit. For example, chemical doping of polymers has been proved to have a disastrous effect on the nonlinear optical properties [12]. Instead consideration must be given to the strong correlation of the electronic and lattice configurations. Modification of the electronic configuration of a polymer backbone must be accommodated by the lattice to avoid charge localisation and a break up of any backbone configuration.

6. CONCLUSIONS.

A review of a range of studies of the nonlinear optical properties of organic conjugated systems enables the identification of some of the important influencing factors. Interchain effects are seen to greatly reduce the bulk nonlinearity. Planar conjugated systems, which have a tendency to aggregate are seen to be strongly influenced by such effects. In the design of a molecular system for nonlinear optics, therefore, it is important to consider not only the intrinsic molecular properties but also the extrapolation of those properties towards bulk. In the region of the near infrared, the nonlinearity of conjugated polymers is seen to be strongly influenced by the proximity of multiphoton absorptions. A complete understanding of the relationship between the positioning of such resonances and the monomeric electron density and structure should render their influence tailorable, enabling more control over the dispersion of the material nonlinearity in the near infrared region. In all systems investigated, a strong or dominant contribution from imaginary components of the nonlinearity is observed and a purely real, nonresonant nonlinear susceptibility appears elusive. This imaginary contribution may be linked with the strong electron-lattice coupling interaction characteristic of these systems. Modification the electron distribution for the purposes of an increased nonlinearity must take account of this interaction.

REFERENCES

[1]   W.M. Dennis, W.Blau and D.J. Bradley, Appl. Phys. Lett., 47, 200 (1985)
[2]   G.M. Carter, J.V. Hryniewicz, M.K. Thakur, Y.J. Chen and S.E. Meyler, Appl. Phys. Lett., 49, 998 (1986)
[3]   H.J. Eichler, P. Günter and D.W. Pohl, "Laser Induced Dynamic Gratings", Springer Series in Optical Sciences, Springer Verlag (1986)
[4]   H.J. Byrne and W. Blau, Synth. Metals, 37, 231 (1990)
[5]   J.M. Nunzi and D. Grec, J. Appl. Phys., 62, 2198 (1987)
[6]   P. D. Maker, R. W. Terhune and C. M. Savage, Phys. Rev. Lett., 12, 507 (1964).
[7]   See for example "Polydiacetylenes", D. Bloor and R. R. Chance eds., Nato ASI series No 102, Martinus Nijhorf, (1985).
[8]   H. Kuhn, Fortschr. Chem. Org. Naturst., 16, 169 (1958); 17, 404, (1959).
[9]   K. C. Rustagi and J. Ducuing, Opt. Comm., 10, 258, (1974)
[10   H.J. Byrne, W. Blau, R. Giesa and R.C. Shulz, Chem. Phys. Lett., 167, 484 (1990)
[11] C. Grossman, J. R. Heflin, K. Y. Wong, O. Zamani-Khamiri and A. F. Garito, in "Nonlinear optical effects in Organic Polymers', J. Messier, F. Kajzar, P. Prasad and D. Ulrich eds., Nato ASI series, Vol. 162, Kluwer Academic. (1989).
[12] P. N. Prasad, in "Nonlinear optical effects in Organic Polymers', J. Messier, F. Kajzar, P. Prasad and D. Ulrich eds., Nato ASI series, Vol. 162, Kluwer Academic. (1989).
[13] W. P. Su, J. R. Schrieffer and A. J. Heeger, Phys. Rev. B, 22, 2099 (1980).
[14] T.C. Chung, J.H. Kaufman, A.J. Heeger and F. Wudl, Phys. Rev. B, 30,702 (1984)
[15] G.N. Patel, Polym. Prepr., Am. Chem. Soc., Div. Polym. Chem., 19,154 (1978)
[16] H.J. Byrne and W. Blau, Synth. Metals, 37, 231 (1990)
[17] F. Kajzar and J. Messier, in "Polydiacetylenes", D. Bloor and R.R. Chance eds., NATO ASI Series, Martinus Nijhoff (1985)
[18] B.S. Hudsen and B.E. Kohler, J. Chem. Phys., 64, 4422 (1976)
[19] Z.G. Soos and S. Ramasesha, Phys. Rev. B29, 5410 (1984)
[20] P. Tavan and K. Schulten, Phys. Rev. B36, 4337 (1987)

Two Photon Enhanced Non-linearity of Polydiacetylene

J.M. Nunzi and F Charra

CEA/DTA/LETI/DEIN/CEN-SACLAY, LABORATOIRE DE PHYSIQUE
ELECTRONIQUE DES MATÉRIAUX, 91191 GIF SUR YVETTE, CEDEX, FRANCE

Polydiacetylenes exhibit a large two photon absorption at 1064 nm[1]. In high intensity ($GW.cm^{-2}$) picosecond degenerate four wave mixing conditions, it leads to enhanced intensity dependent nonlinearities formally described in terms of one-dimensional $\chi^{(5)}I$ and $\chi^{(7)}I^2$ susceptibilities[2,3]. This is also the case of poly(3-butylthio-phene)[4] and we recently demonstrated the possibility of a purely $\chi^{(5)}$ effect: phase conjugation with frequency doubling[5]. Such enhanced nonlinearities raise the question of their origin in terms of: virtual excitation of quantum levels (optical Stark shifts)[6,7] or real population of excited species(photoinduced carriers, triplet states, polarons)[8-10]. The answer indeed determines ultimate response speed of this large and potentially useful [3,5,6] effect. We thus first discuss a simple model accounting for the way 3 level molecules pumped at high intensity have enhanced nonlinearities. We then report on intensity and time dependence of real and imaginary parts of picosecond phase conjugate response measured at 1064 nm in a polydiacetylene red form.

1 MODEL FOR HIGH INTENSITIES

1.1 Two Photon Resonance in Three Level Systems

Account of the large two photon absorption and of connected third order susceptibilities ($\chi^{(3)}$) exhibited by polydiacetylenes in the near-infrared region requires at least a three level system description[11]. At high fields, high order terms ($\chi^{(5)}$) play a significant role and with some care regarding artificially resonant secular terms[12], perturbative expansions can be extended to their calculation[5]. $\chi^{(5)}$ terms however generally appear when optical pump fields E are such that $E^2 \geq \chi^{(3)}/\chi^{(5)}$, leading to divergence of perturbative expansions. Fortunately, nearly exact non-perturbative calculations are then tractable[13]. In an attempt to find a simple analytical solution to our problem, let us consider the optical Bloch equations for the density matrix of the system[14] submitted to one strong electric field E at frequency ω close to two photon resonance *(Figure 1a)*. In the rotating field approximation where only the most ω and 2ω resonant terms are retained, we write the density matrix as:

$$[\ell] = \begin{pmatrix} \ell_{00} & \ell_{01}e^{-i\omega t} & \ell_{02}e^{-2i\omega t} \\ \ell_{01}^{*}e^{i\omega t} & 0 & \ell_{12}e^{-i\omega t} \\ \ell_{02}^{*}e^{2i\omega t} & \ell_{12}^{*}e^{i\omega t} & \ell_{22} \end{pmatrix} \tag{1}$$

Assuming that 2g population and coherence relaxation times are shorter than pulse duration (γ^{-1} and $\Gamma^{-1} \ll \tau_p$), the system is in a stationary state and coefficients ℓ_{ij} are time independent. As population is conserved:

$$\ell_{00} + \ell_{22} = 1 \tag{2}$$

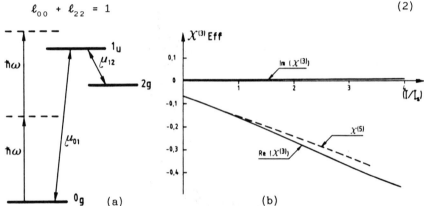

Figure 1 *Two photon resonant three level system (a). For polydiacetylene red form, we choose* $\hbar\omega_{01}$ = 2.28 eV, $\hbar\omega_{02}$ = 2.19 eV, μ_{01} = 6 Åe, μ_{02} = 23 Åe,[6] $\gamma^{-1} \simeq$ 60 fs[15] and $\Gamma^{-1} \simeq$ 800 fs[16]. *Effective phase conjugation* $\chi^{(3)}$ *in arbitrary units vs strong pump intensity using the above parameters (b).*

A solution for $[\ell]$ is readily obtained[17] and with dipolar transition matrix $[\mu]$ (Figure 1a), we get the average ω-polarization of the system:

$$P_w = \frac{1}{2} N \, Tr \, (\mu\ell) \tag{3}$$

where N is the number of elementary molecules per unit volume, close to the inverse volume of an exciton[18] (c.a. 10 monomer units). Notice that relevant local field factor is 1 for conjugated polymers[19]. In the international system, P_ω is related to the intensity dependent first order susceptibility via:

$$P_\omega^{NL} = \varepsilon_0 \; \chi_\omega^{(1)}(I)E$$

and $\hspace{8cm}$ (4)

$$\chi^{(1)}(I) = \frac{N \; \hbar \; \gamma\Gamma}{8\varepsilon_0 I_s} \; \frac{(\Delta_2 + i\gamma)I/I_s + (1 - \gamma/\Gamma)\omega_s (I/I_s)^2}{\gamma^2 + (\Delta_2 + \omega_s I/I_s)^2 + \gamma^2 (I/I_s)^2}$$ (5)

where we define pump intensity as $I = EE^*/4$, Stark shift frequency $\omega_s = \sqrt{\gamma\Gamma} \; (\mu_{12}^2 - \mu_{10}^2)/2\mu_{01}\mu_{12}$, two photon saturation intensity $I_s = |\Delta_1| \hbar^2 \sqrt{\gamma\Gamma}/2\mu_{01}\mu_{12}$, one photon detunings $\Delta_1 = \omega_1 - \omega_0 - \omega \approx -(\omega_2 - \omega_1 - \omega)$ (Figure 1a), and two photon detuning $\Delta_2 = \omega_2 - \omega_0 - 2\omega$. We recognize in (5) the regular two photon resonant term[13] of $\chi^{(3)}$ (first order in I/I_s) as well as higher order terms, intensity power series expansion diverging when $I \geqslant I_s$. In the case of a polydiacetylene red form (4BCMU) pumped at $\lambda = 1064$ nm (1.17 eV Nd-YAG laser frequency), a reasonable set of molecular parameters involved in (5) is given on *Figure 1a*. We get $\Delta_2 = -0.15$ eV, $\gamma \simeq 10^{-2}$ eV, $\Gamma \simeq 8 \; 10^{-4}$ eV, $\omega_s = 5 \; 10^{-3}$ eV and $I_s = 10^{15}$ V^2/m^2 ($\leftrightarrow$ 200 MW.cm^{-2}). We thus see that: optical Stark effect indeed tunes two photon resonance onto 2ω laser frequency (ω_s and Δ_2 have opposite signs in the denominator of (5)) and power series expansions of $\chi^{(1)}(I)$ diverge at low optical fields (in pulsed picosecond experiments).

1.2 Phase Conjugate Response

When nonlinear process is degenerate four wave mixing (inset of *Figure 2*), two pump beams 1 and 2 are counter propagating into the active medium (polymer). A third beam (probe 3) incoming at a small angle θ to 1 is reflected (time reversed) after its interaction with 1 and 2. Conjugate signal 4 is radiated by the source polarization:

$$P_4 = 2 \cdot \frac{3}{4} \varepsilon_0 \; \chi^{(3)}(-\omega;\omega,\omega,-\omega)E_1 E_2 E_3^*$$ (6)

with a reflectivity defined as $\mathfrak{R} = I_4/I_3$, in the small signal limit and in non-absorbing media (thickness L, index n)[14]:

$$\mathfrak{R} = \left| \frac{3\pi L}{2n\lambda} \cdot \chi^{(3)} E_1 E_2 \right|^2$$ (7)

When the polymer is resonantly excited by a strong laser field (intensity $I = EE^*/4$), we can always define an effective phase conjugate $\chi^{(3)}$ by identification in (4) of the component radiated with frequency ω and spatial phase of E_3^*.[20] As a matter of simplicity, we consider the case of an intense forward pump 1 and weak probe 3 and backward pump 2:

$$I_1 \gg I_2 \text{ and } I_3 (\gg I_4) \tag{8}$$

We thus expand total input intensity as:

$$I = I_1 + (E_1^* E_2 + E_1 E_3^*)/4 + \text{2nd order terms} \tag{9}$$

Developing (4) up to second order in (9), we get the effective nonlinearity:

$$\chi_{eff}^{(3)}(I_1) = 2 \frac{\partial \chi_{(I_1)}^{(1)}}{\partial I} + I_1 \frac{\partial^2 \chi_{(I_1)}^{(1)}}{\partial I^2} \tag{10}$$

The phase conjugate source polarization becomes:

$$P_4 = \frac{\varepsilon_0}{4} \chi_{eff}^{(3)} E_1 E_2 E_3^* \tag{11}$$

where $\chi_{eff}^{(3)}$ includes factor 6 considered in (6) for perturbational susceptibilities [12]. The effective nonlinearity derived from (5) is pictured on *Figure 1b*. At $I = 0$, we have the regular negative real part of a system pumped above two photon absorption[11]. We see that the imaginary part (linked with nonlinear absorption)[2] is weakly intensity dependent in this domain while real part increases (X5) with a negative slope ($\chi^{(5)}$ linked with Stark effect)[3] and a curvature ($\chi^{(7)}$ linked with 2g population)[5]. At higher intensities, nonlinearity saturates as for resonant two level systems,[18,20,21] but enhanced intensity dependent nonlinearity is characteristic of systems pumped above two photon resonance[13,17]. Except for vertical and horizontal scales on Figure 1b, $\chi_{eff}^{(3)}$ intensity dependence is rather insensitive to the choice of molecular parameters (*Figure 1a*) as long as $\Delta_2 < 0$ and $\omega_s > 0$. Consequently, account of inhomogeneous broadening and 3-dimensional isotropy in solutions[3] have little effect on $\chi_{eff}^{(3)}$ general behavior.

1.3 Time Dependent Effects and Coherent Artifact

The above presented three level molecule description holds as long as (2) is verified. However, as real electronic excitations take place in semiconducting polymers, π-bonds are broken and within the fraction of picosecond following excitation[22], the excited state electronic wave function faces a forbidden lattice which rapidly tends to reorganize via electron-phonon coupling (formation of polaronic states)[10]. In degenerate four wave mixing experiments, we phenomenologically describe such effects by their additional nonlinear polarization P_{exc}^{NL} resulting from a number $N_{exc}(I,t)$ (intensity and time dependent) of induced excited species (per unit volume) with individual (molecular) polarizability $\alpha_{exc}(I)$ (intensity dependent or not):[8]

$$P_{exc}^{NL} = \varepsilon_0 N_{exc}(I,t) \, \alpha_{exc}(I)E \tag{12}$$

In order to treat uniformly the case of mono- and multiphoton excitations (heating for example)[3], it is convenient to consider α_{exc} as the induced polarizability per absorbed photon. In the approximation of weak absorptions, exponential decays (lifetime τ_e) and impulsive excitation (pulse duration τ_p), the number of excited species is related to the absorption (imaginary part of $\chi_\omega^{(1)}$) by:[2]

$$N_{exc}(I,t) = \frac{\Phi \, \tau_p \, e^{-t/\tau_e}}{n \, c \, \hbar} \, I \, \Im m(\chi_\omega^{(1)}(I)) \tag{13}$$

where Φ is the creation quantum efficiency. In a more general description, different excited species with different lifetimes can be treated, excited species can evolve with time ($\alpha_{exc}(t)$) and as pulse duration is finite, α_{exc} depends on τ_p which operates as a filter selecting longer lifetime effects $(\tau_e \geqslant \tau_p)$[23]. We then apply derivation (10) to the induced $\chi_{exc}^{(1)}(I) = N_{exc}\alpha_{exc}$ to obtain phase conjugate response for real processes involving excited species:

$$\chi_{eff}^{(3)}(I_1, t=0) = \left[2 \frac{\partial}{\partial I} + I_1 \frac{\partial^2}{\partial I^2} \right] (N_{exc}(I_1, 0)\alpha_{exc}(I_1)) \tag{14}$$

This equation holds at zero delay, when beams 1, 2 and 3 are synchronized and mutually coherent (derived form the same laser source). Meanwhile, probing the excited state laser induced grating at different delays gives access to relaxation dynamics[24]. In our experimental configuration (eq. 8), read pulse is backward pump beam 2 (*inset of Figure 4*). At $t \neq 0$, E_2 is no longer involved in expansion (9), and since it is weak, the nonlinear phase conjugate susceptibility radiating signal via (11) is then:

$$\chi_{eff}^{(3)}(I_1, t \neq 0) = \alpha_{exc}(0) \frac{\partial N_{exc}}{\partial I}(I_1, t) \tag{15}$$

We thus see that delayed conjugate signal undergoes a large reduction from (14) to (15) (at least by a factor 2 when absorption is not saturated). It appears as a fast instantaneous effect called coherent artifact and it is a universal feature of time resolved pump-probe experiments in degenerate multi-wave mixing configurations, when only local effects are involved. For example, assuming a N-photon generation of linear and permanent excited species:

$$\chi_{exc}^{(1)}(I,t) = a \ I^N \ \alpha_{exc} \tag{16}$$

and

$$\frac{\chi_{eff}^{(3)}(I_1, t = 0)}{\chi_{eff}^{(3)}(I_1, t \neq 0)} = N + 1 \tag{17}$$

This new result implies that without excited state decay, signal is reduced by a factor $(N+1)^2$ outside coherent artifact, what severely limits time resolution of degenerate pump-probe experiments in multi-photon absorption conditions.

2 PICOSECOND PHASE CONJUGATION IN A RED 4BCMU SOLUTION
2.1 Intensity Dependent Effects

Experimental set-up is described elsewhere.[2] We use 8 mJ, 33 ps pulses at 1064 nm and 1 Hz rep.rate. Samples are 4BCMU red solutions in 1,4 dichlorobutane contained in 1 mm thick cells.[25] For BCMU concentrations ranging between 0 and 20 g/ℓ, modulus of $\chi_{xyxy}^{(3)}(I)$ is reported on *Figure* 2 vs strong pump intensity I_1 at zero delay. Calibration is realized with (7) on carbon disulfide.[2] Polarization of four beams is depicted on the inset. Straight line indicates a slope 2 characteristic of $\chi^{(7)}$ effects. Solutions have a damage threshold higher than 5 GW.cm^{-2} at 1064 nm. Each curve averaged for the display on Figure 2 is composed at least of 5 intensity scans, 100-200 laser shots each, and no hysteresis appeared during the experiment. Above 5g/ℓ, signal evolution reflects solvatochromic transformations, such concentrations are indeed solid while below 5g/ℓ, they are liquid at room temperature. Nonlinearity of pure solvent is 10 times weaker than that of carbon disulfide:

$$\chi_{xyxy}^{(3)} \ solvent = +3.2 \ 10^{-14} esu(4.4 \ 10^{-22} m^2/V^2) \tag{18}$$

It is real and positive as checked in z-scan[26] as well as in Kerr gate experiments.[25] Interferences between solvent and polymer give $\chi^{(3)}$ phase and modulus:

$$\chi_{xyxy}^{(3)}(I)^2 = \left(\chi_{solv}^{(3)} + c.\chi_{r.pol}^{(3)}(I)\right)^2 + \left(c.\chi_{i.pol}^{(3)}(I)\right)^2 \tag{19}$$

where c is polymer volumic fraction identified (with a small error when c $\ll$ 1, 1-c $\approx$ 1) to concentration in g.cm^{-3}.[27]

We propose here an original and straightforward resolution method for (19) at all intensities. An order of magnitude of $\chi_{xyxy}^{(3)}$ for red amorphous 4BCMU is[2] 10^{-11} esu, at 0.25 g/ℓ it is thus a perturbation to (18):

$$\chi_{0.25g/\ell}^{(3)} \approx \chi_{solv}^{(3)} + 25 \ 10^{-5} \ \chi_{r.pol}^{(3)}(I) \tag{20}$$

Xe get $\chi^{(3)}$ real part with a weak error (%) by substraction of the two lowest curves (0.25 g/ℓ and 0 g/ℓ) on *Figure 2*.

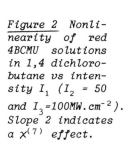

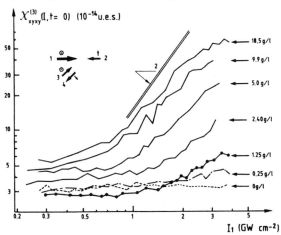

Figure 2 Nonlinearity of red 4BCMU solutions in 1,4 dichlorobutane vs intensity I_1 (I_2 = 50 and I_3 =100MW.cm^{-2}). Slope 2 indicates a $\chi^{(7)}$ effect.

A polynome fitting the result on *Figure 3a* is reported in (19) to extract $\chi^{(3)}$ imaginary part point by point (Figure 3b). Operation is performed on 5g/ℓ data where imaginary part has a large contribution and diluted regime seems to hold. Polymer transparency at 1064 nm implies: $\mathfrak{Im}(\chi^{(3)}(I)) > 0$.

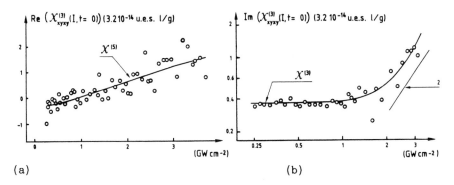

(a)   (b)

Figure 3 $\chi^{(3)}_{xyxy}$ real (a) and imaginary parts (b) deduced from (19) and (20) for 1 g/ℓ red 4BCMU polymer in 1,4 dichlorobutane. 5th order real part appears as a straight line in linear scales (a), 7th order imaginary part as a slope 2 in logarithmic scales (b).

Real part starts with a negative value as predicted on *Figure 1b*, and it becomes positive via a $\chi^{(5)}$ effect above 1 GW.cm^{-2}. More surprisingly, imaginary part which should be regularily constant undergoes a large $\chi^{(7)}$ increase above 1 GW.cm^{-2}, unlike rigid energy levels description of Figure 1b. This indicates that nonlinear contribution of real excitations may be large above 1 GW.cm^{-2} (with failure of (2)).

With ca.20% accuracy, nonlinearities of an ideally pure amorphous red 4BCMU are (1 esu = 7.2 10^7 m^2/V^2):

$$\begin{cases} \text{at 0 GW.cm}^{-2}, \ \Re(\chi^{(3)}_{xyxy}) = -1.6 \ 10^{-11} \text{esu and} \\[4pt] \qquad\qquad\qquad \Im(\chi^{(3)}_{xyxy}) = 1.2 \ 10^{-11} \text{ esu} \\[8pt] \text{at 3 GW.cm}^{-2}, \ \Re(\chi^{(3)}_{xyxy}) = 3.8 \ 10^{-11} \text{ esu and} \\[4pt] \qquad\qquad\qquad \Im(\chi^{(3)}_{xyxy}) = 3.8 \ 10^{-11} \text{ esu} \end{cases} \qquad (21)$$

With account of one dimensionality[3] (factor a) and coherent effects (factor b), we can obtain an estimation of nonlinear index change (n_2) and nonlinear absorption coefficient (α_2) for single beam experiments such as self lensing[28] and self limiting[29] effects at 1064 nm:

$$n_2 = 3ab\Re(\chi^{(3)}_{xyxy})/4\varepsilon_0 n^2 c \text{ and } \alpha_2 = 3\pi ab\Im(\chi^{(3)}_{xyxy})/\lambda\varepsilon_0 n^2 c \qquad (22)$$

where a = $\chi^{(3)}_{xxxx}/\chi^{(3)}_{xyxy}$ is either 3,5 or 7 for $\chi^{(3)}$, $\chi^{(5)}$ or $\chi^{(7)}$ effects (Figure 3) and b is respectively 1, 2/3 and 1/2. We get (ε_0 = 8.9 10^{-12} F/m, n $\approx$ 2 and $\chi^{(3)}$ in m^2/V^2):

$$\begin{cases} \text{at low intensity, } n_2 = -4.7 \ 10^{-7} \text{cm}^2/\text{MW and } \alpha_2 = 4.2 \ 10^{-2} \text{cm/MW} \\[8pt] \text{at 3 GW.cm}^{-2}, \qquad n_2 = 7.4 \ 10^{-7} \text{cm}^2/\text{MW and } \alpha_2 = 6.7 \ 10^{-2} \text{cm/MW} \end{cases} \qquad (23)$$

Those results are consistent with our previous measurements on 4BCMU red gels in chlorobenzene (α_2 = 4.5 10^2 cm/MW)^2, as well as with other groups results: $|n_2' + in_2''|$ = 6.5 10^{-2} cm^2/MW in polymer solid matrices[30] and $n_2 \approx -1.5 \ 10^{-7}$ cm^2/MW at 700 MW.cm^{-2} in spin coated layers[28]. From a practical point of view, some predictible consequences of this enhanced nonlinearity are an efficient power limiting effect[29] (large α_2) and wave front instabilities of focused beams[28] (n_2 sign inversion).

2.2 Life Time Effect

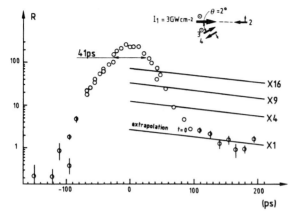

Figure 4 Time dependence of reflectivity R (arbitrary units) with pump 2 delay. Straight lines extrapolate memory at zero delay (ref.5) for 3rd, 5th and 7th order effects. Life time has little influence on half maximum width. Vertical bars are 70% confidence limits.

Figure 4 depicts evolution of laser induced dynamic grating in the 5 g/ℓ solution at 3 GW.cm^{-2} with pump beam 2 time delay. Two orders of magnitude below coherent peak, there is a slowly decaying memory ($\tau_e \approx$ 300 ps) consistent with the multiexponential decay of one dimensional charged excitations in red gels[8] and with two photons state recovery time[5]. For purely $\chi^{(3)}$ effects, memory extrapolation to zero delay is 4 times larger (17), but as it appears on *Figure 2*, extrapolation can be 16 times larger than remaining signal. We thus see that a large contribution to zero delay signal of *Figure 4* is induced by excited species. This points out intensity dependence knowledge necessary to address questions about response speed. However, all the response is not well accounted for by extrapolations based on (17) and some possible causes are:

1) coherent effects (§ I.1) may have a large contribution[13] ;
2) excited species may decay via multiexponential or multistep processes with initial life time shorter than 300 ps;[26]
3) excited species may induce an intensity dependent polarization $\alpha_{exc}(I)$ in (12).[31]

First cause does not account for increased absorption and we will not address second cause here,[5,8,16,22,23,25,26]. At those intensities, excitation ratio can be as large as one absorbed photon pair for 10 polymer repeat units parallel to intense pump field (if $\Phi \approx 1$ and $\tau_e > \tau_p$). It becomes thus possible to probe excited species nonlinearity whose number is close to excitons number.[18] In such conditions, mid gap (1.5 eV) absorbing species induced by two photon absorption (recently observed in the same polymer[25]) may act as intermediate states increasing two photon absorption cross-section: it induces a two step fully resonant two photon absorption, as suggested before by Lequime and Hermann.[1] Phenomenologically, such an effect is described by an imaginary $\chi^{(7)} \propto \Im(\chi^{(3)} . \chi^{(3)}_{exc})$ accounting for our observations.

In summary, after presentation of an analytical model accounting for the way three level systems become highly nonlinear when pumped above two photon absorption, we have addressed coherent artifact specifities for multiphoton degenerate processes. We have then presented experiments allowing straightforward determination of intensity dependent phase and modulus of degenerate four wave mixing nonlinearity in the case of a soluble polydiacetylene in picosecond two photon absorption regime and we have deduced some relevant data for practical applications involving n_2 and α_2. Analysis of time dependent effects with a large signal resolution reveals that excited species can have a large contribution to zero delay response and a possible scheme for excited state induced nonlinearity $\chi^{(7)}$ is proposed.

REFERENCES

1. M. Lequime and J.P. Hermann, Chem. Phys., 26; 431, 1977.
2. J.M. Nunzi and D. Grec, J. Appl. Phys., 62, 2198, 1987.
3. F. Charra and J.M. Nunzi, SPIE Proc. 1127, 173, 1989.
4. H. Byrne, W. Blau and K.Y. Jen, Synth. Metals, 32, 229, 1989.
5. F. Charra and J.M. Nunzi, J. Opt. Soc. Am. B, in press.
6. J.M. Nunzi and F. Charra, Opt. Com., 73, 357, 1989.
7. G.J. Blanchard, J.P. Heritage, A.C. Von Lehman, M.K. Kelly, G.L. Baker and S. Etemad, Phys. Rev. Lett., 63, 887, 1989.
8. J.M. Nunzi and F. Charra in "Organic materials for nonlinear optics", R.A. Hann and D. Bloor eds., Roy. Soc. Chem., 301, 1989.
9. L. Robins, J. Orenstein and R. Superfine, Phys. Rev. Lett., 56, 1850, 1986.
10. K. Fesser, A.R. Bishop and D.K. Campbell, Phys. Rev. B, 27, 4804, 1983.
11. F. Kajzar and J. Messier, Polymer J., 19, 275, 1987.
12. B.J. Orr and J.F. Ward, Mol. Phys., 20, 513, 1971.
13. F. Charra and J.M. Nunzi in "Organic materials for nonlinear optics", R.A. Hann and D. Bloor eds., Roy. Soc. Chem. 40, 1989.
14. Y.R. Shen: "The principles of nonlinear optics", Wiley, 1984.
15. T. Hattori and T. Kobayashi, Chem. Phys. Lett., 133, 230, 1987.
16. B.I. Greene, J. Orenstein, R.R. Millard and L.R. Williams, Chem. Phys. Lett., 139, 381, 1987.
17. T.Y. Fu and M. Sargent III, Opt. Lett., 5, 433, 1980.

18. B.I. Greene, J. Orenstein, R.R. Millard and L.R. Williams, Phys. Rev. Lett., 58, 2750, 1987.
19. C. Cojan, G.P. Agrawal and C. Flytzanis, Phys. Rev. B., 15, 909, 1977.
20. M. Ducloy and D. Bloch, Opt. Com., 47, 351, 1983.
21. B.P. Singh, M. Samoc, H.S. Nalwa and P.N. Prasad, J. Chem. Phys., 92, 2756, 1990.
22. M. Yoshizawa, M. Taiji and T. Kobayhashi, IEEE, J. Quant. Electr., 25, 2532, 1989.
23. T. Kobayashi, M. Yoshizawa, K. Ichimura and M. Taiji in "Ultrafast Phenomena VI", T. Yajima et al., eds., Springer, 277, 1988.
24. H.J. Eichler, P. Gunter and D.W. Pohl, "Laser-induced dynamic gratings", Springer, 1986.
25. F. Charra and J.M. Nunzi, Proceeding NATO Workshop, La Rochelle, France, August 27-31, 1990.
26. F. Charra, Thesis, Université Paris Sud, Orsay, 1990.
27. F. Kajzar in "Nonlinear optical effects in organic polymers", J. Messier et al., Eds., Kluwer, 225, 1989.
28. J. Valera, A. Darzi, A.C. Walker, W. Krug, E. Miao, M. Derstine and J.N. Polky, Electron. Lett., 26, 222, 1990.
29. R.C. Hoffmann, K.A. Stetyick, R.S. Potember and D.G. McLean, J. Opt. Soc. Am. B, 6, 772, 1989.
30. P.P. Ho, N.L. Yang, T. Jimbo, Q.Z. Wang and R.R. Alfano, J. Opt. Soc. Am. B, 4, 1025, 1989.
31. M. Sinclair, D. Moses, D. Mc Branch, A.J. Heeger, J. Yu and W.P. Su, Synth. Metals, 28, D655, 1989.

The Design of New Copolymers for $\chi^{(3)}$ Applications: Copolymers Incorporating Ladder Subunits

C.W. Spangler, T.J. Hall, and P.-K. Liu

DEPARTMENT OF CHEMISTRY, NORTHERN ILLINOIS UNIVERSITY, DEKALB, ILLINOIS 60115, USA

L.R. Dalton, D.W. Polis, and L.S. Sapochak

DEPARTMENT OF CHEMISTRY, UNIVERSITY OF SOUTHERN CALIFORNIA, LOS ANGELES, CALIFORNIA 90089, USA

1 INTRODUCTION

During the past five years organic materials with extended pi-conjugation sequences have been investigated as potential candidates for electro-optic device design due to their large optical nonlinearities and extremely fast switching times.[1,2] In order to design new materials with enhanced nonlinear properties, however, it is essential that we understand how the structural identity of the organic pi-system is related to the molecular level hyperpolarizability or bulk susceptibility. Although our understanding of how structure relates to ß or $\chi^{(2)}$ has advanced dramatically in the past two years, the same relationships for γ and $\chi^{(3)}$ are much more obscure. In our recent work we have attempted to establish design criteria for polymers with enhanced $\chi^{(3)}$ properties, and in this paper we will describe how incorporation of oligomeric ladder subunits in formal copolymer structures with saturated spacer subunits leads to new classes of $\chi^{(3)}$ materials with enhanced physical and nonlinear optical properties.

2 DESIGN OF $\chi^{(3)}$ COPOLYMERS

Several electroactive polymers, such as polyacetylene, polythiophene and poly[p-phenylene vinylene] have shown promise as NLO-active materials over the past few years[3]. However, these materials often have broad absorption bands and processibility problems which make them un-

suitable for device design. However, as several theoretical and experimental research groups have pointed
out in recent publications and symposia, it is not evident that long conjugation sequences are a necessary
criterion for high $\chi^{(3)}$ activity. Flytzanis and
coworkers[4,5] initially suggested that third order
susceptibilities might be proportional to the sixth
power of the electron delocalization length. Beratan,
et al.[6] have shown that the third order hyperpolarizability γ increases rapidly for trans polyenes as conjugation increases up to 10-15 repeat units. Increasing the polyene length beyond this did not lead to
additional enhancement of γ. This suggests that long
conjugation sequences may not be a necessary precondi-
tion for high $\chi^{(3)}$, and that oligomeric segments could
be interspersed with non-active spacers to maximize
both NLO activity and increase processibility. Hurst
and coworkers[7] have also calculated second hyperpolarizability tensors <u>via</u> <u>ab-initio</u> coupled-perturbed
Hartree-Fock theory for a series of polyenes up to
$C_{22}H_{24}$. They found that γ_{xxxx} was proportional to
chain length, with a power dependence of 4.0, but that
this dependence tapered off as N increased. More
recently, Garito and coworkers[8] have calculated the
power law dependence of γ_{xxxx} with chain length L is of
the order 4.6 ± 0.2. Prasad concurs with the prediction that γ/N falls off with increasing N, and has
shown <u>via</u> degenerate four wave mixing experiments that
γ/N reaches a limiting value for poly(p-phenylene)
oligomers after terphenyl, and that effective conjugation does not extend much beyond 3 or 4 rings.[9] Thus it
would appear that long conjugation sequences are not
necessary for enhanced third-order activity. With this
in mind we have proposed[10-12] that polymer structures
in which high-NLO activity oligomeric segments alternate with various low-NLO activity spacers whose
purpose is to enhance solubility, processibility and
film-forming capability. A general formulation for
these copolymers is illustrated in Figure 1.

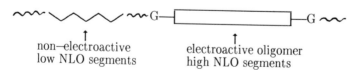

       ↑   ↑
non—electroactive   electroactive oligomer
low NLO segments   high NLO segments

G = Mesomerically interactive functional group

Figure 1   General formulation for NLO-active copolymers

3   COPOLYMERS INCORPORATING LADDER SUBUNITS

Dalton and coworkers[13,14] have discussed the conse-
quences of enhanced electron delocalization in ladder
polymers and have shown by electron nuclear double
resonance (ENDOR) and electron spin echo (ESE) spectroscopies that delocalization does not extend much beyond
one ladder repeat unit. This is illustrated below in
Figure 2 for the ladder polymers POL and PTL.

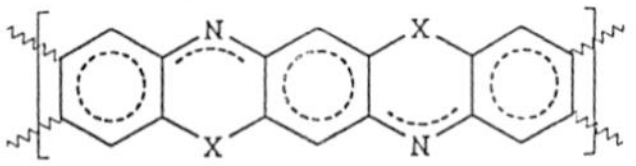

Figure 2   Schematic representation of hypothetical
electron delocalization in POL and PTL
ladder polymers.

Although ladder polymers normally have extremely limited solubility which makes optical film casting from
solution quite difficult, Dalton and Yu[15] have described the formation of ladder copolymers from open-chain
prepolymers, the last step being a double ring closure
forming a five-ring repeat unit:

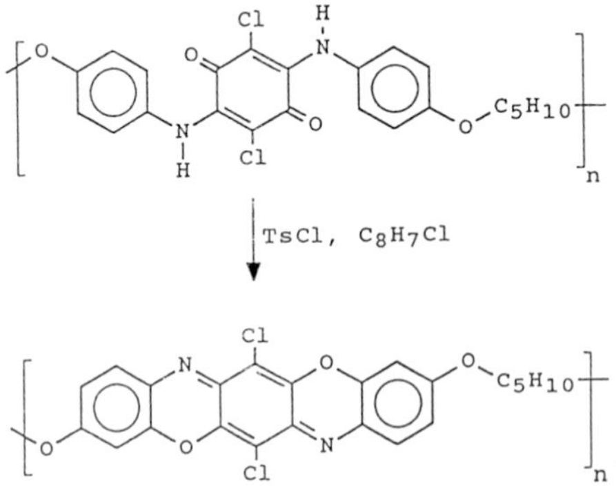

Although the final product displayed large optical nonlinearities, it was evident from elemental analysis and
other spectroscopic characterization that ring closure
was incomplete. In keeping with our previous success
in preparing copolyamides incorporating either diphenyl-
polyene or poly(2,5-thienylene vinylene) (PTV) oligomeric subunits for which good values of $\chi^{(3)}/\alpha$ have
been obtained, we decided to prepare ladder copolyamides by this approach which would ensure that the

ladder units would be fully ring closed. This approach
is illustrated below for Compound I which was prepared
by modifying the procedure of Mital and Jain[16].

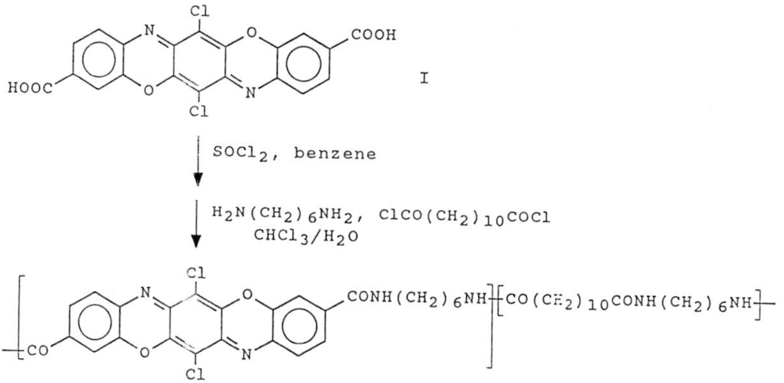

Scheme 1 Synthesis ladder subunits.

Ladder subunits can also be attached as pendant groups
to a polyvinylphenol backbone. Although not copolymers
in the same sense as the backbone incorporated structures, they are of interest in determining the active
unit orientation effect on $\chi^{(3)}$.

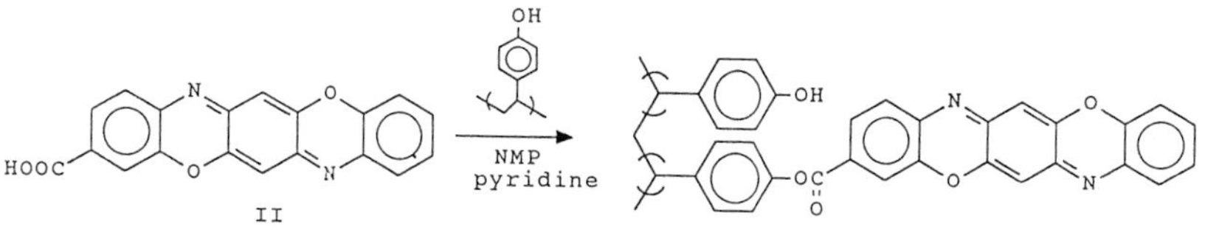

10-40% incorporation

Scheme 2 Synthesis of pendant ladder polymers

4 EXPERIMENTAL

<u>Backbone Copolymers</u>: The 10% incorporated copolymer
was prepared by first dissolving I-diacyl dichloride
(0.048 g, 1.0 x 10^{-4} mole) in CHCl$_3$ (50 mL) followed by
addition of dodecanedioyl chloride (0.24 g, 9 x 10^{-4}
mole). To this vigorously stirred solution was added
all at once a solution of hexamethylenediamine (0.08 g,
1.0 x 10^{-3} mole) in water (50mL) containing 0.2 g NaOH.
The polymer precipitated immediately. After stirring
an additional 12 hours, the aqueous layer was carefully

decanted and the polymer collected on a fine fitted
glass funnel, washed with ca. 100 mL water and finally
with acetone yielding the 10% incorporated copolyamide
(0.29 g, 71%).

Pendant Copolymer: Attachment of the ladder 5-ring
pendant group was accomplished as follows: for a 25%
incorporation on polyvinylphenol, II-acyl chloride
(0.60 g, 1.4 x 10^{-4} mole) was dissolved in NMP (15 mL)
and the solution filtered before adding to a solution
of polyvinylphenol (0.067 g, 5.6 x 10^{-4} mole) in NMP
(2 mL) containing pyridine (ca. 0.1 mL). After stir-
ing overnight the solution was poured into water and
finely divided polymer allowed to settle. After filtration, the polymer was stirred in methanol overnight to
remove NMP, filtered and dried in vacuo at 25°C. The
yield of 25% incorporated pendant polymer was essentially quantitative.

5 CHARACTERIZATION

The polymers were characterized by FT-IR and UV-VIS
spectroscopies, as well as thermal gravimetric analysis. As we have described previously[11,12], the conversion of the acyl halides to the copolyamides can be
followed by observing the loss of C=O stretching at
1755 cm^{-1} (COCl), and the appearance of the amide
stretching frequency in the same general location as
Nylon 6-12. The UV-VIS absorption of the bis-acyl
chloride and the incorporated polyamide are essentially
identical, which implies that the absorption characteristics of the polymer can be predetermined by
measurements on the monomer in solution. This highly
desirable characteristic is shown in Figure 3.
Preliminary NLO characterization was carried out by
degenerate four wave mixing utilizing phase- conjugation
geometry at 532 and 598 nm and employing pulses varying
from 6-25 ps, as has been described in our previous
copolymer studies[11,12,17]. $\chi^{(3)}/\alpha$ for the ladder
incorporated and pendant polymers varies from 10^{-11} to
10^{-12} esu-cm. Up to the present time, however, these
measurements have been carried out for hand-cast films
wherein the film thickness is not uniform. More
detailed measurements, including a complete frequency
profile at several different wavelengths on spin-coated
films of uniform thickness are currently underway which
will allow complete separation of real and imaginary
components of these $\chi^{(3)}$ values, as reported recently
by Cao, et al.[18], and will be reported at a later time.

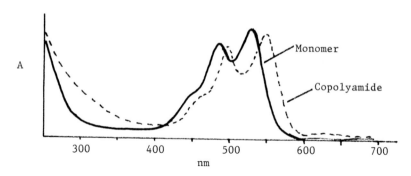

<u>**Figure 3**</u>   UV-Visible spectra of ladder copolyamide and ladder oligomer precursor (monomer)

TGA of the backbone incorporated copolyamide shows essentially the same thermal characteristics as a Nylon 6-12 standard with decomposition beginning around 425-50°. The 25% incorporated pendant ladder shows a lower onset of decomposition than does the backbone polyvinylphenol (300 vs 420°) however the polymer still maintains 60% of its weight at 700° which we interpret as indicative of thermal cross-linking that does not occur in polyvinylphenol itself.

6   CONCLUSIONS

Ladder subunits can be incorporated in a formal copolymer backbone (copolyamide) or as pendant subunits attached to a polyvinylphenol backbone. These polymers show substantial nonlinear susceptibilities, with $\chi^{(3)}/\alpha$ in the 10^{-11} to 10^{-12} esu-cm range, even though the active subunit is diluted (10-25%) incorporation. Full NLO characterization of these promising new materials and their device potential is currently underway, as are efforts to improve their percent incorporation, solubility, thermal stability and film quality.

7   ACKNOWLEDGEMENTS

This work was supported by Air Force Office of Scientific Research under contracts F49620-87-C-0100 and F49620-88-0071 as well as AFOSR grant #90-0060. The authors would also like to thank Drs. J. P. Jiang and X. F. Cao (USC) and Dr. Robert Norwood (Hoechst-Cleanese) for preliminary NLO characterization of the polymers.

8 REFERENCES

1. "Nonlinear Optical Properties of Organic Molecules and Crystals", D.S. Chemla and J. Zyss, Eds, Academic Press, Orlando, FL 1987, Vol. 1 and 2.

2. "Organic Materials for Non-Linear Optics", R. A.Hann and D. Bloor, Eds., Roy. Soc. Chem., London, 1989.

3. For example, see L. R. Dalton, D.W. Polis, M. R. McLean and L. Yu, "Handbook of Conducting Polymers," T. A. Skotheim, Ed., Marcel Dekker, New York, 1990 (In Press) for a review of the status of electroactive polymers for nonlinear optics.

4. G. P. Agrawal and C. Flytzanis, *Chem. Phys. Lett.*, 1976, **44**, 366.

5. G. P. Agrawal, C. Cojan and C. Flytzanis, *Phys. Rev. B*, 1978, **17**, 776.

6. D. N. Beratan, J.N. Onuchic and J. W. Perry, *J. Phys. Chem.*, 1987, **91**, 2696.

7. G. J. Hurst, M. Duplis and E. Clementi, *J. Chem. Phys.*, 1988, **89**, 385.

8. A. F. Garito, J. R. Heflin, K. Y. Wong and O. Zamani-Khamiri, in Ref. 2, pp. 16-17.

9. P. N. Prasad, in Ref. 2, pp. 264-274.

10. C. W. Spangler, T. J. Hall, K. O. Havelka, M. Bader, M. R. McLean and L. R. Dalton, *Proc. SPIE*, 1989, **1147**, 149.

11. C. W. Spangler, T. J. Hall, K. O. Havelka, D. W. Polis, L. S. Sapochak, and L. R. Dalton, *Proc. SPIE*, 1990, **1337** (In Press).

12. C. W. Spangler, P.-K. Liu and T. J. Hall, D. W. Polis, L. S. Sapochak and L. R. Dalton, *Polymer*, 1991 (Submitted).

13. L. R. Dalton, J. Thomson, and H. S. Nalwa, *Polymer*, 1987, **28**, 543.

14. L. R. Dalton, "Nonlinear Optical and Electroactive Polymers", P. N. Prasad and D.R. Ulrich, Eds., Plenum, New York, 1988, pp. 243-273.

15. L.-P. Yu and L. R. Dalton, *J. Amer. Chem. Soc.*, 1989, **111**, 8699.

16. R. L. Mital and S. K. Jain, *J. Chem. Soc. (C)*, 1971, 1875.

17. L. Yu, D. W. Polis, F. Xiao, L. S. Sapochak, M.R. McLean, L. R. Dalton, C. W. Spangler, T. J. Hall and K. O. Havelka, *Polymer*, 1991 (In Press).

18. X. F. Cao, J. P. Jiang, R. W. Hellwarth, M. Chen, L.-P. Yu, and L. R. Dalton, *Proc. SPIE*, 1990, **1337** (In Press).

The Electronic Processes in Carbazolyl Containing Polymers Sensitized by Pyrylium and Rhodamine 6G

A. Tamulis and L. Bazan

INSTITUTE OF THEORETICAL PHYSICS AND ASTRONOMY, ACADEMY OF SCIENCES OF LITHUANIA, K. POZELOS 54, VILNIUS 232600, THE REPUBLIC OF LITHUANIA

A. Undzenas

SCIENTIFIC RESEARCH INSTITUTE OF ELECTROGRAPHY, P. VILEISIO 18, VILNIUS 232055, THE REPUBLIC OF LITHUANIA

1 INTRODUCTION

Nanotechnology development and molecular computers architecture design technology is demanding the investigation of the elementary electronic processes in organic molecules and polymeric compounds. The understanding of these processes is also important for the interpretation of charge carrier photogeneration in sensitized organic electrophotographic layers incorporating polymeric carbazolyl containing photoconductors. The necessary condition in such systems is the formation of donor-acceptor complex between electrondonating carbazolyl radical and electronaccepting sensitizer molecule.

We have earlier investigated the electronic structure of carbazole (Cz), 2,4,7 - trinitro-9-fluorenone (TNF) and 2,4,5,7 - tetranitro-9-fluorenone (TeNF) [1]. Cz acts as electrondonor and polynitrofluorenones act as electronacceptors in polyvinylcarbazole (PVK) and poly-epoxypropylcarbazole (PEPK) films. Electronic structure of PVK and PEPK, as well as of more complex sensitizers have been also investigated [2].

2 THE ELECTRONIC STRUCTURE OF THE PYRYLIUM AND RHODAMINE CATIONS AND THEIR SALTS WITH Cl⁻ ANION

The electronic structure of various pyrylium (Pyr^+) is presented in Table 1. The data were obtained by the MNDO method [3]. The geometry of Pyr^+ and Rhodamine $6G^+$ ($R-6G^+$) was taken from [4]. Pyr^+ with more complicated radicals (R^1, R^2, R^3) have lower values of ionization potentials ($I_k = E_{HOMO}$) and electron affinity (E_{LUMO}) (compare with [5]). The dimethyl amino fragment - $N(CH_3)_2$ in $DMeDMeAPhPyr^+$ dramatically changes the ionization potential and electron

affinity of Pyr^+ (compare with [5]).
    In sensitized organic electrophotographic layers
based on PVK and PEPK this leads to high values of the
residual potential. The sensitizer molecule due to the
small ionization potentials is acting as a neutral trap
for mobile holes.
    Atomic charges on three atoms 2,4 and 6 in Pyr^+
ring make up the positive triangle (Table 1). One can
see the ranges of atomic charges of the various Pyr^+ on
these atoms. Various salts of $DMeDMeAPhPyr^+$ and Cl^-
anions have been investigated using MNDO method. Cl^-
anions were placed in different positions. The distances
from Cl^- to the oxygen atom in the ring we change in
the range 3.2 - 3.5 Å. The values of ionization potentials of $[DMeDMeAPhPyr^+ + Cl^-]^o$ are respectively in the
range - 6.93 - -6.42 eV.
    The ionization potentials of $[DMeDMeAPhPyr^+ + Cl^-]^o$
salts are lower than I_k of Cz. It means that these salts
can be traps for holes. Electrophotographic parameters
of PEPK layers sensitized with chloride of 2,4,6 -tri-
phenylpyrylium confirm this fact. The sensitizing efficiency of this pyrylium salt is lower than the efficiency of tetrafluoroborate or perchlorate of 2,4,6 -triphe-
nylpyrylium salts towards PEPK.
    The ionization potentials of BF_4^- and ClO_4^- are -8.56
eV and -7.51 eV and they do not form hole traps in these
salts.
    The triangle of positive atomic charges one can
find also in the xanthene ring of the $R-6G^+$ cation and
in its salt $[R-6G^+ + Cl^-]$ (Fig.1). We can anticipate
that positive atomic charge triangle in the pyrylium
and Rhodamine salts plays an essential role in the photogeneration processes taking place in sensitized carbazolyl containing polymeric photoconductors.
    We have found that the position of the fragment
$-COOC_2H_5$ in $R-6G^+$ is most favourable when the carbonyl
group $>C = O$ is placed near the xanthene ring: E_{total} =
-5390.757 eV (E_{total} = -5389.771 eV when the $>C = O$ is
placed far from xanthene ring).
    The calculated values of ionization potentials are
in the range -6.05 - -7.83 eV for various $[R-6G^+ + Cl^-]^o$
(the distance from the oxygen atom in the rings to the
Cl^- was taken as 2.0 - 3.5 Å). It means, that these
salts can be traps for mobile holes. Electrophotographic
layers with Rhodamine (Cl^-) sensitizers confirm this
theoretical finding.
The localization of the positive charge on various fragments of the $R-6G^+$ molecule is as follows: $q = +0.545$ on
the xanthene ring, $+0.215$ on the $-C_2H_5$, $+0.112$ on the
$-CH_3$, $+0.041$ on the $-PhCOOC_2H_5$, -0.121 on the NH.

Table 1   The values of atomic charges and I_k(HOMO), A_k
(LUMO) (in eV) for various Pyr$^+$

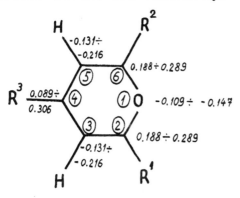

Abrevations of various Pyr$^+$	R^1	R^2	R^3	E HOMO I_k	E LUMO A_k
Pyr$^+$	H	H	H	-15.768	-6.711
DMePyr$^+$	CH_3	CH_3	H	-15.12	-6.39
TMePyr$^+$	CH_3	CH_3	CH_3	-15.015	-6.291
DMeMeOPyr$^+$	CH_3	CH_3	CH_3O	-15.08	-6.19
DMePhPyr$^+$	CH_3	CH_3	C_6H_5	-12.91	-5.76
DMeDMeAPhPyr$^+$	CH_3	CH_3	$C_6H_4N(CH_3)_2$	-11.52	-5.65
DPhPyr$^+$	C_6H_5	C_6H_5	H	-12.63	-5.69
DPhMePyr$^+$	C_6H_5	C_6H_5	CH_3	-12.58	-5.64
TPhPyr$^+$	C_6H_5	C_6H_5	C_6H_5	-12.42	-5.45
					-6.3[6]
DNafMeOPhPyr$^+$	$C_{10}H_7$	$C_{10}H_7$	$C_6H_5OCH_3$	-11.27	-5.24
					-6.2[6]

3 THE TOPOLOGY OF THE ELECTRON DISTRIBUTION AFTER HOPPING FROM CARBAZOLYL TO PYRYLIUM AND RHODAMINE 6G SALTS WITH Cl$^-$

During the photoexcitation process the electron from the carbazolyl radical in PVK or PEPK jumps onto the acceptor molecule.

The largest fraction of electron falls down on the xanthene ring of R-6G$^+$ and 36 % on the C atom which   is

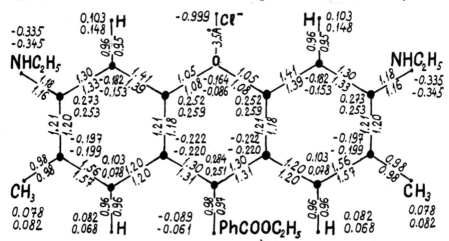

Fig.1 The atomic charges of R-6G⁺ (upper values) and
[R-6G⁺ + Cl⁻]° (lower values). The interatomic bond
orders show that the xanthene ring has slightly
hinoidal structure.

opposite to O atom. The electron hopping leads to some
change in interatomic Wiberg indexes and HOMO, LUMO energies. The electron distribution after hopping on
[R-6G⁺ +Cl⁻]° salt is analogues to the [R-6G]° case (fig.2).
Electron distribution after hopping on the positive
triangle in the case of pyrylium is more homogeneous. The
electron hopping on [DMeDMeAPhPyr]⁺ leads to changes in
LUMO and HOMO. The electron distribution after hopping
on the positive triangle of [DMeDMeAPhPyr⁺+Cl⁻]° salt is
analogues to the [DMeDMeAPhPyr]° case (fig.3).

4. CONCLUSIONS

1. There is a positive atomic charge triangle in
the pyrylium cation, pyrylium salts with Cl⁻ anion, Rhodamine 6G cation, and [R-6G⁺+Cl⁻]° salts.
2. The topology of atomic charges does not change when
Cl⁻ anion is added to Pyr⁺ and R-6G⁺ . The alternating
of atomic charges in this case decreases.
3. The most probable configuration of R-6G⁺ is when
the carbonyl group in -COOC₂H₅ is situated near the
xanthene ring.
4. The calculated values of ionization potentials
are in the range -6.42 - -6.93 eV for various[Pyr⁺+Cl⁻]°
and -6.05 - -7.14 eV for various [R-6G⁺+Cl⁻]° (the distance from the oxygen atom in the rings to the Cl⁻ was
taken as 3.5 Å). It means, that these salts can be traps

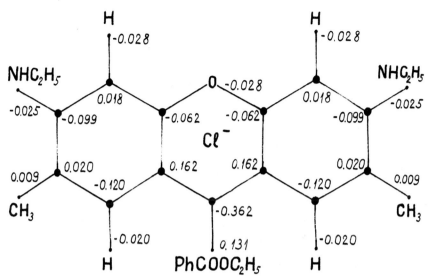

Fig.2. The electron distribution after hopping on [R-6G⁺+Cl⁻]ᶜ salt.

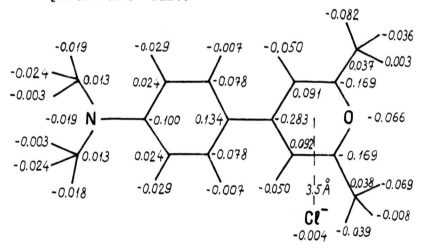

Fig.3. The electron distribution after hopping on the positive triangle of [DMeDMeAPhPyr⁺+Cl⁻]ᶜ salt.

for mobile holes. Electrophotographic parameters of
PEPK layers sensitized by chloride of 2,4,6-triphenylpy-
rylium confirm this fact. The sensitizing efficiency of
this pyrylium salt is lower than the efficiency of
tetrafluoroboride or perchlorate of 2,4,6-triphenylpy-
rylium salts towards PEPK.
5. The calculations show, that during the photogeneration process the electron most probably hopping from
the carbazolyl radical to the positive triangle of sensitizer molecule. The electron hopping is the same   in
various acceptors: Pyr^+, $[Pyr^{+}+Cl^-]^\circ$, $R-6G^+$ and
$[R-6G^{+}+Cl^-]^\circ$.

REFERENCES

1. A.Tamulis and Š.Kudžmauskas, Investigations of electronic structure of carbazole, 2,4,7-trinitro-9-fluo-
   renone, 2,4,5,7-tetranitro-9-fluorenone molecules
   and its single charge ions by using the NDO methods,
   Deposited in the Lithuanian research institute   for
   scientific and technical information, 232659, Vilnius,
   Kalvarijų 3, The Republic of Lithuania, No 1908. L I.
   1987.07.11,(In Russian).
2. Š.Kudžmauskas, G.Vektaris, A.Tamulis, V.Liuolia,
   V.Gaidelis and A.Undzėnas, The modeling of the processes of the photogeneration, photosensibilization
   and charge carrier transport in the carbazolyl containing photoconductors, Deposited in the Soviet
   Union Research Institute for scientific and technical
   information, Moscow, Inventor No 0289003737345.
   1988.12.12., (in Russian).
3. M.J.S.Dewar and W.Thiel, Ground states of molecules.
   38. The MNDO method. Approximations and parameters,
   J.Amer.Chem.Soc., 1977, 99, 4899.
4. A.V.Vilkov, B.C.Mastriukov and N.I.Sadova,'Determination of the geometry structure of the free molecules', Khimija, Leningrad, 1978 (in Russian).
5. R.W.Bigelow, Counterion and solvent perturbations to
   the electronic structure of substituted pyrylium
   cations: a CNDO/S study, J.Chem.Phys., 1980, 73,
   3964.
6. V.E.Kampar, Charge transfer complexes of neutral
   donors with acceptors-organic cations,   Uspekhi
   Khimii, 1982, 51, 185, (in Russian).

Study of a Novel Class of Second-order Non-linear Optical (NLO) Polyurethanes

R. Meyrueix, G. Mignani, and G. Tapolsky

RHÔNE-POULENC RECHERCHES, CENTRE DE RECHERCHES DES
CARRIÈRES, 85, AVENUE DES FRÈRES PERRET. BP 62. 69192 SAINT-FONS.
CEDEX, FRANCE

ABSTRACT. We introduce a novel class of second order non
linear optical (NLO) polymers. Each monomer contains a NLO
unit covalently incorporated into the main chain with its
dipole moment perpendicular to the main chain. With the aid
of a Fabry Perot interferometric technique we measure the
two independent components of the second order electro-
optical (E/O) susceptibility as well as the two order parameters of the chromophores <P1> and <P3>. (P1 : Legendre
polynomial). The very good stability of the chromophore
orientation and the high value of the second order
susceptibility show that these polymers are promising
candidates for NLO applications.

1. INTRODUCTION

A very high concentration of chromophores as well as a
good orientational stability are now recognized as two
important goals. The importance of improving these two
characteristics is illustrated by the amount of work done in
this area [1].

The guest/host ensembles (azo dye/PMMA or PS) were
the first reported quadratic NLO films [2]. In these systems
residual motion of the dye is related to the free volume in
the host matrix and cancels the dye polar order over a
period of weeks to months [2 - 3]. Several routes were
followed in order to stabilize the dopant orientation :
increase of the length of the chromophore [4], physical
ageing [5], use of a remnant field [6-8] and dipolar
interactions [9-10].

Recently, covalent attachment of the dopant to the
polymeric backbone in a side chain position has been used
to increase both the NLO units concentration and the
stability [11, 12]. However, deorientation is not completly
prevented after either contact poling [11] or Corona

poling. The electrooptical susceptibilities of these
polymers are quite large : $\chi^{(2)}(-\omega, \omega, o) \sim$ 40-50 pm V^{-1} after
Corona poling under high fields [12]. Covalent attachment
of the NLO units within the main chain in a "head to tail"
configuration would increase the orientational efficiency of
the poling field [14, 15]. Finally, very stable $\chi^{(2)}$ values,
even at 83°C, have been obtained by M. Eich et al. [16] with
a cross linked epoxy with NLO units covalently incorporated
into the rigid network.

We present in this paper a novel class of main chain
highly functionalized linear polymers. Each monomer contains
a chromophore whose donor group is in the main chain and
whose dipole moment lies perpendicular to the back bone. The
glass transition temperature can be varied by ajusting the
stiffness of the main chain (see fig. 1).

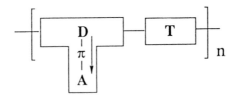

Figure 1 : Schematic representation of NLO main chain
functionalized thermoplastic D - Π - A stands for donor and
acceptor substituted Π system as covalently incorporated NLO
unit. The arrow illustrates the ground state dipole.

The main caracteristics of two of these polymers are
presented. The two independent components of $\chi^{(2)}(-\omega, \omega, o)$
are measured at λ = 830 nm. The orientational stability of
the NLO units is evaluated from the decay with time of their
two order parameters <P1> and <P3>(P1 = Legendre polynomial)

2. EXPERIMENTAL

2.1. Synthesis

Under nitrogen, Disperse Red 17 was dissolved in freshly
distilled dimethyl acetamide and a small excess of
diisocyanate was added. The mixture was stirred overnight at
60°C. The solution was filtered and water was slowly added
until precipitation of a red/purple solid. PU1 was the
polymer prepared with the 1,6 hexamethylene diisocyanate and
PU2 was prepared with the 4,4'methylenebis(phenylisocyanate)

2.2. Film preparation and poling

Films were prepared from a concentrated solution of approximately 28 wt% in dimethylacetamide (DMAC). The poling at T ≃ T_g was achieved in situ on the E/O experimental set up. The ratio [poling field/temperature] was maintained constant during the cooling in order not to disturb the thermodynamic equilibrium as the temperature was decreased [5].

2.3. Electrooptical measurement of the Non Linear suscepti- bility $\chi^2(-\omega, \omega, o)$.

The two independent χ_{113} and χ_{333} coefficients were simultaneously measured by Fabry Perot under Oblique Incidence (PFOI) interferometry as described previously [9, 13].

3. RESULTS AND DISCUSSION

3.1. Thermal, dielectrical and linear optical properties

DSC studies were performed on PU1 and PU2. The glass transition temperature T_g increases noticeably with the stiffness of the main chain : T_g = 75° and 115°C for PU1 and PU2 respectively. Mesogenic transition was not detected either by thermal analysis or by direct microscopic inspection between crossed polarizers. This was confirmed by the very small difference between the refractive indexes. The residual anisotropy nx-nz ~ $1-10^{-3}$ is commonly observed in solvent cast polymeric films [18].

3.2. Electrooptic experiments

The χ_{113} and χ_{333} components were measured after the chromophore orientation under a poling field of 25 V/μm at 100°C. It appears from the results shown in Table 1 that the ratio R = χ_{333}/χ_{113} stays close to the theoretical value R = 3 predicted for a film doped with unidimensional NLO units [20].

	χ_{113} (pm.V^{-1})	χ_{333} (pm.V^{-1})	χ_{333}/χ_{113}
PU 1	5.0	14.8	2.94
PU 2	3.01	10.3	3.3

Table 1: E/O quadratic susceptibilities of PU 1 and PU2 measured at 830 nm under a poling field of 25 V/μm at 100°C.

The chromatic dispersion of $\chi^{(2)}{}_{113}$ $(-\omega,\omega,o)$ between 830 and 670 nm is reported fig 2. Several authors have shown that the wavelength dependence of $\chi^{(2)}$ in a SHG experiment is in good agreement with a two level model [19, 21].
In the electrooptical domain we find that the accordance is qualitatively good however the experimental dispersion is sharper than the theoretical one.

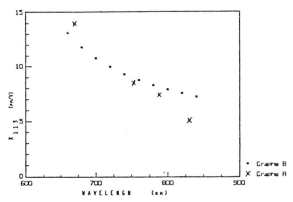

Figure 2 : Chromatic dispersion of $\chi^{(2)}{}_{113}$ $(-\omega,\ \omega,o)$
          A : experimental points
          B : theoretical value derived from a two level model.

The static value βs of the hyperpolarisability can be related to the E/O effect by the relation [20] :

$$\chi^{(2)}{}_{333} = \frac{N\beta sQF^2(\omega)F(o)\ \mu oF(p)E(p)}{5kT(p)} \tag{1}$$

where Q describes the chromatic dispersion of β [18] where $F(\omega)$ and $F(o)$ are the Lorentz local field factors and where $E(p)$ is the poling field. The Onsager local field factor

$F(p) = \dfrac{\varepsilon(p)(n^2+2)}{2\varepsilon(p)+n^2}$ has to be calculated at the poling

temperature $T(p)$ at which $\varepsilon(p) \sim 23$ and $n \sim 1,7$.
With a measured absorption wavelength λ max = 471 nm and a dipole moment μ = 8D, eq (1) gives $\beta s \sim 45.10^{-30}$ esu.
By comparison with the known βs value for the Disperse Red one - $\beta s \sim 47.10^{-30}$ esu [19] - that we use as reference, it appears that the orientational process in PU1 is correctly

described by a mean field approximation with Lorentz and
Onsager local field factors. Thus, strong dipole/dipole
interactions do not perturb greatly the chromophores'
orientation.

3.3. Residual mobility of the NLO units in the glassy state

After the removal of the poling field, the oriented chromophores are out of thermodynamic equilibrium and tend to
disorder. The decrease with time of the order parameters
<P1> and <P3> are calculated from the χ_{113} and χ_{333} values
[9, 13] assuming a dipole moment $\mu \sim$ 8D and a static
hyperpolarisability $\beta s \sim 45.10^{-30}$ esu.
At room temperature, the orientation in PU2 is extremely
stable over periods of months (fig 3). A greater segmental
mobility and a lower T_g value of PU1 make the rotational
diffusion faster in this latter polymer, although it remains
slower than in previously reported side chain NLO functionalized polymers after contact poling [11].

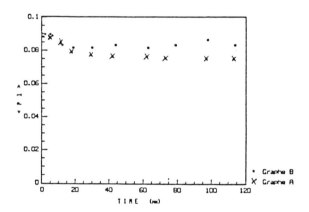

Figure 3 : Long term relaxation of the order parameter <P1>
           at room temperature. A : PU1   B : PU2

The short term deorientation in PU1 should be compared with
those of Disperse Red One guest molecules inserted in the
PMMA host matrix (fig. 4) at equivalent (T_g - T)
temperatures. The stabilisation by the covalent linkage of
the NLO unit within the main chain is clearly demonstrated.

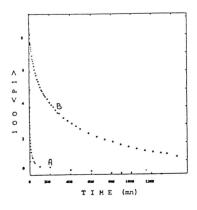

<u>Figure 4</u> : Comparison of the short term relaxation of the
order parameter <P1> near T_g in a guest/host
ensemble and in PU1. $T_g - T = 16°C$
A : PMMA/DR1 (7,5 W%)   B : PU1

It should be noted that <P3>(t) remains close to zero during
the deorientation (fig.5). This result indicates that the
orientation function f(Θ) does not differ greatly from its
initial Boltzman distribution after the removal of the
poling field.

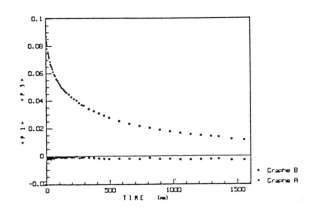

<u>Figure 5</u>   : Short term decay at 60°C of the order parameters
<P1> and <P3> in PU1.
Poling field at 100°C : 20 V/μm
A : <P1>   B : <P3>

4. CONCLUSION

In this paper, we introduced a novel class of second order NLO polymers. Each monomer contains a NLO unit covalently incorporated into the main chain with its charge transfer axis perpendicular to the main chain. The two independent components of $\chi^{(2)}(-\omega,\omega,o)$ are simultaneously measured by oblique incidence Fabry Perot interferometry. Even under moderate poling field (Ep ~ 50 V/μm) the nonresonant nonlinearity is fairly high ($\chi^{(2)}_{333}$ ~ 30 pm V^{-1}). The theoretical ratio $\chi_{333}/\chi_{113} = 3$ is approached and maintained during the disordering of the NLO units. Our first results show that the chromatic dispersion of $\chi^{(2)}$ is slightly sharper than predicted with a two level model. At 23°C, the deorientation in the polymer of highest T$_g$ is extremely stable over a period of months. These results show that main chain NLO functionalized polymers appear to be very promising for integrated optics applications.

References

[1]   G.T. BOYD J.Opt.Soc.Am.B, 6,(4),685, (1989)

[2]   L.A. Blumenfeld, F.P. Chernyakovskii, V.A. Gribanov, I.M. Kanerskii, J. Macromol. Sci.Chem.A, 6, (7), 1201, (1972)

[3]   H.L. Hampsch, J. Yang, G.K. Wong, Macromol. 21, 526, (1988).

[4]   H.L. Hampsch, J. Yang, G.K. Wong, J.M. Torkelson Polym. Com. 30, 40, (1989)

[5]   R. Meyrueix, G. Mignani. SPIE vol. 1127, 160, (1989)

[6]   M.A. Mortazavi, A. Knoesen, S.T. Kowel, B.G. Higgins et al. J. Opt.Soc - Am.B, 6,(14),733, (1989)

[7]   H.L. Hampsch, J. Yang, G.K. Wong, J.M. Torkelson. Mater. Res. Soc. Symp.Proc.Ser. Vol 173, paper Q 13.6 (1989)

[8]   J.R. Hill, P. Pantelis, G.J. Davies, Ferroelectrics 76, 435, (1987)

[9]   R. Meyrueix, G. Mignani, Mater.Res. Soc. Symp.
           Proc.Ser. Vol 173, paper Q 13.5 (1989)

[10]   M.A. Hubbard, T.J. Marks, J. Yang, G.K. Wong,
           Chem. Mater. 1, (2), 167, (1989).

[11]   C. Ye, N. Minami, T.J. Marks, J. Yang, G.K. Wong.
           Macromol. 21, 2899, (1988).

[12]   K.D. Singer, M.G. Kuzyk, W.R. Holland, J.E. Sohn,
           S.J. Lalama et al. Appl.Phys.Lett. 53, (19),
           1800, (1988)

[13]   R. Meyrueix, G. Mignani, G. Tapolsky "Organic
           molecules for Nonlinear Optics and Photonics"
           NATO ASI Series Ed. J.Messier, D.Ulrich, 1991
           (to be published)

[14]   C.S. Willand, D.J. Williams Ber. Bunsenges Phys.
           Chem. 91, 1304, (1987)

[15]   M.L. Schilling, H.E. Katz, Chem.Mater. 1, 668,
           (1989)

[16]   M. Eich, B. Reck, D.Y. Yoon, C.G. Wilson,
           G.C. Bjorklund, J.Appl.Phys. 66, (7), 3241,
           (1989)

[17]   J.L. Oudar, J. Zyss, Phys.Rev., A 26, 2076,
           (1982)

[18]   W.M. Prest, Jr and D.J. Luca, J.Appl.Phys.
           50, (10), 6067, (1979)

[19]   K.D. Singer, J.E. Sohn, L.A. King, H.M. Gordon,
           J. Opt. Soc. Am.B., 6, (7), 1339, (1989)

[20]   D.S. Chemla, J. Zyss, Non linear Opt. Prop. of
           Org.Molecules and Cryst. Acad. Press. (1987)

[21]   C.C. Teng, A.F. Garito, Phys. Rev. B 28, 6766,
           (1983)

Stabilization of Polaronic and Bipolaronic States in Electro-active Oligomeric Segments of Non-linear Optic-active Copolymers

C.W. Spangler and K.O. Havelka

DEPARTMENT OF CHEMISTRY, NORTHERN ILLINOIS UNIVERSITY, DEKALB, ILLINOIS 60115, USA

1 INTRODUCTION

Electroactive polymers are now generally recognized as multifunctional materials capable of having such diverse properties as metallic-type conductivity as well as enhanced nonlinear optical susceptibilities. We have previously demonstrated[1-5] that polaronic and bipolaronic charge states similar to those formed upon chemical or electrochemical doping of various electroactive polymers can be formed and stabilized in small oligomeric model compounds in solution, and during the past year several of these oligomeric segments related to polyacetylene, poly(p-phenylene vinylene) (PPV) and poly(2,5-thienlyene vinylene) (PTV) and ladder polymers such as POL and PTL have been incorporated into formal copolymers[6-9]. In these copolymers the optical characteristics of the oligomeric segment can be controlled to give predictable narrow absorption windows. In this paper we would like to demonstrate how polaron and/or bipolaron charge states can be formed and stabilized in various extended pi-conjugation sequences by chemical or protonic doping, and how the formation of these charge states in formal polymers incorporating these segments might lead to enhanced nonlinear optical characteristics.

2 FORMATION OF POLARONIC (P) AND BIPOLARONIC (BP) CHARGE STATES IN MODEL OLIGOMERS

In the previous OMNO Conference (1988) we described how polaronic and bipolaronic charge states could be formed and the bipolaronic charge state stabilized[1]. In general, bipolaron formation during oxidative doping of

a diphenyl polyene series is facilitated by increased
conjugation length and strong electron donating substituents. Thus one can control the change in absorption characteristics by either varying the size of the
oligomer or the identity of the substituent:

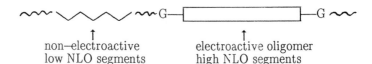

In general, then, one can design a series of copolymers
in which (+)(+) or (-)(-) bipolarons could be stabilized by varying the functionality joining an NLO-active
oligomeric segment to an inactive spacer. This is
illustrated in Figure 1.

non–electroactive   electroactive oligomer
low NLO segments   high NLO segments

G = Mesomerically interactive functional group
    (a) electron donor stabilizes (+)(+)
        bipolaron
    (b) electron withdrawer stabilizes (-)(-)
        bipolaron

    Figure 1. General formulation for NLO-active
               copolymer capable of supporting BP
               charge states

The phenyl polyenes can be considered as phenyl end
capped oligomers of polyacetylene. Other electroactive polymers whose charge states are dominated by
bipolarons and which exhibit large optical nonlinearities include poly[p-phenylene vinylene] (PPV)   and
poly[2,5-thienylene vinylene] (PTV). These polymers
can also be modeled accurately with short oligomeric
segments[3,10].

3 TYPICAL PREPARATIONS

The alkoxy substituted-diphenylpolyenes and PPV dimers
reported here are new compounds prepared by Wittig

methodology developed in our laboratory[11] whose detailed syntheses will be published elsewhere. However the
two preparations detailed below may be considered
typical.

3',3'',4',4'', 5', 5''-hexamethoxy diphenyldodeca-
1,3,5,7,9,11-hexaene

3,4,5-Trimethoxycinnamaldehyde (4.44 g, 0.02 mole) and
E,E-hexa-2,4-diene-1,6-diyl bis(tributylphosphonium)
dibromide (4.84 g, 0.01 mole) were dissolved in DMF
(100 mL) at 90°. Sodium ethoxide (30 mL, 1 M soln,
0.03 mole) in ethanol was added dropwise over a period
of 0.5 hour and the resulting mixture stirred an
additional 16 hours. Water (ca. 50 mL) was then
added to initiate precipitation and the mixture cooled
to room temperature. The product was isolated by
filtration, dried and recrystallized from toluene/DMF
yielding the title compound (67%), m.p. 254-7°, λ_{max}
(CH_2Cl_2) 403, 428, 454 nm; calcd. for $C_{30}H_{34}O_6$; C,
73.44; H, 7.0; found C, 73.46; H, 7.0.

2,5-Dimethoxy-1,4-bis(4'-methoxystyryl) benzene

4-Methoxybenzaldehyde (2.72 g, 0.02 mole) dissolved
in ethanol (100 mL) was added to a solution of 2,5-
dimethoxy-1-4-xylylene bis(tributylphosphonium)
dibromide (0.01 mole) in DMF (100 mL) at 90°. Sodium
ethoxide (30 mL, 1 M soln, 0.03 mole) was added and
the reaction worked up as described above. The crude
product was recrystallized from toluene yielding the
title compound (69%), m.p. 216-217°, λ_{max} (CH_2Cl_2);
392 nm; calcd. for $C_{26}H_{26}O_4$; C, 77.59; H, 7.51; found:
C, 77.58; H, 6.58.

4 PROTONIC VS. OXIDATIVE DOPING

Although strong oxidizing or reducing agents are often
used to "dope" conjugated polymers to conductive states
incorporating delocalized polaron or bipolaron charge
sites, Han and Elsenbaumer[12] have recently described a
non-oxidizing protonic acid doping procedure. The technique involves in-situ doping of the polymer as formed
in acid solution (e.g. CF_3COOH). We have recently compared oxidative and protonic doping and their compara-
tive charge states in a series of methoxy-substituted
polyenes[6]. Protonic doping has the advantage of the
dopant and solvent being the same, whereas in $SbCl_5$
doping, the oxidant must be added to the organic substrate. The changes in optical absorption were moni-

tored as a function of time utilizing a Guided Wave Model 200-25 VIS-NIR Spectrometer <u>via</u> fiberoptic cable to the remote sample cell at a scan rate of 2 nm/sec (300-1600 nm). Whereas $SbCl_5$ doping (10^{-2} M $SbCl_5$ added to 10^{-5} M polyene) is essentially complete in less than a minute, protonic doping may require up to four hours for bipolaron formation (2-4 hours is typical).

5 RESULTS AND DISCUSSION

The results of our doping studies for diphenyl polyene and PPV oligomers are shown in Tables 1 and 2. Han and Elsenbaumer[12] suggest a doping mechanism that involves initial protonation ortho to the vinyl repeat unit in PPV-type polymers:

However, they then propose that an intermolecular redox occurs between neutral polymer and the BP to form two dominant polaron states:

$$N \xrightarrow{\ 2H^+\ } BP \xrightarrow{\ N\ } P_1 + P_2$$

Their conclusion is based primarily on the VIS spectrum and ESR activity. In contrast to 2,5-dimethoxy-PPV, we do not observe a typical polaronic absorption pattern, but one almost identical to the bipolaron obtained from $SbCl_5$ doping, except blue shifted. We have suggested[6] that P and BP states can exist in dynamic equilibrium with the BP dominating the optical absorption.' In order to determine the relative contributions of P states, quantitative ESR measurements were carried out on 10^{-4} M solutions of the polyenes and PPV dimers in CF_3COOH for 2,4 and 24 hours. In no case are polarons present to an extent greater than 1%. Thus it is not surprising that the dominant optical absorption is bipolaronic. The blue shift in comparison to $SbCl_5$-- generated bipolarons can be explained on the basis of which site undergoes protonation. If protonation in the PPV dimers occurs at positions 4,4' rather than positions 2,2', the delocalization length will be two C atoms shorter than for oxidative doping. We believe

this position may be sterically favored in the polyenes, but not in the polymer. We are currently at-
tempting to solve the question as to which site
protonates by utilizing CF_3COOD as protonic doping
agent. One disturbing feature of our study is that
only two PPV oligomers undergo protonic doping. The
others appear to be inert. At the current time there
does not appear to be an easy answer for this lack of
reactivity, since these oligomers do undergo oxidative
doping easily. At this stage of our investigation we
favor the following overall mechanism for protonic
doping:

$$N \xrightarrow{\;2H^+\;} BP \rightleftarrows P$$

Further studies on PPV oligomers are continuing in
order to further elucidate the protonic doping mechanism.

6 CONCLUSIONS

Recent studies by Cao, _et al._[13], have indicated that
third-order optical nonlinearity in ladder copolymers
can be significantly enhanced by bipolaronic charge
states. Such enhancement can be initiated either by
photo-excitation or by chemical doping ($SbCl_5$). Our
current studies show how such states can be formed in
polymer oligomer units and stabilized by electron-
donating substituents. In addition we have shown that
protonic doping also generates stable bipolaronic
states. This is of particular importance in terms of
long-term stability since CF_3COOH-doped solutions are
stable for many months as compared to days for $SbCl_5$-
doped samples. In addition, composite films (substrate-
polycarbonate) can be spin-cast from these solutions
more readily than from $SbCl_5$ - containing solutions.
We are now in the process of evaluating $\Delta\gamma$ (N   -->
BP) for the compounds reported here, and $\Delta\chi^{(3)}$ for
the polycarbonate composites to determine if bipolaronic enhancement in the small oligomers is similar to
that found in ladder polymers.

7 ACKNOWLEDGEMENT

Acknowledgement is made to the Donors of the Petroleum
Research fund, administered by the American Chemical
Society, and the Air Force Office of Scientific Research
Grant #90-0060, for partial support of this research,

to Paul Bryson (USC) for ESR measurements of the protonically doped samples and to Pei-Kang Liu (NIU) for help in preparing this manuscript.

<u>Table 1</u>   Protonic and Oxidative Doping: Substituted Polyene

Substituents	n	H+ Doping λ_{max}^a BP (nm)	SbCl5 Doping λ_{max}^b BP
4,4'-(OMe)2[d]	5	664,727	627,692,755
4,4'-(OMe)2	6	713,776	680,741,818
2,2',5,5'-(OMe)4	5	653	c
2,2',5,5'-(OMe)4	6	713	755
2,2',4,4',5,5'-(OMe)6	5	691,726	762
2,2',4,4',5,5'-(OMe)6	6	707,767	797
3,3',4,4',5,5'-(OMe)6	5	804,881	653,768
3,3',4,4',5,5'-(OMe)6	6	839,937	693,796
4,4'-(OC8H17)2	5[e]	672,740	706,769
4,4'-(OC8H17)2	6[f]	726,792	755,825

[a] CF$_3$COOH solvent, absorption maxima after 24 h.   [b] CH$_2$Cl$_2$, solvent, absorption after 1 h.   [c] broad absorption.   [d] the first six entries were first reported in Ref. 6, the others are reported here for the first time.   [e] C$_8$H$_{17}$ = 1-octyl and 2-ethyl-1-hexyl.   [f] C$_8$H$_{17}$ = 2-ethyl-1-hexyl.

<u>Table 2</u>   Protonic and Oxidative Doping Results: PPV Dimers

Substituents	π-π*	SbCl5 Doping λ_{max}^a BP(nm)	H+ Doping λ_{max}^b (nm)
Unsubstituted	390	580,951	c
4,4'-(OMe)2	392	657,1196	620,1152,1398
2,2',5,5'-(OMe)4	400	594,1399	c
2,2',4,4',5,5'-(OMe)6	408	895	c
3,3',4,4',5,5'-(OMe)6	387	671,1399	c
4,4'-(OC8H17(2	394	671,1244	640,1196,1404

[a] CH$_2$Cl$_2$ solvent, absorption after 1 h,   [b] CF$_3$COOH solvent absorption maxima after 2 h.   [c] no bipolaron formed.

8   REFERENCES

1. C. W. Spangler, L. S. Sapochak and B. D. Gates,
    "Organic Materials for Nonlinear Opitcs", R. A.
    Hann and D. Bloor, Eds., Roy. Soc. Chem., London,
    1989.
2. C. W. Spangler and R. Rathunde, J. Chem. Soc.
    Chem. Commun. 1989, 26.
3. C. W. Spangler, T. J. Hall, and P.-K. Liu,
    Polymer, 1989, 30, 1166.
4. C. W. Spangler, T. J. Hall, K. O. Havelka, M.
    Brader, M. R. McLean and L. R. Dalton, Proc.
    SPIE, 1989, 1147, 149.
5. C. W. Spangler and K. O. Havelka, Polymer
    Preprints, 1990 31(1), 396.
6. C. W. Spangler and K. O. Havelka, "New Materials
    for Nonlinear Optics", ACS Symposium Series, G.
    Stucky, J. Sohn and S. Marder, Eds. (In Press).
7. C. W. Spangler, T. J. Hall, K. O. Havelka, D. W.
    Polis, L. S. Sapochak and L. R. Dalton, Proc.
    SPIE, 1990, 1337, (In Press).
8. C. W. Spangler, P.-K. Liu, T. J. Hall, D. W.
    Polis, L. S. Sapochak and L. R. Dalton,
    Polymer, 1991, (In Press).
9. L. Yu, D. W. Polis, F. Xiao, L. S. Sapochak, M.
    R. McLean, L. R. Dalton, C. W. Spangler, T. J.
    Hall and K. O. Havelka, Polymer, 1991, (In Press).
10. C. W. Spangler, P.-K. Liu and K. O. Halveka,
    Polymer Preprints, 1990, 31(1), 394.
11. C. W. Spangler, R. K. McCoy, A. A. Dembek, L. S.
    Sapochak and B.D. Gates, J. Chem. Soc. Perkin
    Trans. 1, 1989, 151.
12. C. C. Han and R. L. Elsenbaumer, Synthetic Metals,
    1989, 30, 123.
13. X. F. Cao, J. P. Jiang, R. W. Hellwarth, M. Chem,
    L. P. Yu and L. R. Dalton, Proc. SPIE, 1990,
    1337, (In Press).

Non-linear Optical Properties of Conjugated Polymers and Dye Systems

A. Grund, A. Mathy. A. Kaltbeitzel, D. Neher, C. Bubeck, and G. Wegner

MAX-PLANCK-INSTITUT FÜR POLYMERFORSCHUNG. POSTFACH 3148. D-6500 MAINZ. FEDERAL REPUBLIC OF GERMANY

1. Introduction

Nonlinear optical investigations on conjugated polymeric systems have shown strong nonlinear effects [1]. Systematic analysis seems to be necessary to resolve the relationship between chemical structure and the nonlinear optical behaviour. To study the influence of the dimensionality of the π-electron system we compared a simple dye (rhodamine 6G) and phthalocyanine with different conjugated polymers that we described recently [2-6]. Third harmonic generation (THG) experiments are used to determine the value of $\chi^{(3)}(-3\omega;\omega,\omega,\omega)$ at a wavelength of 1064nm. Degenerate Four Wave Mixing (DFWM) of picosecond laser pulses is further used to study the wavelength dependence of $\chi^{(3)}(-\omega;\omega,\omega,-\omega)$.

2. Experimental Setup and Investigated Systems

The experimental setup for THG measurements [2,3] was based on the Maker fringe method. Infrared light pulses of 0.4mJ and 30ps length generated by an active/passive mode-locked Nd-YAG laser were focused on the sample which was mounted on a rotation stage in an evacuated chamber. The third harmonic intensity was analyzed via a formalism including all bound waves [4]. $\chi^{(3)}(-3\omega;\omega,\omega,\omega)$ was evaluated with respect to a quartz reference.

DFWM experiments were performed using a folded BOXCAR configuration [3,7,8]. The output of a synchronously pumped and cavity dumped dye laser system was split into three beams with variable delay and focused on the sample. The intensity of the nonlinear signal was used to evaluate $\chi^{(3)}(-\omega;\omega,\omega,-\omega)$ with reference to CS2 [9].

Thin films of polymers and dyes were prepared by spin-coating and casting on glass or fused silica substrates. The polymers are substituted poly (para-phenylene vinylene) (PPV) [2,3], poly phenyl acetylene (PPA) [3,4,5] and polythiophene (PT) [5]. Langmuir Blodgett films were prepared with a polymeric phthalocyanine (Pc), where a flexible spacer connects the Pc-rings [7]. The rhodamine 6G (R6G) films were prepared by casting from solution.

3. Results

3.1. Third Harmonic Generation

Fig.1 shows a survey of the measured $\chi^{(3)}(-3\omega;\omega,\omega,\omega)$ values for all investigated systems in dependence of the absorption maximum [5,6]. To take the density of the nonlinear active groups into consideration, $\chi^{(3)}$ was divided by the absorption coefficient α_{max}. The dye R6G shows a $\chi^{(3)}$ value one order of magnitude smaller than the conjugated polymers. For a 45nm thick film of rhodamine 6G we obtained $\chi^{(3)}(-3\omega;\omega,\omega,\omega)$ $=(3.4+/-1)*10^{-12}$ esu and a phase of 241 deg.

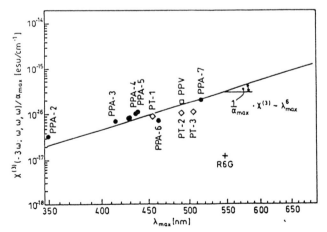

Figure 1: Survey of the THG result with conjugated polymers and dye systems measured at λ(laser)=1064nm (double logarithmic plot).

3.2. Degenerate Four Wave Mixing

Fig.2 shows the absorption spectrum and the measured $\chi^{(3)}(-\omega;\omega,\omega,-\omega)$ values for a 600nm thick R6G film. The decay time of the transient grating for this film was determined to be 50 ps (see Fig.3).

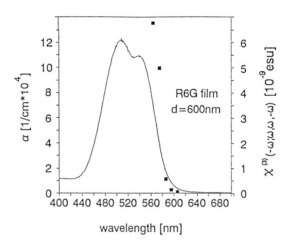

wavelength [nm]

Figure 2:   Absorption coefficient and $\chi^{(3)}(-\omega;\omega,\omega,-\omega)$ (squares) as function of the wavelength for a R6G film

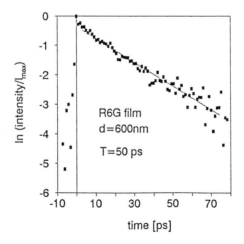

time [ps]

Figure 3: Decay of the transient grating of a thin film of R6G (semilogarithmic plot)

To gain an overview on the DFWM results in the range of the one-photon absorption bands of PPV, PPA, PT, Pc with flexible spacer and R6G, Fig.4 shows $\chi^{(3)}(-\omega;\omega,\omega,-\omega)$ as a function of the absorption coefficient α. Surprisingly, the resonant $\chi^{(3)}(-\omega;\omega,\omega,-\omega)$ values for chemically very different systems show a strong correlation with α. Even with a relatively "simple" dye like R6G, $\chi^{(3)}(-\omega;\omega,\omega,-\omega)$ values similar or even superior to those of conjugated polymers are obtained, if the laser wavelength is tuned across the one-photon absorption band.

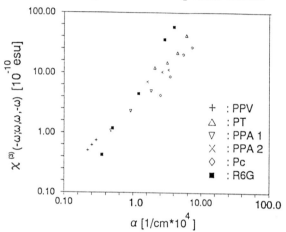

Figure 4: Survey of the DFWM results with conjugated polymers and dye systems described in the text (double logarithmic plot)

4. Discussion

The high values of $\chi^{(3)}(-\omega;\omega,\omega,-\omega)$ for PPA, PT, Pc and R6G are obtained only in the case of strong absorption of the incident laser light. The responsible process is saturable absorption [10].

Further results [11] indicate that resonant $\chi^{(3)}(-\omega;\omega,\omega,-\omega)$ scales with α^2 for the dyes R6G and Pc and linearly with α in the case of polymers with a one dimensional π-electron system. Conjugated polymers such as PPV, PPA, PT show a very fast response time in the range of 1ps or less [2-6], the dye systems Pc [7] and R6G show a decay time in the range of 5-50ps.

In contrast to DFWM the $\chi^{(3)}$ values obtained from THG differ very significantly for dye systems and conjugated polymers. Clearly, it is necessary to distinguish between $\chi^{(3)}$ values obtained from DFWM and THG - a fact that is quite often confused in the literature.

Although the $\chi^{(3)}(-3\omega;\omega,\omega,\omega)$ data contain various resonance contributions, a clear trend to higher values with increasing λ_{max} is observed in the case of the polymers with a one dimensional π-electron system. This is in qualitative accordance with the scaling law derived by Flytzanis [12]. Further experiments, especially the determination of the nonresonant values of $\chi^{(3)}(-3\omega;\omega,\omega,\omega)$ are necessary to provide sufficient evidence to establish this theory. To obtain materials with high third order nonlinearities and useful figures of merit $\chi^{(3)}/\alpha$, the conjugated polymers still seem to be very promising.

Acknowledgement

We thank H.Menges for considerable technical support. Financial support by the BMFT under the project number 03M4008E9 is gratefully acknowledged.

References

[1] D.S.Chemla, J.Zyss, Ed. "Nonlinear Optical Properties of Organic Molecules and Crystals", Academic Press, New York, 1987

[2] C.Bubeck, A.Kaltbeitzel, R.W.Lenz, D.Neher, J.D.Stenger-Smith, G.Wegner, in J.Messier, F.Kajzar, P.Prasad, D.Ulrich (ed.): Nonlinear Optical Effects in Organic Polymers, NATO ASI Series E 162, Kluwer Acad. Publ., Dordrecht, 1989, p.143

[3] C.Bubeck, A.Kaltbeitzel, D.Neher, J.D.Stenger-Smith, G.Wegner, A.Wolf, in H.Kuzmany, M.Mehring, S.Roth (ed.): "Electronic Properties of Conjugated Polymers III", Springer Series in Solid State Science, 1989, 91, p.214

[4] D.Neher, A.Wolf, C.Bubeck, G.Wegner, Chem.Phys. Lett, 1989, 163, 116

[5]   D.Neher,   A.Wolf,   M.Leclerc,   A.Kaltbeitzel,
      C.Bubeck, G.Wegner, Synth.Met., 1990, 37, 249
[6]   D.Neher,   A.Kaltbeitzel,   A.Wolf,   C.Bubeck,
      G.Wegner, in J.L.Bredas, R.R.Chance (ed.):
      Conjugated Polymeric Materials: Opportunities in
      Electronics, Kluwer Acad. Publ., Dordrecht, 1990,
      p.387
[7]   A.Kaltbeitzel,   D.Neher,   C.Bubeck,   T.Sauer,
      G.Wegner, W.Caseri, in H.Kuzmany, M.Mehring,
      S.Roth   (ed.):   "Electronic   Properties   of
      Conjugated Polymers III", Springer Series in
      Solid State Science, 1989, 91, p.220
[8]   G.M.Carter, J.Opt.Soc.Am., 1987, B4, 1018
[9]   N.P.Xuan,   J.-L.Ferrier,   J.Gazengel,   G.Rivoire,
      Opt.Commun., 1984, 51, 433
[10] R.L.Abrams, R.C.Lind, Opt.Lett., 1978, 2(4), 94
[11] A.Grund, A.Kaltbeitzel, C.Bubeck in preparation
[12] G.P.Agrawal, C.Cojan, C.Flytzanis, Phys.Rev.B,
      1978, 17, 776

Comparison of Electro-optic Coefficients of Amorphous and Liquid Crystalline Non-linear Optical Copolymers

M.S. Griffith and M.R. Worboys

GEC-MARCONI RESEARCH CENTRE, GREAT BADDOW, ESSEX CM2 8HN, UK

N.A. Davies

GEC-HIRST RESEARCH CENTRE, WEMBLEY, MIDDLESEX HA9 7PP, UK

D.G. Mcdonnell

ROYAL SIGNALS AND RADAR ESTABLISHMENT, GREAT MALVERN, WORCESTERSHIRE WR14 3PS, UK

1 INTRODUCTION

This work was undertaken to determine the magnitude of any enhancement in electro-optic coefficients which may occur in liquid crystalline systems relative to those of amorphous systems and to compare this to the expected theoretical enhancement[1]. Two materials were prepared which contained nominally the same concentration of non-linear optical (NLO) molecules, one exhibiting liquid crystalline properties and the other not. The approach used was to copolymerise a liquid crystalline monomer with a NLO dye monomer[2], thus overcoming the problem of low dye concentration usually encountered in guest-host systems. To obtain high electro-optic coefficients it is necessary to d.c. pole these materials using very high fields. It has been shown that the optimum poling conditions are approximately 10-20°C below the glass transition temperature T_g in amorphous systems[3]. At this temperature breakdown due to impurities is minimised whilst sufficient mobility·of the side group is retained to enable poling to occur within reasonable time scales.

2 DISCUSSION

The monomers which were used are depicted in Figure 1. The monomer A/2ANAS is the acrylate ester of Disperse Red 1 and is a readily obtainable material possessing a high β coefficient[3]. Copolymerisations were carried out using AIBN initiated polymerisation in dry dimethylsulphoxide at 85°C for 40 hours in a nitrogen atmosphere. Repeated precipitation with methanol gave copolymers which were found to be free of monomers by TLC and GPC. The phase diagrams for the copolymers are shown in Figure 2. Similar phase behaviour has been reported[4] for an anthraquinone dye system where it was found that the clearing temperature T_c was depressed at higher concentrations of dye. It is expected that Disperse Red 1 exhibits some mesogenic character and that T_c is obscured at high concentrations of the NLO co-monomer by the presence of the glassy phase. This eutectic type behaviour has been observed in other copolymer

A/2ANAS

A/2OCPB

A/2OCBB

A/2OCBAB

A/6OMBB

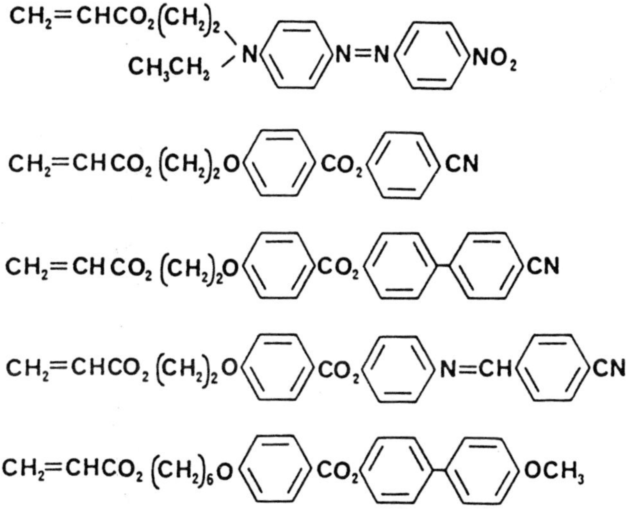

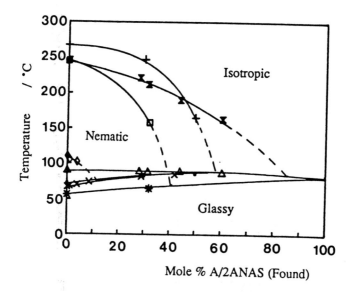

Figure 1 Chemical structure of the monomers

Figure 2 Phase diagram for the NLO liquid crystal copolymers. Symbols; T_g, A/2OCBAB(■), A/6OMBB(✱), A/2OCPB(×), A/2OCBB(△); T_{NI}, A/2OCBAB(+), A/6OMBB(□), A/2OCPB(◇), A/2OCBB(⧗).

systems[5]. Glass transition temperatures, as measured by DSC, are not significantly affected by high concentrations of the NLO group and are primarily determined by the type of backbone and length of the methylene spacer chain.

Molecular weights were ~4,000 for each material as measured by GPC, indicating ~10 acrylate units per copolymer chain, and so the materials must be regarded as oligomers. However, with ten repeat units in the backbone the values of T_g and T_c do not significantly change for higher molecular weight materials[6]. Phase transition temperatures and types were determined by optical microscopy, number densities by UV/visible spectroscopy and refractive indices by an optical interferometric technique. The optical order parameters (S_2) were estimated using a similar method to that described in the literature[7].

Figures 3 and 4 show field dependent axial ordering above T_c in 3.5 Mole% A/2ANAS in A/2OCPB (liquid crystalline) and 10 Mole% A/2ANAS in A/2OCPB (amorphous) respectively and the data shows a striking similarity to that predicted by the MSVP model[8]. Switch-on and switch-off times (not shown here) at temperatures above T_c were found to be equal and did not depend on the applied field, but were found to be very long at temperatures near T_g. Switching times were field dependent in the nematic region of the liquid crystalline copolymer, with shorter times at higher applied fields.

The r_{eff} coefficient was measured as a function of poling field for dipped thin films of 30 Mole% A/2ANAS in A/2OCPB (amorphous) and 30 Mole% A/2ANAS in A/2OCBB (liquid crystalline) copolymers, as shown in Figures 5 and 6 respectively. Measurements were made at 633 nm on a tilted, poled cell. The magnitude of electro-optic modulation was found to be independent of frequency up to 60 kHz which was the limit of the detector used. Measurements were made immediately after poling to avoid thermal degradation of the electro-optic coefficients. The electro-optic stability as a function of storage temperature will be reported elsewhere but early results suggest a marked increase in stability is obtained in liquid crystalline systems.

The electro-optic coefficient measured in the amorphous polymer (where S_2 was 0 at 0 V and 0.54 at 180 V) became non-linear with field at poling fields in the region of 70 V.μm^{-1}, but was linear with applied field at lower poling temperatures. Poling times as short as 15 minutes did not affect the results. For the liquid crystalline material (S_2=0.76), a linear field dependence was observed for films which had not been a.c. aligned, giving an enhancement relative to the amorphous material of 2.8 and 1.4 at low and high poling fields respectively. Dipped films of the liquid crystalline polymer appeared isotropic and only gave a scattering texture when heated above the glass transition temperature. Films of liquid crystalline polymer which were initially a.c. aligned using 10V.μm^{-1} at 50 Hz and at 108°C, exhibited little or no enhancement of r_{eff} relative to the amorphous material when poled at low fields, suggesting that poling is inhibited by the presence of liquid crystalline order. The a.c. aligned films which were then poled at low field did not appear scattering in contrast to the initially unaligned films. In general, poled liquid crystalline films were found to be more scattering than poled amorphous films, probably due in part to impurity related electro-hydrodynamic instabilities[7].

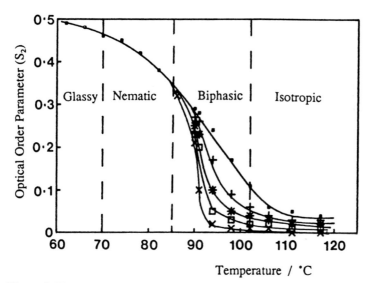

Figure 3 Field dependent optical order parameter for 3.5 Mole% A/2ANAS - A/2OCPB liquid crystal copolymer. Poling field; 12 V.μm^{-1} (✗), 18 V.μm^{-1} (□), 23 V.μm^{-1} (✳), 30 V.μm^{-1} (+) and 38 V.μm^{-1} (■).

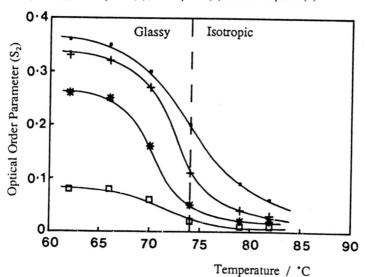

Figure 4 Field dependent optical order parameter for 10 Mole% A/2ANAS - A/2OCPB amorphous copolymer. Poling field; 16 V.μm^{-1} (□), 27 V.μm^{-1} (✳), 38 V.μm^{-1} (+) and 60 V.μm^{-1} (■).

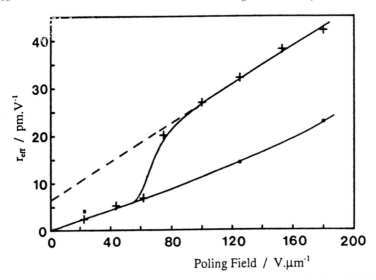

<u>Figure 5</u> Electro-optic measurements on 30 Mole% A/2ANAS in A/2OCPB (amorphous copolymer) as a function of poling field when poling at 20°C(■) and 10°C(+) below T_g. Samples poled for one hour and cooled at 5°C.min^{-1} to 25°C.

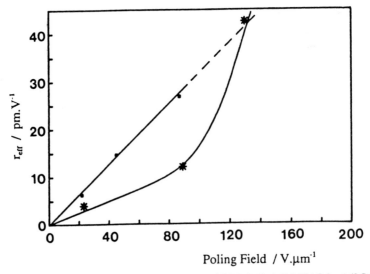

<u>Figure 6</u> Electro-optic measurements on 30 Mole% A/2ANAS in A/2OCBB (liquid crystalline copolymer) as a function of poling field at 20°C below T_g, ac aligned(✳) and not ac aligned(■) before dc poling. Samples poled for one hour and cooled at 5°C.min^{-1} to 25°C.

To obtain a meaningful comparison of the experimental data with that predicted theoretically, it is necessary to estimate the value of $\chi^2_{(333)}$ for these materials. Figure 7 shows the estimated values as a function of poling field for the copolymers of A/2OCPB (amorphous) and A/2OCBB (liquid crystalline) containing 30 Mole% of A/2ANAS. As the electro-optic measurements made here did not measure r_{33} direct we have estimated $\chi^2_{(333)}$ from the values of r_{eff}, n_o, n_e and S_2. It was found that for the amorphous copolymer at low field, $n_o = n_e = 2.27$, whereas for the liquid crystalline copolymer $n_o = 1.85$ and $n_e = 2.71$. It was assumed that S_2 approximated to $<P_2>$ and the value of $<P_4>$ was estimated from microscopic order parameters published for the low molar mass liquid crystal, BBOA[9]. Using these values it was possible to calculate the theoretical ratio $\chi^2_{(333)}/\chi^2_{(113)}$ using the SKS model[10]. The estimated enhancement (calculated from the slopes in Figure 7) was found to be 5.0 at low fields which compares to a theoretically predicted enhancement of 3.7. The discrepancy may occur as a result of the different molecular properties of the liquid crystalline monomers used in each material.

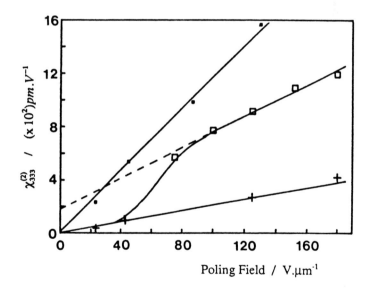

Figure 7 Variation in $\chi^{(2)}_{333}$ with poling field for copolymers containing 30 Mole% of A/2ANAS. Co-monomers were A/2OCBB(■) and A/2OCPB poled at 20°C(+) and 10°C(□) below T_g.

3 CONCLUSIONS

Electro-optic coefficients in non-linear optical copolymers are enhanced by liquid crystalline order relative to those obtained in an isotropic material. However, amorphous copolymers have been found to exhibit a field dependent axial ordering at high poling fields, leading to higher than expected electro-optic coefficients. During the transition from an isotropic state to an axially ordered state the electro-optic coefficient exhibits a non-linear response with poling field. The definitive comparison of experimental results with theoretical predicted enhancement of electro-optic coefficients requires further work, although the results obtained here suggest a reasonably close agreement might be expected.

REFERENCES

1. G.R. Meredith, J.G. Van Dusen and D.J. Williams in "Non-Linear Optical Properties of Organic and Polymeric Materials", Ed. D.J. Williams, ACS Symposium No 233, pp109-133, American Chemical Society, Washington DC, (1982).

2. C. Noël, C. Friedrich, V. Leonard, P. Le Barny, G. Ravaux and J.C. Dubois, Makromol. Chem., Makromol. Symp., 1989, 24, 238.

3. S. Esselin, P. LeBarny, P. Robin, D. Broussoux, J.C. Dubois, J. Raffy and P.J. Pocholle, SPIE, 1988, 971, 120.

4. H. Ringsdorf and H.W. Schmidt in "Recent Advances in Liquid Crystalline Polymers". Ed. L.L. Chapoy, Elsevier Applied Science Publishers Ltd., 1983, p253.

5. F. Cser, J. Horváth, K. Nyitrai and G. Hardy, Israel J. Chem., 1985, 25, 252.

6. V. Percec and C. Pugh in "Sidechain Liquid Crystal Polymers", Ed. C.B. McArdle, Blackie (London), 1989, pp30-105.

7. T.V. Talroze, V.P. Shibaev, V.V. Sinitzyn and N.A. Platé in "Polymeric Liquid Crystals, Proc. 2nd Symp.", Plenum, Washington DC, 1985, pp331-344.

8. G.R. Möhlmann and C.P.J.M. Van Der Vorst in "Sidechain Liquid Crystal Polymers" Ed. C.B. McArdle, Blackie (London), 1989, pp330-356.

9. S. Jen, N.A. Clark, P.S. Persham and E.B. Priestly, J. Chem. Phys., 1977, 66(10), 4635.

10. K.D. Singer, M.G. Kuzyk and J.E. Sohn, J. Opt. Soc. Am., 1987, B4(6), 968.

ACKNOWLEDGEMENTS

This work has been carried out with the support of the Procurement Executive, Ministry of Defence.

Stability of Poled Non-linear Optical Polymers and Glasses

F.C. Zumsteg, R.P. Foss, and R. Beckerbauer

CENTRAL RESEARCH AND DEVELOPMENT DEPARTMENT, E.I. DU PONT DE
NEMOURS AND CO., WILMINGTON, DELAWARE 19880-0356, USA

1 Introduction

A primary obstacle to the realization of the
technological potential of poled polymers has been the lack
of long term temporal stability of the polar molecular
orientation necessary for second order nonlinear optical
effects. In developing new materials for this purpose,
researchers have progressed from the original concept of
dye dissolved in polymer[1,2] to systems in which the oriented
dye molecules are chemically incorporated into a highly
crosslinked matrix while it is being poled[3]. We report
here on a new synthetic approach to a stable poled organic
system and an environmental effect which can cause a
substantial decrease on stability.

2 Poled monomeric glasses

A new approach to creating a stable poled stucture
utilizes nonlinearly polarizable molecules which form a
glass with a glass transition temperature, T_g, substantially
above room temperature. Since such glasses do not have
spacer groups and polymer backbones, which do not contribute
to the nonlinearity, higher concentrations of nonlinear
molecules and, hence, higher bulk nonlinearities should be
possible . Organic molecules have long been known to form
glasses when rapidly cooled to very low temperatures. The
application to nonlinear optics was made by Eich et al.[4] who
showed that it was possible to prepare thin glassy monomeric
films which could be poled to create an acentric structure.
T_g of their materials, however, is below room temperature.

We have found several nonlinearly active materials
which form stable monomeric glasses (Table 1) with T_g's

substantially above room temperature. All of the materials
shown can be prepared in the glassy state by melting and
then quickly cooling to room temperature. Typical cooling
rates are approximately 10 °C/sec. Although the viscosity
is much less than that of typical melted polymer, the large

		T_g	T_c^*	T_m
1	*p*-Nitroanalinoglutarimide	66	120	204
2	N-(*p*-Cyanophenyl)aminomethyl-succinimide	43	99	194
3	N-(*p*-Carboxymethylphenyl)amino methylsuccinimide	39	92	203
4	N-(4'-nitro-4-stilbenyl)amino methylsuccinimide	71	131	210
5	3,5-Dihydroxyacetophenone, p-nitrophenylhydrazone	123		235

Table 1   Glass, crystallization, and melting temperatures
           of monomeric glasses. (Note: Crystallization
           temperatures at cooling rates of 20 °C/min.)

intermolecular forces are sufficient to cause the viscosity
to be large and modify other physical factors enough to
prevent the formation of single crystals while being
quenched to the glassy state. One of these materials, *p*-
Nitroanalinoglutarimide (*p*-NAG) is particularly attractive
in that thin glassy films can be formed directly by spin
coating from solution.

The *p*-Nitroanalinoglutarimide is conveniently prepared
by the ring opening reaction of cyclic anhydrides with
hydrazines to form amide/acids which, in turn, can be
cyclized to the imide by heating with acetic anhydride in
the presence of an acetate catalyst. The structure was
confirmed by elemental analysis, IR and NMR

The differential scanning calorimetry, DSC, trace of
p-NAG shown in fig. 1 illustrates the thermodynamic
features of the monomeric glass which make it useful as
poled nonlinear optical material. T_g is sufficiently above
room temperature so that long term stability can be
achieved for room temperature operation. The crystallization temperature, T_c, at 120 °C is sufficiently above
T_g that the material can be poled without inducing

crystallization. Finally the melting temperature, T_m, at
204 °C occurs below the decompositon temperature.

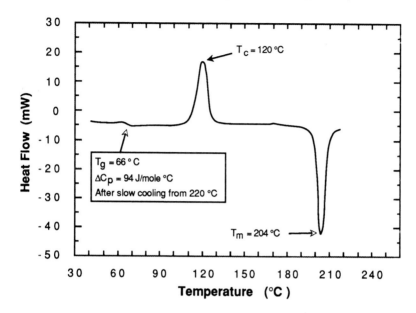

<u>Figure 1</u>   Differential scanning calorimeter trace of
             p-Nitroanalinoglutarimide   (20 °C/min)

    Films of *p*-NAG were prepared by spin coating from a
solution of tetrahydrofuran on a substrate of indium tin
oxide coated glass. The resulting films were approximately
1.0 μm thick and showed no birefringence when viewed
through crossed polarizers. To ensure that no
microcrystals existed which could act as nucleation
centers, the films were heated to slightly above the melt
temperature and then rapidly cooled to room temperature.

    Samples were poled using a corona discharge to create
the electric field. *p*-NAG was poled by rapidly heating it
to 85 °C and maintaining it there for 10 min. It was then
cooled to room temperature at which time the electric field
was removed. The degree of molecular orientation was
monitored by measuring the second harmonic intensity
generated when placed in the beam of a Q-switched Nd-YAG
(1.06μm) laser. The resulting material has an initial
value of d_{33} = 4.2 pm/V. Figure 2 shows its exceptional

stability. The estimated half-life is greater than three
years.

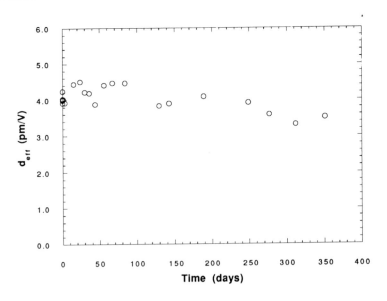

<u>Figure 2</u> Nonlinear optical coefficient d33 of
            p-Nitroanalinoglutarimide as a function of time

     Extension of these results to other systems has
proved to be difficult. Although we have found a large
number of other materials which have T_g appreciably above
room temperature and have $T_c > T_g$, material, processing is
very material specific. The materials reported on here as
well as other materials are under study to determine the
factors important in orientation and stability as well as
appropriate processing techniques.

     3 Effect of relative humidity on poled polymer stability

     The orientational stability of some poled polymers is
sensitive to not only the processing conditions used in
poling, but is also extremely sensitive to environmental
conditions such as humidity. To test the effect of
humidity on stability, we measured effect of relative
humidity on the stability of the two side chain polymers
shown in fig. 3. These two polymers were chosen because of

well known differences in solubility of water in
polymethacrylate and polystyrene based polymers.

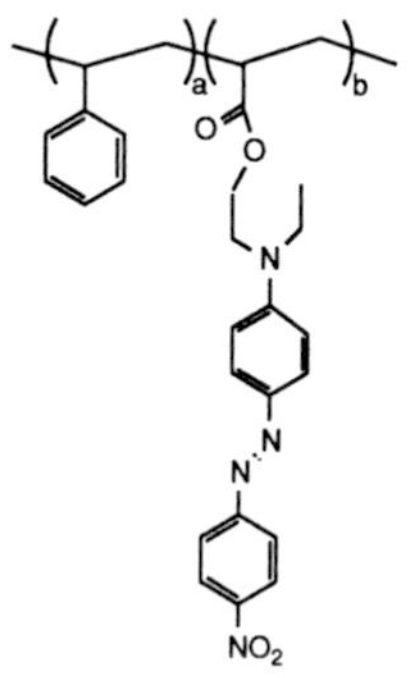

Styrene - Disperse Red 1   Methylmethacrylate -
Copolymer   Disperse Red 1 Copolymer

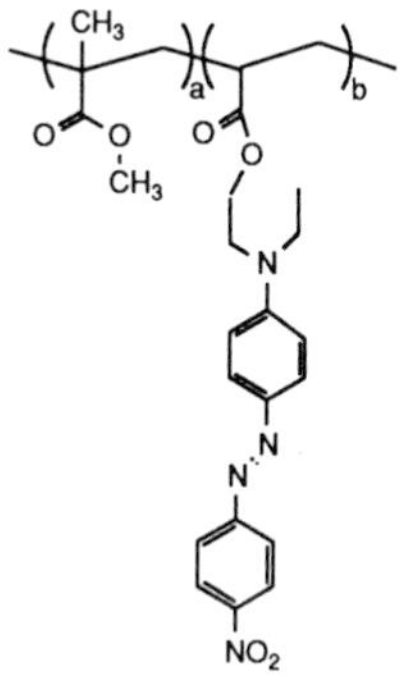

Figure 3   Side chain copolymers used to study the effect
of relative humidity on orientational stability

The side-chain polymers were prepared by reacting the
dye, Disperse Red 1, with poly-acryloyl chloride / polymethylmethacrylate copolymers and polyacryloyl chloride /
polystyrene copolymers. In both materials the ratio of
styrene or methylmethacrylate to acryloyl chloride was
20:1. The reaction was carried out in an excess of dye so
that minimal unreacted acryloyl chloride was left in the
final product.

Thin films of these polymers were prepared and poled
using the same techniques as described above for monomeric
glasses. The only differences were that the polymers were
spin coated from methylene chloride and samples were heated
to 120 °C during the poling process. After poling the
samples were kept in sealed containers maintained at
constant relative humidity by the presence of saturated
aqueous solutions of various salts[5]. The samples were
exposed to ambient atmosphere only during the time required
to make second harmonic generation measurements.

The results are shown in Figs.4 and 5. The effect of
increasing relative humidity on the temporal stability of
the PMMA based polymers is particularly dramatic for the
sample exposed to 100% relative humidity with the second
harmonic intensity decreasing to nearly the unpoled value
within the first day. Other PMMA samples became

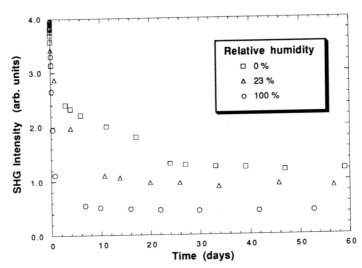

<u>Figure 4</u>   Second harmonic intensity from PMMA / Disperse
Red 1 side-chain copolymer as a function of time
for various relative humidities. Normalized to
start at the same value.

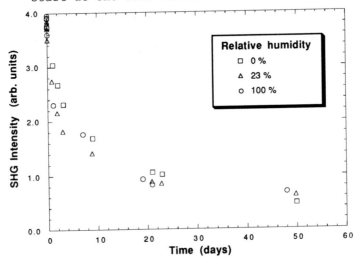

<u>Figure 5</u>   Second harmonic intensity from Styrene / Disperse
Red 1 side-chain copolymer as a function of time
for various relative humidities. Normalized to
start at the same value.

increasingly stable as the relative humidity was decreased. The most stable material was the one maintained at 0% relative humidity. On the otherhand the styrene based polymers in which only small amounts of water can be dissolved show essentially no effects of changing relative humidity.

4 Conclusions

We have focused on material properties which affect the temporal stability of poled organic systems. We have found that the monomeric glass consisting only of nonlinear optically active molecules can be poled and has exceptionally good temporal stability. Also we have found that environmental effects such as humidity can dramatically reduce temporal stability in some poled polymers.

5 Acknowledgements

The authors wish to thank W. Tam for providing some of the monomeric glasses used in this study.

REFERENCES

1.   G. R. Meredith, J. G. Vandusen and D. J. Williams, 'Nonlinear Optical Properties of Polymeric Materials', ed. D. J. Williams, American Chemical Society, Washington, D.C., 1983, p. 109.

2.   K.D. Singer, J.E. Sohn, and S.J. Lalama, <u>Appl. Phys. Lett.</u>, 1986, <u>49</u>(5), 248

3.   M. Eich, B. Reck, Do Y. Yoon, C.G. Willson, and G.C. Bjorklund, <u>J. Appl. Phys.</u>, 1989, <u>66</u>(7), 3241.

4.   M. Eich, H. Looser, Do Y. Yoon, R. Twieg, G. Bjorklund, and J. C. Baumert, <u>J. Opt. Soc. Am. B</u>, 1989, <u>6</u>(6), 1590.

5.   ed. J.A. Dean, 'Lange's Handbook of Chemistry', McGraw-Hill Book Co., New York, N.Y., 1979, p. 10-84.

Devices

Non-linear Optical Devices: Current Status of Organics

George I. Stegeman

CENTER FOR RESEARCH IN ELECTRO-OPTICS AND LASERS, UNIVERSITY OF
CENTRAL FLORIDA, ORLANDO, FLORIDA 32826, USA

Summary: We compare nonlinear optical devices based on organic materials
with those made from other materials.

1. INTRODUCTION

Nonlinear organic materials have been investigated since the mid 1970s.
However, their application to devices is very recent, and in this sense they are
at a disadvantage because other materials such as inorganic dielectrics and
semiconductors have been investigated in the context of nonlinear devices for
a number of years now.   Nevertheless, progress in the development and
application of nonlinear organics has been rapid, especially in the area of second
harmonic generation.   In this paper we will assess the state-of-the-art in
nonlinear devices based on organic materials, and compare it with devices made
from other nonlinear materials.

2. SECOND ORDER DEVICES

Until the mid 1980s the principal application of doublers was in frequency
shifting lasers to shorter wavelengths.   Since then it was realized that using
waveguide[1] or resonator[2] geometries, or a combination of the two[3], it
would be possible to convert $\simeq 100$ mW of laser diode power to at least a few
mW of blue light for applications in optical data storage.   In fact lithium niobate
channel waveguides with end faces polished to produce a resonator at both the
fundamental and harmonic wavelengths have been shown to have more than
sufficient conversion efficiency for this application.[3]   Phase-matching was
achieved by using the material birefringence with an output polarization
orthogonal to the input.   The problem is that mass production of such devices
would be difficult.

Interest in using organic materials has been spurred on by discoveries that the appropriate nonlinear coefficients can be factors of 20-60 larger than those of dielectric materials such as lithium niobate, potassium niobate etc. [for example, 4-12]. Over the last few years there have been two distinct approaches to making efficient doublers out of organic materials. The first has been to use single crystal materials in waveguide form.[for example 7-10] The second has been to take very nonlinear molecules which would normally crystallize into a centrosymmetric crystal structure, and deposit them as a LB film,[for example 13] or partially orient them in a polymer host.[for example 11,12]

As shown in Table 1, the inherent advantage of organic materials are primarily their very large second order susceptibilities d_{ijk}. Relative to lithium niobate, organic diagonal elements are almost one order of magnitude larger, and up to 25x larger for the off-diagonal elements used in Type I and II phase-matching. Also listed is the wavelength beyond which the doubled light is strongly absorbed. A major problem is clear, namely that the stronger the activity, the higher the cut-off wavelength. Therefore for doubling of semiconductor diode lasers into the 410 nm region, many of these materials are not adequate and it is on solving this problem that effort needs to be concentrated.

Table 1   Nonlinear coefficients d_{ij} for some representative
            inorganic and organic materials

MATERIAL	d_{ii} pm/V	d_{ij} pm/V	Transparency nm
LiNbO$_3$	41	5.8	400-2500
β-BaB$_2$O$_4$	2.0	0.3	190-3500
KTiOPO$_4$	4	7	350-4500
MMONS [4]	184	71	530-1600
MNA [5]	165	25	480-2000
NPP [6]	30	84	480-1800
DMNP [7]	30	90	450-
DAN [8]	5	50	485-2270
DCV/MNA [12]	19	6	
DCANP [11]	8	2	400->2000

MNONS	3methyl-4-methoxy-4-nitrostilbene
NPP	N-(4-Nitrophenyl)-(L)-prolinol
DAN	4-(N,N-Dimethylamino)-3-Acetamidonitrobenzene
DCANP	2-docosylamino-5-nitropyridine (Y-herringbone)
MNA	2-Methyl-4-nitroaniline
DCV/PMMA	dicyanovinyl azo dye in copolymer
DMNP	3,5dimethyl-1-(4-nitrophenyl) pyrazole

Optical waveguides are the optimum geometry for efficient doublers. To date, single crystal organic materials have been formed into fibers,[7,8] planar[9] and channels[10] waveguides. Poling and Langmuir-Blodgett deposition techniques have been used to orient highly SHG active molecules to form SHG active planar and channel waveguides with promising propagation losses.[11,13] A few cases are also listed in Table 1.

One of the advantages to using waveguide geometries is that there are a number of techniques available for phase-matching in addition to the traditional Type I and II birefringent methods used for bulk (unguided) SHG devices. Two are shown in Figure 1. "Cerenkov" radiation is the simplest to implement. It was first demonstrated by Tien and coworkers in 1970.[14] The refractive index, n_b (2ω), of at least one of the bounding media must be large enough so that $2k_0 n_b(2\omega) > 2\beta(\omega)$ where $\beta(\omega)$ is the wavevector associated with the guided wave at the fundamental frequency. In this case, the second harmonic is radiated into the bounding medium at an angle θ (relative to the propagation direction) given cos $\theta = \beta(\omega)/k_0 n_b(2\omega)$. This approach was first applied to lithium niobate waveguides to be able to use the large d_{33} coefficient in that material which is 7x the d_{13} coefficient used in Type I phase-matching.[15] For organic materials it has been used in both fiber and channel waveguide geometries.[7,8,10,16]

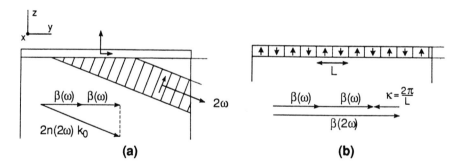

(a)　　　　　　　　　　　　　**(b)**

Figure 1　　Two new recently implemented schemes for second harmonic generation in waveguide structures. (a) "Cerenkov" radiation in which the second harmonic signal is radiated into one or more of the waveguide bounding media. (b) Quasi-phase-matching in which a periodic modulation of the nonlinear susceptibility (or refractive index) is used to ensure phase matching.

A second approach is to spatially modulate either the refractive index, the waveguide structure or the nonlinear susceptibility in a periodic fashion.[17] The required periodicity L is given by $\beta(2\omega) - 2\beta(\omega) = \pm \kappa_p$ where $\kappa_p = p\pi/L$

and p is an integer. The optimum situation is to use $p=1$ and to modulate the nonlinearity so that interference effects associated with the coherence length $[\beta(2\omega - 2\beta(\omega)]^{-1}$ are minimized. This approach was first used again in lithium niobate to be able to phase-match the large d_{33} coefficient, and in fact progress in this area has been dramatic.[18,19] For organic material, Khanarian and coworkers have produced partial modulation of the second order susceptibility (as well as an index modulation) during the poling of oxy-nitrostilbene molecules.[13] Photobleaching has also been implemented to make index gratings in poled polymers.[20]

A comparison of device figures of merit is given in Table 2. The goal is to produce a few milliwatts of blue light for up to 100 mW near infrared input. Normalizing to one centimeter devices, a figure of merit of a least 10% W^{-1} is required. (Note that the efficiency increases only linearly with device length for Cerenkov, and quadratically for all other cases.)

Clearly from Table 2, these target numbers have been reached and exceeded. Amongst actual devices reported to date, the Cerenkov ones are most advanced in organic materials. However, there are difficulties in easily producing environmentally insensitive diffraction limited spots. For the quasi-phase-matching, the results are still preliminary but the approach appears very promising.

Table 2 Conversion efficiencies, extrapolated to a 1 cm long device for virion, material and phase-matching in waveguides, $M = P(2\omega)/P^2(\omega)$

Reference	Waveguide	Phase Matching Technique	η %W^{-1}
Sohler [1]	$LiNbO_3$ (1)	Birefringence	12
Sohler [3]	$LiNbO_3$ (2)	Birefringence	45
Taniuchi [15]	$LiNbO_3$	Cerenkov	42
Uemiya [7]	DMNP (3)	Cerenkov	40
Fejer [18]	$LiNbO_3$ (4)	Quasi-Phase Matching	37
Khanarian [13]	Poled Polymer (5)	Quasi-Phase Matching	0.04

(1) Resonator at fundamental wavelength
(2) Resonator at both ω and 2ω
(3) Crystal core fiber
(4) Calculated efficiency of 370
(5) 4-oxy, 4-nitrostilbene, poled with periodic electrodes,
    projected channel waveguide $\simeq 8\%$ W^{-1} cm^{-2}

Highly nonlinear organic materials will also find applications in plane wave devices. For example, NPP has been used in parametric amplified sampling spectroscopy to time-resolve luminescence signals.[21] Another interesting example is autocorrelators for measuring femtosecond pulses.[22]

Organic materials have certain advantages and disadvantages relative to their inorganic counterparts. The large coefficients make very short doubling devices possible, thus reducing the requirements on material homogeneity, attenuation coefficients and acceptance angles. There are also many features which facilitate phase-matching, for example the thickness control afforded by L-B films, photobleaching to tune index or form quasi-phase-matching gratings, reverse poling to modulate the susceptibility etc. Finally, poling techniques obviate the necessity of producing single crystal materials.

But, there are still problems to be solved. The losses are of the order of cm^{-1} and in some cases there are impurities which lead to damage and high powers. However, the biggest challenge is the trade-off between spectral transmission and nonlinearity. Clearly what is needed is a material with a cut-off wavelength below 400 nm which still maintains the large nonlinearity associated with organics.

3. THIRD ORDER DEVICES

The investigation and application of third order nonlinearities in organic materials is not as advanced as it is for the second order nonlinearities. Here there are basically two types of devices possible. Near or on resonance, the nonlinearity is large, but so also is the attenuation coefficient. Here phenomena such as bistability and bleaching of the absorption are of principal interest. The non-resonant response is most useful in waveguide configurations for switching devices etc.

The bleaching of the absorption, which can be useful for optical limiting, has been studied for silicon naphthalocyanine.[23] The results indicate a nonlinearity relaxation time of 5 ns and a saturation intensity of 200 KW/cm^2 which begins to compare favorably with values of 550 W/cm^2 for exciton bleaching and 4.4 KW/cm^2 for the bandfilling mechanism in MQW GaAlAs.[24] An attractive feature of this organics result is that the absorption can be almost completely bleached out, in contrast to previous studies of semiconductors.

Bistability has also been reported for organic materials placed in a Fabry-Perot cavity.[23,25,26] Hysteresis loops observed for a thin film of silicon naphthalocyanine have been interpreted as due to absorptive bistability.[23]

Other reports of etalon bistability have been due to thermal effects.[25, 26] Bistability has also been measured in waveguide resonators based on MNA,[27] and again the origin of the effect was thermal nonlinearities.

There has been more activity in the measurement and application of third order nonlinearities in waveguides, specifically in grating couplers, distributed feedback structures and nonlinear directional couplers. The geometries are shown schematically in Figure 2.

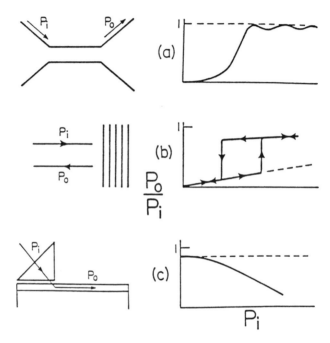

Figure 2   Three all-optical waveguide structures investigated in organic materials to date and their theoretical responses. (a)The nonlinear directional coupler. (b) The nonlinear distributed feedback grating. (c) The nonlinear prism (or grating) input coupler.

The efficient coupling of radiation into a planar (usually thin film) waveguide via a grating (or a prism) requires that the sum of the wavevector components of the grating, incident light and guided mode parallel to the surface be zero. For nonlinear waveguides, the guided wavevector changes with guided wave power leading to a power-dependent reduction in coupling efficiency and change in the optimum coupling angle. Experience has shown that both a thermal and electronic nonlinearity can lead to qualitatively and quantitatively similar results. Optical limiting via electronic nonlinearities have now been reported in MNA,[28] polimic acid,[29] poly 4BCMU[30] and DANS based side chain polymers[31]. To date, most nonlinear coupler experiments in non-organic materials have been based on thermal nonlinearities, with the exception of work on GaAs waveguides.[32]

Another example of a grating device is nonlinear distributed feedback grating. The Bragg condition for efficient reflection involves conservation of guided mode and grating wavevector, $2\beta(\omega) = \kappa$. When the guided mode wavevector changes with power, so also does the reflectivity of the grating structure. Here the only report on polydiacetylene films appears to be thermal in nature.[33] In fact, similar results in InSb films were shown to be due to a combination of nonlinear input grating coupler and a distributed feedback grating using a thermal nonlinearity.[34,35] Most recently, however, there has been a report of a nonlinear distributed feedback grating experiment in an optical fiber which is clearly due to an electronic nonlinearity.[36]

The final device studied to date in organic materials has been the nonlinear direction coupler. This was implemented in poly-4BCMU channel waveguides and studied at 1.06 μm.[37] Limited switching was obtained both due to thermal effects and due to two photon absorption, and switching due to non-resonant refractive nonlinearities is still to be demonstrated. The best results to date have been obtained in fiber waveguides in which electronic (albeit very weak) nonlinearities were used.[38,39]

The most common problem experienced with implementing third order guided wave devices in organic (as well as other) materials is thermal effects. That is, absorption causes a local temperature rise and hence index change via the thermooptic effect. There are two regimes. For times longer than the thermal relaxation time, typically microseconds, a thermooptic nonlinearity exhibits most of the characteristics of an electronic nonlinearity. For times shorter than the thermal relaxation time, usually achieved with pulsed lasers, the index change accumulates over the duration of the pulse. Therefore, the shorter the pulse (at the same peak power), the smaller the net index change and the smaller the effective nonlinearity. For reference, a typical effective thermal nonlinearity n_2 is tabulated (Table 3) versus the absorption (not total loss) coefficient and pulse duration time, using material parameters typical of polymers. The key point is that when studying and using non-resonant nonlinearities of the order of 10^{-12} cm^2/W or less, it is necessary to use picosecond pulses! This fact is not widely appreciated.

Table 3   **Effective thermal nonlinearitites for various absorption coefficients and pulse durations Δt**

n_2 cm^2/W (single pulse)

α (cm^{-1})	Δt = 1s	10 ns	10 ps
10^5	$10^{-(4\pm1)}$	$10^{-(6\pm1)}$	$10^{-(9\pm1)}$
1	$10^{-(9\pm1)}$	$10^{-(11\pm1)}$	$10^{-(14\pm1)}$

There are well-defined figures of merit by which the suitability of materials for all-optical switching devices can be judged.[40,41] They are both based on the nonlinear phase shift which can be obtained from a given material.   The maximum optically induced index change can be limited either by the physics of the material in the form of saturation or damage.   The effective propagation distance or device length is limited by the attenuation coefficient to $\simeq \alpha_t^{-1}$. Thus the maximum nonlinear phase shift is $\Delta n_{sat} \, 2\pi/\lambda \, \alpha_t^{-1}$ leading to the figure of merit W - $\Delta n_{sat}/\alpha_t \, \lambda$ (normalized to 2π.)   Similarly, when two photon absorption is large (i.e. $\alpha = \alpha_0 + \gamma$ I), the effective distance is limited by $(\gamma I)^{-1}$, and writing $\Delta n = n_2$ I, this leads to a figure of merit $T = 2\gamma\lambda/n_2$.   For reasonable device performance, $W > 1$ and $T < 1$ are needed.[40,42]

It is necessary to know the spectral dispersion of n_2 and γ to find wavelength regions in which the values of W and T are suitable.   Unfortunately the spectral dispersion in n_2 has not been measured or calculated for most materials and measurements of γ are even more rare.   The only complete measurement of γ in any material is reproduced for ply-4BCMU in Figure 3, reaching maximum values of γ can reach 100's cm/GW.[43]   It is clearly important to find spectral windows in which γ (and therefore T) is small.

Table 4 shows W and T at the few wavelengths where they are known for ultrafast nonlinearities.   (We do not include charge excitation nonlinearities due to their relatively long relaxation times.)   Although both W and T are promising for AlGaAs near the band gap, the absorption limits, the useful device length, the large dispersion in the index there leads to large pulse spreading, and charge excitation dominates the response after just a few pulses.[44]   Glasses at the moment are the most promising, but the nonlinearity is sufficiently small that long device lengths are still required.[45]   For organics, the numbers are promising but the spectral variation in the relevant parameters is just not available for useful assessment.[31, 46-48] This problem needs to be remedied over the next few years.

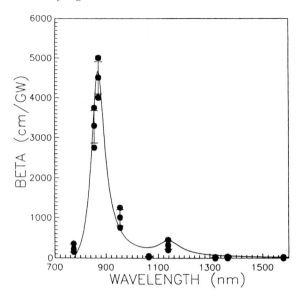

Figure 3   Two photon absorption spectrum (β instead of γ as in text) of poly-4BCMU. The solid line gives the best fit of this and third harmonic generation data (not shown) to a four-level model.[43]

Table 4   Material with ultrafast nonlinears and their all-optical switching figures of merit. An incident intensity of 1 GW/cm^2 was assumed.

Material	n_2 cm^2/W	α cm^{-1}	W	T	λ Microns
AlGaAs [44]	-4×10^{-12}	18	2.5	0.9	0.81
λ_g = 790 nm	-3×10^{-13}	0.5	7.1	11	0.85
PTS [46]	-3×10^{-11}	0.8	350	0.4	1.06
PTS [47]	-10^{-12}	0.8	13	23	1.06
poly 4BCMU[43]	-10^{-13}	0.2	3.5	127	1.06
	5×10^{-14}	1.7	0.2	1.3	1.3
DANS [48]	2×10^{-13}	< 1	> 1.4	$\cong$ 1	1.06
RN Glass [45]	1.3×10^{-14}	0.01	13	< 0.1	1.06

4. SUMMARY

There has been very rapid progress in the application of nonlinear organics to all-optical devices in the last two years. For doublers, organic based devices have overtaken those made from inorganic materials, primarily because of the very large nonlinearities available. The first devices based on third order nonlinearities in organics have been reported in the last two years and in the case of nonlinear grating couplers were superior to those made from other materials. It is expected that, within 2-5 years, devices based on intensity dependent refractive indices in organics will rival those in other materials.

This research was supported by AFOSR [88-0317] and NSF [ECS-8911960]

5. REFERENCES

1. W. Sohler and H. Suche, in Integrated Optics III, L.D. Hutcheson and G. Hall, eds, SPIE Proceedings, 1983, 408, 163.

2. G.J. Dixon, C.E. Tanner and C.E. Wieman, Opt. Lett., 1989, 14, 731.

3. R. Regener and W. Sohler, J. Opt. Am. B, 1988, 5, 267.

4. J.D. Bierlein, L.K. Cheng, Y. Wang and W. Tam, Appl. Phys. Lett., 1990, 56, 423.

5. B.F. Levine, C.G. Bethea, C.D. Thurmond, R.T. Lunch and J.L. Bernstein, J. Appl. Phys., 1979, 50, 2523.

6. I. Ledoux, D. Josse, P. Vidakovic and J. Zyss, Opt. Engin., 1986, 25, 202.

7. T. Uemiya, U. Uenishi, Y. Shimizu, S. Okamoto, K. Chikuma, T. Tohma and S. Umegaki, SPIE Proceedings , 1989, 1148, 207.

8. P. Kerkoc, Ch. Bosshard, H. Arend and P. Gunter, Appl. Phys. Lett., 1989, 54, 487.

9. K. Sasaki, T. Kinoshita and N. Karasawa, Appl. Phys. Lett., 1984, 45, 333.

10. T. Kondo, N. Hashizume, S. Miyoshi, R. Morita, N. Ogasawara, S. Umegaki and R. Ito, SPIE Proceedings, in press, 1337.

11. Ch. Bosshard, M. Kupfer, P. Gunter, C. Pasquier, S. Zahir and M. Seifert, Appl. Phys. Lett., 1990, 56, 1204.

12. J.E. Sohn, K.D. Singer, M.G. Kuzyk, W.R. Holland, H.E. Katz, C.W. Dirk, M.L. Schilling and R.B. Comizzoli, Proceedings of NATO Advanced Research Workshops on "Nonlinear Optical Effects in Organic Polymers," Series E: Applied Sciences, (Kluwer Acad. Pub., London, 1989), 291-7.

13. G. Khanarian, R.A. Norwood, D. Haas, B. Feuer and D. Karim, Appl. Phys. Lett.,1990, 57, 977.

14. P.K. Tien, R. Ulrich and R.J. Martin, Appl. Phys. Lett., 1970, 17, 447.

15. T. Tanuichi and K. Yamamoto, Oyo Buturi, 1987, 56, 1637.

16. K. Chikuma and S. Umegaki, J. Opt. Soc. Am. B, 1990, 7, 768.

17. T. Suhara and H. Nishihara, "Theoretical Analysis of Waveguide Second Harmonic Generation Phase-Matched with Uniform and Chirped Gratings", J. Quant. Electron., In press.

18. E.J. Lim, M.M. Fejer and R.L. Byer, Electron. Lett., 1989, 25, 174.

19. J. Webjorn, F. Laurell and G. Arvidsson, IEEE J. Lightwave Techn., 1989, 7, 1597 (1989); J. Webjorn, F. Laurell and G. Arvidsson, CLEO Digest, 1989, paper PD10.

20. G.L.J.A. Rikken, C.J.E. Seppen, S. Nijhuis and E. Staring, SPIE Proceedings,1337, in press.

21. D. Hulin, A. Migus, A. Antonetti, I. Ledoux, J. Badan and J. Zyss, Appl. Phys. Lett., 1986, 25, 3278.

22. M.A. Mortazavi, D. Yankelvich, A. Dienes, A. Knoesen, S.T. Kowel and S. Dijaili, Appl. Optics, 1989, 28, 3278.

23. J.W. Wu, J.R. Helfin, R.A. Norwood, K.Y. Wong, O. Zamani-Kamiri and A.F. Garito, J. Opt. Soc. Am. B, 1989, 4, 707.

24. D.S. Chemla, D.A.B. Miller and P.W. Smith, Opt. Engin., 1985, 24, 556.

25. W. Blau, Opt. Commun., 1987, 64, 85.

26. T.G. Harvey, W. Ji, A.K. Kar, B.S. Wherrett, D. Bloor and P.A. Norman, CLEO Digest, 1990, paper CTUH67, 146-8.

27. S. Ura, Y. Hida, T. Suhara and H. Nishihara, "Bistable Behavior of Fabry-Perot Resonator Constructed with MNA-Diffused ADC Polymer Waveguide", unpublished.

28. M.J. Goodwin, C. Edge, C. Trundle and I. Bennion, J. Opt. Soc. Am., 1988, 5, 419.

29. R. Burzynski, P. Banhu, P. Prasad, R. Zanoni and G.I. Stegeman, Appl. Phys. Lett., 1988, 53, 2011.

30. M. Sinclair, D. McBranch, D. Moses and A.J. Heeger, Appl. Phys. Lett., 1988, 53, 2374.

31. M.B. Marques, G. Assanto, G.I. Stegeman, G.R. Mohlmann, E.W.P Erdhuisen and W.H.G. Horsthuis, "Intensity Dependent Refractive Index of Novel Polymer Materials: Measurements by Nonlinear Grating Coupling", Appl. Phys. Lett., submitted.

32. Y.J. Chen, G.M. Carter, G.J. Sonek and J.M. Ballantyne, Appl. Phys. Lett., 1986, 48, 272.

33. K. Sasaki, K. Fujii, T. Tomioka and T. Kinoshita, J. Opt. Soc. Am., 1988, 5, 457.

34. G. Assanto, J.E. Ehrlich, and G.I. Stegeman, Opt. Lett., 1990, 15, 411.

35. J. Ehrlich, G. Assanto and G.I. Stegeman, Appl. Phys. Lett., 56, 602-4.

36. S. LaRochelle, Y. Hibino, V. Mizrahi and G.I. Stegeman, Electron. Lett., 1990, 26, 1459.

37. P.D. Townsend, J.L. Jackel, G.L. Baker, J.A. Shelbourne III, S. Etemad, Appl. Phys. Lett., 1989, 55, 1829.

38. S.R. Friberg, A.M. Weiner, Y. Silberberg, B.G. Sfez and P.S. Smith, Opt. Lett., 1988, 13, 904.

39. S. Trillo, S. Wabnitz, W.C. Banyai, N. Finlayson, C.T. Seaton, G.I.
     Stegeman and R.H. Stolen, IEEE J. Quant. Electron., 1989, QE-25, 104.

40. G.I. Stegeman, C.T. Seaton, A.C. Walker, C.N. Ironside and T.J. Cullen,
     Optics Commun., 1987, 61, 277-81.

41. V. Mizrahi, K.W. DeLong, G.I. Stegeman, M.A. Saifi and M.J.
     Andrejco, Opt. Lett., 1989, 14, 1140.

42. K.W. DeLong, K.B Rochfort and G.I. Stegeman, Appl. Phys. Lett., 1989,
     55, 1823.

43. W.E. Torruellas, K.B. Rochford, R. Zanoni and G.I. Stegeman, "The
     Cubic Susceptibility Dispersion of poly-4BCMU Thin Films: Third
     Harmonic Generation and Two Photon Absorption Measurements", Opt.
     Comm., submitted.

44. K.K. Anderson, PhD Thesis, M.I.T., 1989; M.J. Lagasse, K.K.
     Anderson, C.A. Wang, H.A. Haus and J.G. Fujimoto, Appl. Phys. Lett.,
     1990, 56, 417.

45. D.W. Hall, M.A. Newhouse, N.F. Borrelli, W.H. Dumbaugh and D.L.
     Weidman, Appl. Phys. Lett., 1989, 54, 1293 (1989); D.L. Weidman, J.C.
     Lapp, and M.A. Newhouse, Digest of the "Nonlinear Optics: Materials,
     Phenomena and Devices", paper MP17, pgs. 45-46.

46. D.M. Krol and M. Thakur, Appl. Phys. Lett., 1990, 56, 1406.

47. S.T. Ho, M. Thakur and A. Laporta, IQEC Digest, 1990, paper QTUB5,
     40-2.

48. K.B. Rochford, R. Zanoni, G.I. Stegeman, W. Krug, E. Miao and M.W.
     Beranek, "Measurement of Nonlinear Refractive Index and Transmission
     in Polydiacetylene 4BCMU Waveguides at 1.319 μm", Appl. Phys. Lett.,
     in press.

Application of Electro-optic Polymer Waveguides to Optical Interconnects

G.F. Lipscomb, R.S. Lytel, A.J. Ticknor, T.E. Van Eck, D.G. Girton, S.E. Ermer, J. Kenney, E. Binkley, and S.L. Kwiatkowski

LOCKHEED RESEARCH AND DEVELOPMENT DIVISION, DEPARTMENT 9720, BUILDING 202, 3251 HANOVER ST., PALO ALTO, CALIFORNIA 94304, USA

1. INTRODUCTION

In this paper we report on our initial progress toward the development of electro-optic polymers and devices for application to optical interconnects. We have selected a concurrent engineering approach, in which materials development and device development are pursued simultaneously with both the starting and ending points being the system application and end-use. Therefore, one specific application, optical interconnects for multi-chip modules, will be discussed. Optical interconnects are of particular interest, because, in addition to the other benefits of optics, optical multi-chip modules also require very high levels of integration, which are likely achievable only with polymer based waveguides. The fabrication and initial test results on two classes of integrated optic devices for optical interconnects based on E-O polymer materials will be described. The first is an optical railtap for the distribution of many different optical signals from a single CW laser diode, and the second is a traveling wave Mach-Zehnder integrated optic modulator, which was tested at modulation frequencies up to 8 GHz using direct detection and up to 20 GHz using heterodyne detection. Electro-optic polymer materials supplied by Akzo Research, BV, were used in the devices described.

Organic and polymeric materials and devices have been the center of intense scientific and engineering investigation for many years due to the extraordinary nonlinear optical and electro-optic properties of certain conjugated π-electron systems and due to the fundamental success of molecular engineering in creating new materials with appropriate linear optical, structural and mechanical properties.[1] Organic electro-optic (E-O) materials offer potentially significant advantages over conventional inorganic electro-optic crystals, such as $LiNbO_3$ and GaAs, in several key areas of integrated optics technology, including primary E-O parameters, processing technology and fabrication technology.[2,3,4] Equally as important as the primary E-O properties, polymeric integrated optic materials offer far greater fabrication flexibility and processing simplicity than current titanium indiffused $LiNbO_3$ waveguide technology, which requires processing at temperatures approaching 1000°C after expensive and difficult crystal growth. In organic electro-optic materials a nonlinear optical moiety is included in a guest/host or polymer system with appropriate linear optical, mechanical and processing properties, and the desired symmetry is artificially created through electric field poling.[5,6] These materials can then be simply and rapidly spin coated into high quality thin films, processed with standard photolithographic techniques and poled quickly and efficiently. In addition, channel waveguides and integrated optic circuits can be defined by the poling process itself[7], by photochemistry of the E-O polymer[8,9], or by a variety of well understood micro-machining techniques.[10] Furthermore, unlike $LiNbO_3$ where the device structure is limited to one side of a single crystal, multi-layer integrated optic waveguide structures can be fabricated in much the same manner as multilayer multi-chip module substrates. The fabrication flexibility of organic E-O polymers, coupled with standard photolithography and fabrication processes, potentially makes possible far higher integration densities than are achievable with any inorganic integrated optic technology.

Finally, a note of caution must always be interjected into the comparison between the "potential" advantages of organic E-O materials and the demonstrated performance of $LiNbO_3$ based devices. $LiNbO_3$ integrated optics is a fairly mature technology, and, after over 25 years of development, the performance levels achievable are well known. The potential advantages of polymer based integrated optic devices still remain to be achieved in actual devices and systems.

2. OPTICAL INTERCONNECTION

The switching speeds of individual electronic circuits in silicon and GaAs now reach well into the hundreds of megahertz and even the gigahertz regimes. The electrical high density interconnect multi-chip module (MCM) is a new electronic interconnect packaging technology that addresses current generation high speed chips and is just now entering systems implementation. As the package sizes become larger, however, electronic interconnects have difficulty handling the high data rates and interconnection densities required by emerging high speed integrated circuit technology, and optical interconnects become attractive. Optical signals can carry information great distances with minimal loss, crosstalk and excess noise. Consequently, the development of passive optical interconnect technologies that are based on passive polymer waveguides and are compatible with current and projected high speed microelectronics is now underway. In passive optical interconnects, fixed optical waveguides capable of very high speeds and high interconnect densities with minimal crosstalk and attenuation replace the conventional metal interconnections.

The incorporation of active electro-optic polymer switching elements with electronic multichip modules (MCM) and passive optical interconnections may enable optical multi-chip modules (OMCM), as shown schematically in Figure 1. The substrate is overlaid with a multi-layer structure of active and passive organic polymer waveguides, metallic control electrodes, signal paths and insulation layers. In addition to electronic ICs, laser diodes and detector arrays are also mounted on the substrate and electrically connected to the optical devices through very short tabs. Each functional element can be fabricated out of the material with the best performance and using the most efficient process technology, and then efficiently integrated together. A key component of OMCM technology, invented at Lockheed, is an optical rail with active electro-optic taps, which greatly simplifies the process of converting the electrical signals to optical form for distribution throughout the package. This conversion step is currently one of the greatest impediments to achieving an optical MCM. The design, initial fabrication and test of a optical railtap is described below. Electro-optic polymer technology may make possible the incorporation of both electronic and optical functionalities in a single highly compact structure, provided the organic materials can meet the thermal and reliability requirements of the electronics industry.

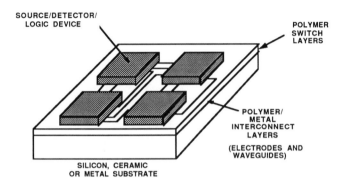

Figure 1. Schematic Diagram of an Optical Multi-Chip Module Based on Electro-Optic Polymers.

We have conducted a detailed systems analysis comparing an electrical MCM with an optical MCM for several specific examples.[11] The analysis included design of the output and input buffers on the ICs, the IC technology, the power required to drive the railtap modulators, the power required to drive the laser diodes, the optical power budget in the interconnect network, and the design of the detector and amplifier to regenerate the logic level. For data rates of 50 MHz and above optical interconnects would require significantly less power than electrical interconnects for either CMOS or ECL devices. Above 500 - 1000 MHz electrical interconnects become increasingly more difficult and will not be possible without expensive custom microwave transmission line design. At what frequency, and even whether, optical MCMs become cost competitive with electrical interconnects will be determined by how readily organic electro-optic materials fit into existing electronic fabrication and packaging processes and end-use environments.

3. MATERIALS REQUIREMENTS

The achievement of optical multi-chip modules will require the hybrid assembly of Silicon and/or GaAs integrated circuits, E-O polymer devices, diode lasers, photodetectors, optical fibers, passive polymer dielectrics, metallic power and ground connections and ceramic packages. The E-O polymer devices and optical interconnects must enhance the performance of the electronic system and not impose additional constraints on the fabrication, assembly or use of the electronic assembly. The product performance levels required of such assemblies are already well known through experience with telecommunications optical receiver and transmitter modules and military hybrid microelectronics. The specifications for military applications are given by Mil. Spec. 883c Level 2, and require use to 125°C and storage to 200°C. The generally accepted requirements for commercial applications are only slightly less stringent than those for military applications. Perhaps the most difficult requirements to meet, however, are those imposed by the standard assembly processes used for electronic modules. While the times are relatively short, on the order of minutes, the temperatures reached in packaging and die attach are quite stressing, above 300°C. While these requirements might be reduced somewhat by using nonstandard assembly procedures, this will greatly increase the cost of the product and significantly reduce the likelihood that organic materials can compete on a cost/performance basis.

Films of electro-optic polymers can be formed quickly and inexpensively by spin coating, spray coating, or dip coating, but are amorphous as produced and exhibit no second-order electro-optic effects. These materials must then be processed by electric-field poling[5] to achieve a macroscopic alignment in order to enable second-order nonlinear optical effects. In brief, the E-O polymer material is heated above its glass transition temperature, a high electric field is applied by electrodes or a corona to partially align the nonlinear molecules in the direction of the field, and the material is cooled back to room temperature under the influence of the electric field. This process "freezes" in the molecular orientation and creates a macroscopic electro-optic coefficient in the material. The process of electric field poling is both the great advantage of E-O polymers and the great unknown as to their ultimate practicality. The only way to induce electro-optic effects in inorganic materials such as $LiNbO_3$ is to find a naturally non-centrosymmetric crystal and engage in expensive single crystal growth. By comparison, electric field poling is quick and inexpensive and allows molecules with extremely large nonlinearities to be used without regard to the capricious way in which they crystallize. The catch is that the poled state is, of course, not permanently "frozen" in, but is thermodynamically unstable and will decay back to the randomly aligned state with no E-O effect after the passage of enough time. This decay time can be very long and not interfere with the device lifetime, but it is also highly temperature dependent. As the temperature approaches the glass transition temperature the polymer diffusion rates increase greatly and the decay becomes rapid. The current generation of E-O materials with glass transition temperatures of 100 - 150°C are clearly inadequate for electronic systems applications. This fact has been widely recognized, and much effort is now directed toward higher temperature stable E-O polymer materials.[12] The thermal stability can be improved without detriment to the E-O response because the E-O effects arise in the NLO moiety while the thermal properties and alignment dynamics are determined mainly by the backbone or host polymer, provided, of course, that the NLO moiety can itself withstand the thermal cycles.

4. DEVICE RESEARCH

Because the factors affecting the long term thermal stability of the E-O polymer materials are effectively decoupled from the E-O response mechanisms, much can be learned by fabricating and testing prototype integrated optic devices using current generation E-O polymers. In a concurrent engineering approach materials development must be intertwined with device development and, of course, always guided by the systems application and ultimate end-use. In this section we report on the fabrication and initial test of two classes of devices for optical interconnects. The first device type, an optical railtap, is a new device invented at Lockheed for the distribution of many optical information channels from a single CW laser diode. Initial proof-of-concept devices were fabricated and tested to demonstrate the optical functionality required for optical interconnection. The second device type is a standard architecture integrated optic Mach-Zehnder interferometric modulator. Sample devices with traveling wave electrodes were fabricated and tested in order to determine the materials performance parameters for E-O polymers and high speed modulation.

RAILTAP

A unique solution to the problem of converting the electrical signal to optical form in an optical interconnect is provided through the use of an optical rail and a sequence of railtaps, as shown schematically in Figure 2. The optical railtap is a key enabling component of optical interconnect technology and is made possible by the inclusion of active electro-optic polymers to make the interconnection network itself active.[13] A passive optical waveguide, or rail, routs optical power around the package and acts as an optical power supply. The optical rail runs near the edge of every IC in the package and an optical tap switch is connected to every output pin. The purpose of the railtap is to convert the electrical data stream coming off of the IC output pin into a series of bursts of light representing "1"s and "0"s and to place the optical data stream onto an optical channel that routes it to a receiver, which drives an input pin of the receiving IC. The railtap is a compact electro-optic modulator that is fabricated using electro-optic polymer materials and is driven directly by the IC output electrical signal.

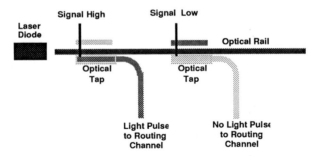

Figure 2.   Schematic View of a Railtap for Optical Interconnects.

Several active and passive railtap devices, with up to five taps on a single rail, were fabricated and tested.[14] The dimensions of one such device are shown schematically in Figure 3. This device was diced such that a single tap remained on the central rail. For these experiments an electro-optic polymer material supplied by Akzo Research BV[8,15] was used, and the waveguides were formed by the technique of photobleaching.[8,9] Independent contact was made to each of the electrodes over the channels. The device was inherently slow because the metal electrodes were far larger than necessary and very thin leading to larger resistance and capacitance and thus a lower RC roll-off frequency.

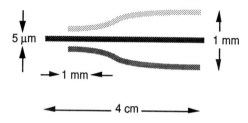

Figure 3. Dimensions of a Sample Single Element Railtap.

The railtap was driven asymmetrically and modulated in a complementary fashion. Shown in Figure 4 are the intensity line scans of the output endface of the 3 port railtap. In each case the central spot is the optical rail, the signal line is on the right and the complementary signal is on the left. In the righthand picture the signal is applied and a strong beam appears in the signal channel, with almost no light in the complementary channel. In the lefthand picture, the complement of the signal is applied and almost no light appears in the signal channel, while a strong peak appears in the complementary channel. In both cases, the optical power left in the central rail is almost unchanged. This is a critical feature for multiple tap rails, since noise depending on the logic state of the upstream taps will not be imparted to the downstream taps. Approximately 6 dB of modulation depth is observed.

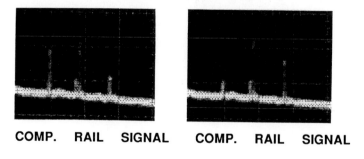

COMP. RAIL SIGNAL   COMP. RAIL SIGNAL

Figure 4. Intensity Line Scan Of the Railtap Endface Showing Complementary Modulation.

The railtap was designed to be only 1mm long for modulation of 1-5% of the light in the central rail and was overdriven to achieve the modulation depth above the leakage background shown in Figure 4. The off-state leakage is due to design, process and fabrication variations leading to deviations from the optimal values of the materials and device parameters, such as waveguide width and index difference. The beam propagation design code was then rerun after fabrication, using the best estimate of the device parameters actually achieved. The results are given in Figure 5, showing a modulation depth close to that actually observed in Figure 4. In addition, this model can be used to estimate the electro-optic coefficient of the polymer. The calculated induced electro-optic coefficient is $r_{33} = 14 \pm 4$ pm/V, in good agreement with that expected from direct measurements on the pure polymer for these poling conditions. This both verifies the predictions of the beam propagation code and indicates that the presence of buffer layers during poling does not greatly alter the induced electro-optic coefficient.

Quantitative measurements of modulation depth and frequency response of the railtap were made by coupling the channel waveguide output to an optical fiber and transmitting it to a photodetector. When a low frequency 20 V voltage swing was applied, a modulation depth of 25% was observed. The frequency response of the railtap was measured with an HP3577A low frequency network analyzer, which has a response from 5 Hz to 200 MHz. Optical fibers were used to endfire couple into and from the channel waveguides and were not permanently attached to the device. The railtap response is flat from 50 Hz all the way out to approximately 20 MHz. Although at this point the RC roll-off of the electrode structure begins to reduce the signal, a clear signal is seen out to 100 MHz.

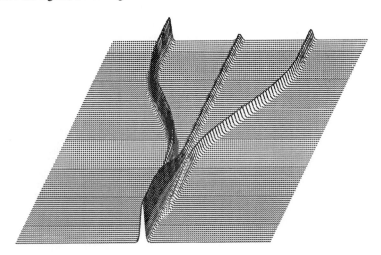

Figure 5. Beam Propagation Model of the Railtap Using Estimated Parameters.

INTEGRATED OPTIC MACH-ZEHNDER MODULATOR

Standard Mach-Zehnder integrated optic modulators were fabricated with impedance matched 50Ω traveling wave electrodes in order to investigate the polymeric E-O materials parameters at multi-GHz frequencies.[16] As before an E-O polymer material supplied by Akzo Research, BV, was used,[8,15] and the waveguides were fabricated by photobleaching[8,9]. Schematic diagrams of the device and microstrip configuration are shown in Figure 6. Since the waveguide stack is 10 μm high, the drive electrodes were 20 μm wide to achieve a 50 Ω impedance. In order to reduce the resistance in the microstrip, gold was electroplated to the target thickness of 10 μm. After the 3" wafer was fabricated, it was diced up and the individual modulators were separated for testing. The final devices were on Silicon pieces 3.5 cm long and 1 cm wide. The modulators were then mounted onto a holder and the center tabs from the input and output electrical SMA connectors were connected to the bonding pads of the electrode structure. The electrical impedance of the microstrip electrode was measured with an HP 8510B network analyzer. For one typical device at 11 GHz the measured impedance was 45.9 Ω + i 3.6 Ω, very close to the nominal 50 Ω, and the electrical throughput was down -6.7 dB. On some devices a second microstrip electrode was placed over the other arm of the Mach-Zehnder for heterodyne detection modulation measurements.

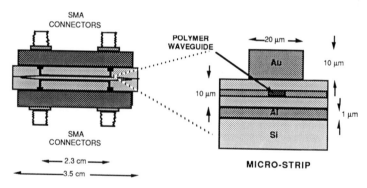

Figure 6. Diagram of Mach-Zehnder Modulator with Cross-section of Microstrip Electrode.

Measurements on a low frequency HP3577A network analyzer showed the device response to be flat to 200 MHz. In order to measure the higher frequency response, the device was driven with an HP8672A/HP86720A frequency generator and the signal was measured with an HP8560B spectrum analyzer. The detector electronics consisted of an Optoelectronics PD-50 photodiode with an Avantek AGT 8235 preamplifier and had an aggregate bandwidth of approximately 2 to 8 GHz. Both the detector electronics and the laser were carefully shielded from rf pickup. The microstrip was terminated off-chip into 50Ω. Shown in Figure 7 are response measurements for one device made at discrete frequencies from 4 to 9 GHz. A clear modulated signal well above the noise floor is observed out to 8

GHz. This frequency represents the extreme limit of the response of the current detector electronics and the roll-off is not due to the E-O polymer material.

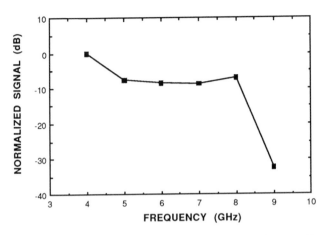

Figure 7. Response of an E-O Polymer Mach-Zehnder Modulator vs. Frequency.

In order to determine the response of the polymer electro-optic material at frequencies above the current detector bandwidth, a heterodyne detection modulation experiment was carried out.[16] One arm of the Mach-Zehnder was modulated at 20 GHz and the second arm was modulated at 18.5 GHz. The difference frequency at 1.5 GHz was then detected using an Antel APD with a 2.0 GHz bandwidth, providing a proof-of-concept demonstration of electro-optic modulation in a polymer material at 20 GHz.

5. SUMMARY AND CONCLUSIONS

We have demonstrated many critical aspects of key components for E-O polymer based optical interconnect systems for electronic applications. A preliminary understanding of the E-O polymer materials parameters that will have to be met in an electronic environment has been achieved, and new devices, made possible by the unique properties of E-O polymers, have been conceived, designed and implemented. Several device demonstrations were carried out using an electro-optic polymer supplied by Akzo Research, B.V.[8,15] Passive railtaps were fabricated and tested, and active railtaps with "slow" electrodes were fabricated and operated in both a symmetric and complementary fashion, providing a proof-of-concept demonstration of the modes critical for optical interconnect applications. Modulation depths of 6 dB (optical) were achieved and the frequency response was observed to be flat to 20 MHz and clear modulation was seen at 100 MHz. Standard Mach-Zehnder interferometer modulators with high speed microstrip electrodes were designed, fabricated and tested. These devices demonstrated flat material and device response out to 200 MHz on a low frequency testset. The devices also showed clear modulation on a high frequency testbed out to 8.0 GHz using direct detection and out to 20 GHz using heterodyne detection. This experiment provides a proof-of-concept demonstration that organic electro-optic materials can be used for modulation in the multi-GHz regime.

6. ACKNOWLEDGEMENTS

The research reported in this paper is due to the Lockheed Photonic Switch and Interconnect Group, and is hereby acknowledged with warmth and gratitude. Special acknowledgement is given to Dr. G.R. Mohlmann, Akzo Research, BV, for his participation in the Lockheed research program.

7. REFERENCES

1 Nonlinear Optical Properties of Organic Molecules and Crystals, Vol. 1 and 2, D. Chemla and J. Zyss, ed. (Academic Press, FLA) 1986.
2 R. Lytel, G.F. Lipscomb, and J.I. Thackara, "Recent Developments in Organic Electro-optic Devices", in Nonlinear Optical Properties of Polymers, A.J. Heeger, J. Orenstein, and D.R. Ulrich, ed., Proc. Materials Research Society Vol. 109, 19 (1988).
3 S.J. Lalama and A.F. Garito, "Origin of the Nonlinear Second-order Optical Susceptibilities of Organic Systems", Phys. Rev. A 20, 1179 (1979)
4 K.D. Singer and A.F. Garito, "Measurements of Molecular Second-order Optical Susceptibilities Using DC Induced Second Harmonic Generation", J. Chem. Phys. 75, 3572 (1981).
5 K. D. Singer, J.E. Sohn, and S.J. Lalama, Appl. Phys. Lett. 49, 248 (1986), and K.D. Singer, M.G. Kuzyk and J.E. Sohn, "Second-Order nonlinear-optical processors in orientationally ordered materials: relationship between molecular and macroscopic properties", J Opt. Soc. Am. B4, 968 (1987).
6 D.J. Williams, "Nonlinear Optical Properties of Guest-Host Polymer Structures", in Nonlinear Optical Properties of Organic Molecules and Crystals, Vol. 1, D. Chemla and J. Zyss, ed., Academic Press, NY (1987), p. 405.
7 J.I. Thackara, G. G. Lipscomb, M.A. Stiller, A.J. Ticknor and R. Lytel, Applied Physics Letters 52, 1031 (1988)
8 G. R. Mohlmann, W.H. Horsthuis, C.P. van der Vorst, "Recent Developments in Optically Nonlinear Polymers and Related Electro-Optic Devices," Proc. SPIE 1177, 67 (1989)
9 J. Yardley, ACS Spring Meeting, Boston (1990) to be published
10 K. D. Singer, W. R. Holland, M.G. Kuzyk, G. L. Wolk, H.E. Katz, M.L. Schilling, "Second Order Nonlinear Optical Devices in Poled Polymers," Proc. SPIE 1147, 233 (1989)
11 R.S. Lytel, J.T. Kenney, E.S. Binkley and G.F. Lipscomb, to be published.
12 See for example R.S. Lytel, "Applications of Electro-optic Polymers to Integrated Optics", Proc. SPIE 1216-04, Los Angeles, Jan. 1990, to be published; and S.E. Ermer et. al., Proc. SPIE, Vol. 1337, San Diego, July 1990, in press.
13 R.S. Lytel , A.J. Ticknor, Patent Pending, and R.S. Lytel, A.J. Ticknor, T.E. Van Eck, G. F. Lipscomb to be published.
14 T.E. Van Eck, A.J. Ticknor, R.S. Lytel and G.F. Lipscomb, to be published.
15 G. R. Mohlmann et al., Proc. SPIE, Vol. 1337, San Diego, July 1990, in press.
16 S.L. Kwiatkowski, D.G. Girton, G.F. Lipscomb and R.S. Lytel, to be published.

Optical Waveguides in Organic Materials: Results and Implications for Third-order Devices

N.E. Schlotter

BELL COMMUNICATIONS RESEARCH, RED BANK, NEW JERSEY, USA

INTRODUCTION

The realization of all-optical switching using organic materials has yet to be achieved. However, significant progress has been made toward the goal. Methods for producing device structures that are compatible with current technologies have been developed. Prototype optically nonlinear organic materials have been tested in these structures. Although the results of the test have shown that further materials development is needed, it is now possible to rapidly test promising materials when they become available. This paper will cover the development of optical waveguides in polydiacetylene materials, the patterning of the materials to form channel guide structures, and the testing of directional couplers in the laboratory. A general review of all-optical switching has been written by Stegeman and Wright[1].

ASYMMETRIC SLAB WAVEGUIDES

An asymmetric slab waveguide (ASW) is a three layer structure where the refractive index of the middle layer is higher than the enclosing layers (for a guiding structure). Should the enclosing layers have the same index, the guide becomes symmetric, which is simply a variant. The asymmetric slab waveguide is of importance because it is one of the best structures for the initial testing of nonlinear optical (NLO) materials and forms the base structure for the formation of devices. Asymmetric slab waveguide structures have been prepared on a variety of substrates. Detailed discussions of the behavior of the ASW can be found in Marcuse[2].

For the purposes of this paper only a few concepts are necessary. First, for guiding layers only a few microns in thickness, only a small number of modes are available for the propagating beam. In geometrical optics these correspond to bounces between the upper and lower interfaces that maintain a constructive interference relationship for the beam. The parameters that control this behavior are the refractive indices of the layers, the "angle" at which the light hits the interfaces, the wavelength of the light, and the thickness of the guiding layer. Finally, when the incident angle of the beam with the interface is less than that for total internal reflection, the guide will no longer confine the beam, limiting the number of guided modes.

Injecting light into the guide is necessary; however, the guiding structure must be defeated locally for this to happen. There are a number of ways to accomplish this trick. First, prism coupling involves placing a high index prism on the guiding layer to reverse the total internal reflection direction. Second, grating coupling requires the fabrication of a transmission grating on the guiding layer. Appropriate grating orders can then be used to inject the beam into the guide. Third, if a sharp edge can be cleaved through the guide, then direct focusing can be used to "edge-fire" into the guide. Finally, the same cleaved geometry can be used with an optical fiber to "butt-couple" light into the guide by the direct contact of the fiber guide with the waveguide. For laboratory applications it is practical to use prism coupling, when possible, since it minimizes the amount of sample preparation and maintains single mode behavior. For devices to be used in applications, fiber coupled geometries will be chosen, since once formed they will be more robust and require minimal skill to install.

POLYDIACETYLENES

The polydiacetylenes (PDAs), poly(3BCMU) and poly(4BCMU), have been found to be processible due to their favorable solvent-side chain interactions. We have been able to spin cast thin films from them, similar to the technique used in preparing photoresist coatings for lithography[3,4].

Films of both PDAs show an anisotropy in their refractive index when measured in the plane versus

out-of-the-plane[4]. This is believed to be due to the
chains aligning in the plane as the film is dried. No
evidence of orientation due to the spin coating process
was observed.

The optical quality of the polymeric waveguide is
often a serious problem. A discussion of the contributing
factors can be found in reference 5. In the PDAs it is
often hard to process them into a useable form without
causing extensive oxidation to occur. With spin coating
the degraded material can be precipitated out just before
coating, thereby reducing this source of contamination.

CHANNEL WAVEGUIDE FABRICATION

Channel waveguides are formed when the guided light
in the ASW is also confined laterally. In polymeric
materials this can be done in several ways. Lithography
and beam writing techniques are natural approaches, but
the resulting sharp, and often ragged, edges may present
problems with optical propagation. Additionally, loss
from scattering and poor interface geometry may occur. An
alternate approach to lateral confinement that has a
smooth, graded index, edge involves patterning via
ion-exchange. The channel pattern in the glass substrate
is formed via diffusion of ions through a metal mask.
This patterned glass substrate is constructed so that the
channels cannot act as light guides independently. By
forming a composite guide structure with the polymer film
coat the complete waveguide is formed[6,7]. Calculation of
the intensity distribution in the composite guide shows
that approximately 80% of the propagating beam is in the
polymer layer[7]. Figure 1 shows a plot of the intensity in
the composite guide.

DIRECTIONAL COUPLERS

A directional coupler is an optical device in which
one guiding channel is placed near another guiding
channel. If the optical parameters are such that the
evanescent fields from the guides overlap, energy can be
transferred. With a properly adjusted interaction length
and spacing, the light in one guide can be nearly
completely transfered to the other guide. This can occur
with no nonlinear behavior involved. If nonlinear

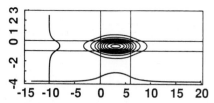

Figure 1 Mode intensity contours calculated by the
effective index method for an inverted rib
waveguide with a constant index rib structure.
The cross-sectional view has the epoxy, polymer,
and glass substrate ordered from top to bottom.
Dimensions are in microns.

materials are present the geometry can be arranged such
that the energy transfer cannot occur until a certain
minimum power is propagated in the channel. Or, if the
guide is designed to transfer at low power it will cease
to transfer at high power. One test design is shown in
Figure 2 where an array of spacings have been prepared to
test the optical behavior of the polymer composite
directional coupler[6,8]. Figure 3 shows the output from the
cleaved guide structure demonstrating the occurrence of
light transfer.

Figure 2 Directional coupler mask image showing the
transition region between 1 to 2 guides.

A figure of merit for the effects of variations in
the refractive index on the directional coupler
performance[9] is defined by the equation

$$B = (L \cdot L_{\sigma})^{\frac{1}{2}} \sigma_n / \lambda \qquad (1)$$

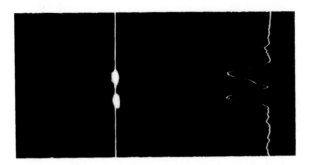

<u>Figure 3</u> Shows the imaged output from a directional
coupler prepared from the mask in Figure 2. The
guides are 6 μm wide and the space between them
is 9 μm.

where L is the device length, L_σ is the index
inhomogeneity correlation length, σ_n is the standard
deviation of the index variation, and λ is the
wavelength. For L = 1 cm, L_σ = 100 μm, λ = 1.3 μm, and
$\sigma_n = 0.5 \times 10^{-4}$, B = 0.038 and the resulting fluctuation in
the switching is seen in Figure 4. The dashed lines in
the plot are for various members of an ensemble of
directional couplers that have been numerically simulated.
If B increases by an order of magnitude to 0.38 the
degradation of the switching performance can be seen in
Figure 5. This can occur from a hundred fold increase in
the device length or correlation length, or from an
increase by 10 of the standard deviation of the index
variations. Note that the above standard deviation
corresponds to about a 1% deviation in the index.

NONLINEAR BEHAVIOR

The relevant device parameter in all optical
switching is the intensity-dependent component of the
refractive index given by

$$n(\omega) = n_0(\omega) + n_2 \cdot I \qquad (2)$$

The nonlinear refractive index term, n_2, is often
calculated from the expression

$$n_2 = 16\pi^2 X^{(3)} / c\epsilon_1 . \qquad (3)$$

which relates n_2 to $X^{(3)}$ (ref. 10) the third order term in the expansion of the polarizability. This is significant

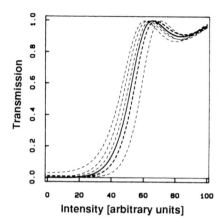

Figure 4 A range of individual responses for single-coupling length directional couplers. Parameters are stated in the text. The solid line is the perfect device response.

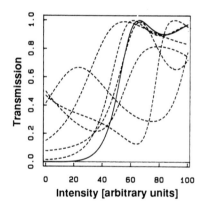

Figure 5 Similar to the plot in Figure 4, but B has been increased to 0.38.

since $X^{(3)}$ is the easiest and, therefore, most often measured parameter. Note, however, that this expression is valid for non-resonant $X^{(3)}$s only. In the PDAs there are a large number of resonances that can give additional contributions to $X^{(3)}$ output. Figure 6 shows a free electron laser study of $X^{(3)}$ from a poly(4BCMU) film in which the non-resonant $X^{(3)}$ is found in measurements beyond 2 μm (from refs. 11,12). Resonance values are 30-40 times the non-resonant $X^{(3)}$ value.

Using the previously developed directional coupler structure, the nonlinear response of poly(4BCMU) films have been tested at 1.06 μm[8]. In this case it was determined that there is a strong two photon absorption (TPA) response. Figure 7 shows the increase in the

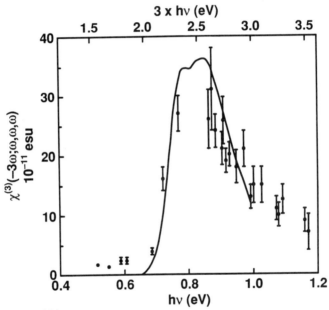

Figure 6 $X^{(3)}$ output from a poly(4BCMU) film irradiated with tunable light from a free electron laser. The non-resonant $X^{(3)}$ is shown near 0.6 eV (about 2.0 μm).

transmission loss as a function of increasing power in the guide. Even more intriguing is that the TPA effect also has an apparent switching behavior. However, this has been satifactorily modeled as an effect of the nonlinear absorption and isn't really switching[8]. Finally, at repetition rates comparable to the thermal relaxation rate thermal switching is observed. This is shown in Figure 8. The pulse energy is below that for observing TPA, but at 82 MHz switching occurs. However, if the repetition rate is lowered to 1 kHz the peak power can be raised above that in the previous experiment with no switching occurring. No switching was observed that could be attributed to the intensity dependent component of the refractive index, n_2. It should be noted that measurements on a PTS single crystal claiming considerably higher values of n_2, have been reported[13,14] as well as electronic switching[15] for the conditions: absorption coefficient of 0.5 cm^{-1}, 100 ps pulses, and a repetition rate of 76 MHz at 1.06 μm, in disagreement with the above results on poly(4BCMU). Note the the absorption loss is approximately the same as for the BCMUs.

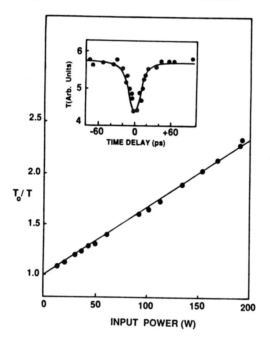

Figure 7 Power-dependent variation in the total transmission of a poly(4BCMU) directional coupler. The inset shows that the decrease in total coupler transmission occurs within the laser cross-correlation time and is therefore a fast electronic response.

CONCLUSIONS

Since the PDAs are typical of the conjugated materials these results suggest that it will be difficult to avoid multi-photon resonances. These can contribute to both loss by absorption and mis-estimation of the size of

n_2 if $X^{(3)}$ measurements are used as the basis of
estimation. Non-resonant measurements of $X^{(3)}$ suggest
that polymers may not be the best materials for all
optical switching if other criteria are included, such as
absorption loss and scattering. In fact, there is reason
to believe that the current improvements in semiconductor
lasers will favor going to inorganic glasses if new
polymeric systems can not be developed. Table 1 shows the
relative position of current materials based on the best
available data. The ratio of n_2 to α, the absorption
constant, is a figure of merit for nonlinear switching
performance.

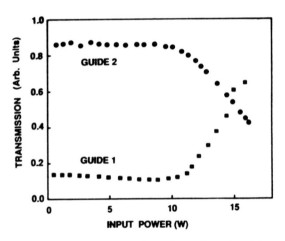

Figure 8 Power-dependent switching due to thermal change
          in the refractive index.

Table 1 Relative Performance of Materials for All Optical
         Switching

Material	Wavelength	Absorption	$n_2 \cdot$	n_2/α
	μm	α/cm	cm^2/MW $(\times 10^{-6})$	$(\times 10^{-6})$
Conjugated Polymers	1.06	1	1	1
GaAs	1.06	1	0.3	0.3
SF-59 Glass	1.06	0.006	0.007	1

ACKNOWLEDGEMENTS

I wish to thank Greg Baker, Shahab Etemad, Wun-Shain Fann, Janet Jackel, John Shelburne III, and Paul Townsend. Working in the NLO field demands a very multidisiplanary group. Their work has made this project much more successful, in the sense that the whole is greater than the sum of the individual parts.

REFERENCES

1. G. I. Stegeman and E. M. Wright, Optical and Quantum Electron., 1990 22, 95.
2. D. Marcuse, 'Theory of Dielectric Optical Waveguides', Chapter 1, Academic Press, New York, 1974, pp. 1-49.
3. P. D. Townsend, G. L. Baker, N. E. Schlotter, C. F. Klausener, and S. Etemad, Appl. Phys. Lett., 1988 53, 1782.
4. P. D. Townsend, G. L. Baker, N. E. Schlotter, and S. Etemad, Syn. Met., 1989 28, D633.
5. A. Tanaka, H. Sawada, T. Takoshima, and N. Wakatsuki, Fiber Integ. Opt., 1988 7, 139.
6. J. L. Jackel, N. E. Schlotter, P. D. Townsend, G. L. Baker, and S. Etemad, SPIE, 1988 971, 239.
7. N. E. Schlotter, J. L. Jackel, P. D. Townsend, and G. L. Baker, Appl. Phys. Lett., 1990 56, 13.
8. P. D. Townsend, J. L. Jackel, G. L. Baker, J. A. Shelburne, III, and S. Etemad, Appl. Phys. Lett., 1989 55, 1829.
9. J. L. Jackel, IEEE J. Quantum Electron., 1990 26, 622.
10. D. J. Sandman, G. M. Carter, Y. J. Chen, B. S. Elman, M. K. Thakur, and S. K. Tripathy, in 'Polydiacetylenes Synthesis, Structure, and Electronic Properties', eds. D. Bloor and R. R. Chance, Martinus Nijhoff Publishers, Boston, 1985, pp. 299-316.
11. W. S. Fann, PhD Thesis, "Applications of the Infrared Free Electron Laser in Nonlinear and Time Resolved Spectroscopy", Stanford University, 1990.
12. W.S. Fann, P. D. Townsend, S. Etemad, G. L. Baker, and J. L. Jackel, private communication, 1990.
13. D. M. Krol and M. Thakur, Appl. Phys. Lett., 1990 56, 1406.
14. M. Thakur, R. C. Frye, and B. I. Greene, Appl. Phys. Lett., 1990 56, 1187.
15. M. Thakur and D. M. Krol, Appl. Phys. Lett., 1990 56, 1213.

Non-linear Optic Polymer Thin Films in Waveguide Device Applications

J.B. Stamatoff, A. Buckley, A.J. East, H.A. Goldberg, G. Khanarian, R.A. Norwood, J.R. Sounik, and C.C. Teng

HOECHST CELANESE RESEARCH DIVISION, 86, MORRIS AVENUE, SUMMIT, NEW JERSEY 07901, USA

1 INTRODUCTION

Synthesis and characterization of NLO polymers is an established area of technology with a very substantial international effort. The focus of this paper will be the issues surrounding the application of these materials to waveguide devices in the near term and the use of similar design principles for the creation of new waveguide materials for the longer term.

NLO polymers essentially have two important near term application areas in waveguides. The first and largest area involves electro-optical applications and the second area involves optical frequency mixing (most notably frequency doubling). There have been a number of works[1-2] describing the electro-optical properties of NLO organic polymers. The advantages of these new materials for waveguides have been clearly defined; however, the difficulties which must be addressed have received less attention. Frequency mixing, on the other hand, has received far less study but this technology potentially has an equal possibility of broad application.[3]

With the same material design criteria, new functionalities may be imparted to organic polymers for use as waveguide materials in the future. For example piezoelectric and pyroelectric effects can be investigated for NLO polymers and new multifunctional polymers may be designed specifically to enhance one or several properties.

2 ELECTRO-OPTICAL WAVEGUIDE APPLICATIONS

The properties of one Hoechst Celanese NLO Polymer are given in Table 1.

Table 1: Properties of a Hoechst Celanese Polymer

Typical E-O Properties at 1.3 μm	Typical NLO Properties at 1.3 μm
n = 1.63	n = 1.63
$\varepsilon = 3.0$ (1-40 GHz)	$\chi^{(2)}(-2\omega;\omega,\omega) = 80 - 165$ pm/V
Slab Waveguide Loss = 0.5 dB/cm	
r = 20 - 40 pm/V	

These properties have been measured for a side chain amorphous polymer of the type shown in Figure 1, which has been specifically designed for NLO applications. In the figure, D represents a donor which may be, for example, sulfur, oxygen, or an amine. For electro-optical applications, these polymers have the advantage that

$$\sqrt{\varepsilon(\omega)} \cong n\,(\omega)$$

(1)

which is a general property of organic materials that permits very high frequency, large bandwidth applications.[4]

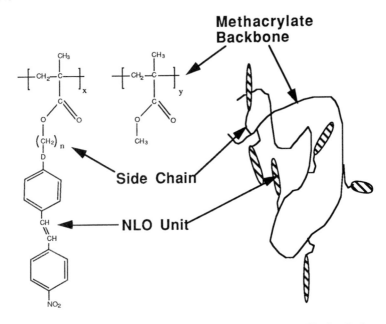

<u>Figure 1</u>: Structure of a Non-linear Optical Side Chain Polymer

For organics, the NLO (and electro-optical) response emanates from the non-linear polarization of delocalized electrons within the NLO chromophore.[5] In contrast to some inorganic materials, this response does not involve atomic motions which give rise to a significant discrepancy between the dielectric constants at optical and microwave frequencies. Accordingly, NLO polymers have only intramolecular polarization based electro-optical activity, and therefore the size of the total NLO effect ($\chi^{(2)}$ or r) is diminished. Currently, quite respectable activities have been achieved (e.g. Table 1), but these activities only represent equality with Pockels coefficients for inorganic materials. The important figure of merit for electro-optical materials is $n^3 r$, and that remains smaller than for $LiNbO_3$, the most widely used E-O material.

Table 2 gives a performance comparison of E-O polymers in a Mach-Zehnder modulator, including material and device relevant figures of merit. From a materials point of view, two fundamental modulator parameters are $V_\pi L$ and Lf which are the modulation voltage required to reach extinction times the length of the modulation region, and the length frequency product, respectively. Longer devices require lower V_π, but must operate at lower frequencies, f.

For a simple Mach-Zehnder design,

$$V_\pi L = \frac{G\lambda}{n^3 r} \tag{2}$$

and,

$$Lf = \frac{c}{\sqrt{\varepsilon} - n} \tag{3}$$

where   G is the effective gap between modulation electrodes
      λ is the wavelength of light
      ε is the dielectric constant of the electro-optical material
      n is the index of refraction of the material and
      c is the speed of light

In the table, G is taken as 12 μm for the organic, with a 1.0 overlap integral between optical and electrical fields for microstrip electrodes directly above the active organic layer while for LiNbO$_3$, G is taken as 16 μm with an 8 μm actual gap and a 0.5 overlap integral assumed between the optical and electrical fields for coplanar electrodes.

Table 2:   Performance   Comparison

	n	r pm/V	n^3r pm/V	$V_\pi L$ $V\text{-}cm$	fL cm/sec	$V_{min}{}^1$ V	$f_{max}{}^1$ GHz
NLO Polymer*	1.63	20	87	17.9	2.9×10^{11}	4.5	145
	1.63	40	173	9.0	2.9×10^{11}	2.3	145
$LiNbO_3$	2.15	31	308	6.75	8.2×10^9	1.7	4

*Projected device performance based upon materials properties.

[1]For a 2 cm active length device.

It can be seen that at present, polymers enjoy a vast advantage over $LiNbO_3$ in frequency performance, but are still slightly inferior in E-O activity. It is important to note that the potential for substantial increases in the Pockels activity exists through improvements in the fundamental E-O activity of the side chain chromophore.

There have been many efforts to improve the electro-optical activity of NLO chromophores. Enormous enhancements of the molecular second order polarizabilities were established in early molecular design studies and improvements in poled polymer films have resulted from the incorporation of these units into the side chain structure. Thus, it is reasonable to suggest that substantial activity improvements can be achieved the near term, so that polymers will permit not only high bandwidth devices but also low power devices as a result of improved Pockels activity.

Most NLO polymer research has concentrated on amorphous polymers. To obtain a Pockels material, the polymer film must be poled, which is accomplished by heating the polymer to the glass transition temperature and applying a high voltage.[4] The film is cooled under the electric field and the noncentric distribution of dipoles is frozen within the structure forming an electret. Figure 2 shows an experimental method developed to measure the Pockels coefficient at Hoechst Celanese. The Pockels material effects polarization rotation of light impinging on the sample at approximately 45°.

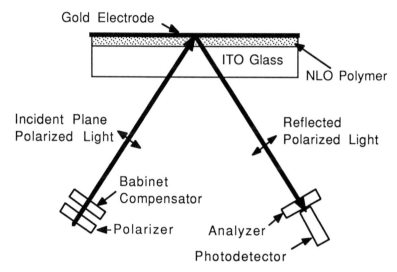

Figure 2: Device to Measure the Pockels Coefficient for Poled Polymer Films

Figure 3 shows results for a side chain polymer at poling voltages exceeding 100V/µm. At low voltages, there is a linear dependance of the Pockels coefficient on poling voltage in agreement with existing theory. In the analysis, poling is modeled by a Boltzmann distribution of molecular dipoles in which µE is taken as the energy of the dipole in the applied poling field E.[6] This analysis does not consider local field effects, birefringence, dielectric anisotropy, intermolecular interactions, or intramolecular conformation effects. In short, it is quite remarkable that the theory (developed for guest-host systems), works at all for side chain polymers.

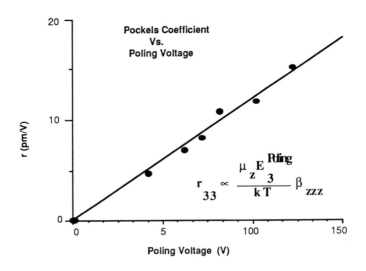

Figure 3:   Development of the Pockels Coefficient for a Poled
Side Chain NLO Polymer as a Function of the Poling
Field

A major issue for poled amorphous polymers is the
stability of the poled state. Because the poled film is in a frozen,
non-equilibrium state, the poled phase is metastable. Studies of
depoling have been previously described.[4] By using polymers
with high glass transition temperatures, T_g, and by annealing
the poled film to reduce free volume,[7] reasonable stability is
achieved. Accelerated testing at elevated temperatures suggests
that the electro-optic polymer described in Table 1 would
retain 80% of its $\chi^{(2)}$ activity at 70°C after 5 years.

Other approaches to achieving improved stability include
crosslinking.[8]   Simple crosslinking results in a polymer with a
T_g equal to the temperature at which the crosslinking is

performed. Even for these systems, the polymer must be heated to a temperature above the T_g for poling so that the advantages of crosslinking are not readily apparent. However, each polymer must be tested individually, and the depolarization behavior will be dependent on the location of crosslinking sites with regard to the NLO chromophore.

All of these approaches are limited by the thermal stability of the NLO chromophore itself. Very high T_g amorphous polymers are well known (e.g. polybenzimidazole) and attachment of side chain NLO chromophores is quite possible. However, degradation processes in stilbene or azo type chromophores are well under way at 200°C. Thus it seems that the thermal stability of the chromophore remains the critical limiting factor for thermal stability of the poled phase.

Optical clarity of polymers was once thought to be a limiting issue; however, for use at 1.3μm, very low losses have been achieved. Figure 4 shows optical loss data for the polymer described in Table 1 in which a loss of 0.53 dB/cm is obtained. Measurement of absorption suggests that this loss is not due to absorption, but rather to scattering processes. In addition, tailoring of the absorption spectrum through the use of specifically designed chromophores is well known. For example, the λ_{max} of 4,4'amino-nitrostilbene occurs at 440 nm whereas the λ_{max} of 4,4'oxy-nitrostilbene occurs at 370 nm. Thus, for electro-optical applications, optical clarity does not appear to be a critical issue. On the other hand, for frequency doubling much more stringent requirements exist.

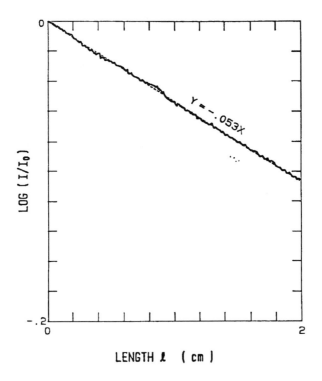

LENGTH *l* (cm)

Figure4:   Loss Measurement of an NLO Polymer Thin Film
Waveguide on SiO_2 (TEO, $\lambda = 1.3$ μm, d = 3.3 μm)

3   Frequency Doubling

Another interesting area of application for NLO polymers is
waveguide devices for frequency mixing. The simplest of this
class of devices is a waveguide frequency doubler. This device
could play a central role in optical storage systems where the
density of information storage increases as the wavelength of
light decreases.   Ideally the device would work with an 800 nm
laser diode. The foremost problem for organic NLO polymers is
therefore optical transmission at the doubled wavelength. For
organics this is a challenge, however there now appear to be
several materials with appropriate transmission windows.

To obtain high conversion efficiencies, phase matching between the fundamental and doubled frequencies must be achieved. At Hoechst Celanese, methods to develop a periodic $\chi^{(2)}$ have been demonstrated and shown to establish a quasi-phasematched condition.[3] The basic concept is shown in Figure 5 and is not new.[9] However, there are several polymer processing methods which appear to make this device concept particularly appropriate to NLO polymer thin films. For example, a periodic $\chi^{(2)}$ can be established by periodic poling as well as other methods.

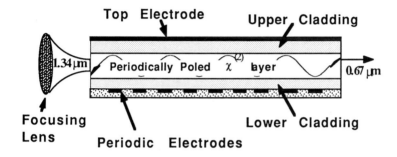

Figure 5: A Periodically Poled NLO Polymer Waveguide Frequency Doubler

Quasi-phasematching has been observed for slab NLO polymer waveguides as shown in Figure 6. In the figure, second harmonic intensity is shown as a function of rotation of a periodically poled slab waveguide about an axis orthogonal to the slab. For this example, 4,4'oxy-nitrostilbene was used as the NLO chromophore ($\lambda_{max}= 370$ nm) and 1.34 μm light was doubled to 670 nm. Using this technology, conversion efficiencies of 1×10^{-2} %/Watt for $P^{2\omega}/(P^{\omega})^2$ have been obtained.[10-11] Significant improvements appear to be quite possible by confining the guided light to a channel, as well as increasing poling efficiency and the interaction length.

Phase Matched Second Harmonic Generation
in a Periodic Polymer Waveguide

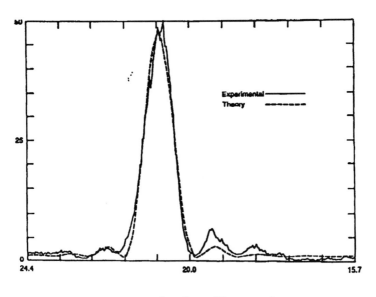

Angle   (Degrees)

Figure 6:   Second Harmonic Generation for a Periodically Poled
             Slab Waveguide as a Function of Angle

4   New Materials

As expected for NLO active materials with uniaxial
symmetry, poled NLO polymer films show both pyro- and
piezo-electric activity in addition to electro-optical activity.[12]
Using   the   4,4'oxy-nitrostilbene   unit   attached   to   the
methacrylate backbone, studies have been performed at
Hoechst Celanese to measure and correlate all three physical
properties. Figure 7 gives the relationship between the electro-
optical and pyroelectric coefficients for these poled polymer
films; significant .activity exists for both effects.

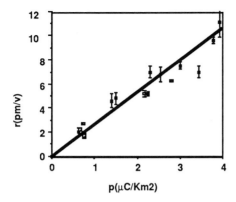

Figure 7: Pyroelectric and Linear Electro-optical Response for a Side Chain NLO Polymer containing 4,4'oxy-nitrostilbene

Perhaps more significant is our ability to tailor the activity of side chain polymers so that new materials may be designed to enhance several types of activities; these materials are then truly multifunctional. Figure 8 shows the structure of a polymer designed to have enhanced pyroelectric and electro-optical activity by the replacement of the inert vinyl acetate comonomer with an NLO chromophore unit. This polymer has a nearly doubled pyroelectric activity, while maintaining the electro-optical activity of the parent NLO polymer.

Vinylidene Dicyanide - Vinyl Acetate Copolymer

Multifunctional Copolymer

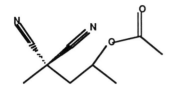

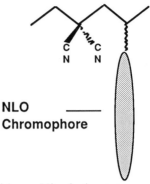

NLO _____
Chromophore

Figure 8:   PolyVinylidene Dicyanide - Vinyl Acetate
Copolymer and a Multifunctional Copolymer
containing the VDCN unit with an E-O Active Unit

There are many other examples of new organic functionalities which may be used to design multifunctional polymers. For example, naphthalocyanines form a class of organometallics demonstrating saturable absorption and enhanced $\chi^{(3)}$ activities.[13] Both properties are useful for the realization of all optical devices. Figure 9 shows a typical structure where functionalizing groups are attached via the central silicon atom.

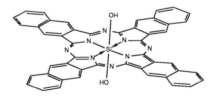

Figure 9:   Naphthalocyanine Species

These materials are soluble in poly methyl-methacrylate and the resulting guest/host structures show narrower spectral responses than neat films, owing to reduced molecular interaction.

Irradiation with monochromatic light at 810nm leads to excitation of the molecules and hence a different absorption spectrum, thus the material becomes much more transparent at the activating wavelength. This excitation process can occur in picoseconds and the decay occurs in nanoseconds. Since the materials are very stable and the saturation intensities relatively low, real device potential exists.

This concept has been further extended by Hoechst Celanese to the covalent incorporation of metallated macrocycles into polymer structures. An example is shown in Figure 10, where the phthalocyanine moiety has been chemically combined into an acrylic based polymer system.

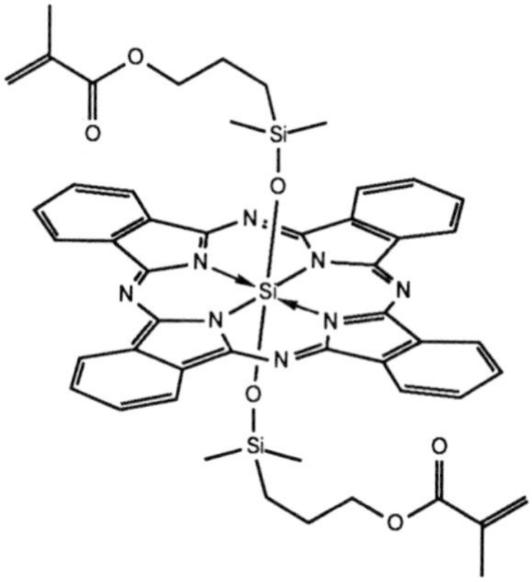

Figure 10: Phthalocyanine Monomer

This material is readily soluble in reasonable solvents and can be processed into thin films for desired device formats. In addition, it shows an even narrower absorption spectrum than its guest/host equivalent and because the moiety is covalently incorporated into the polymer, there is no possibility of slow crystallization of the pthalocyanine species which would lead to undesirable optical scattering effects and lowered optical damage threshold. Obviously, this monomer may be copolymerized with a unit designed for electro-optical or pyroelectric activity to create materials with new multifunctional properties.

5   Conclusion

Organic non-linear optical polymers have been developed in a very short period of time and appear to have high utility in the rapidly growing optical processing, communication, and data storage areas. Polymers with properties better than materials currently used in the emerging optical device industry have been achieved. More importantly, because these properties may be altered through chemical synthesis, it is expected that significant improvements in $\chi^{(2)}$ activities will be achieved.

New waveguide materials are emerging as the molecular design rules for different properties are defined. These properties may be combined into one material by copolymerization leading to new multifunctional materials, for both electro-optical and all-optical device applications.

6   Acknowledgements

In common with most industrial research programs, this work draws on the activities of many people, however, we would be remiss in not recognizing our colleagues at Hoechst Celanese, our collaborators in other Companies and Universities and particular Government Agencies for their support. Dr. D.R. Ulrich of the Air Force Office of Scientific Research has played a particularly significant role.

7   REFERENCES

1. Nonlinear Optical Properties of Organic and Polymeric Materials, D.J. Williams, Ed. (American Chemical Society, Washington, DC 1983).

2. K. D. Singer, J.E. Sohn, L.A. King, H.M. Gordon, H.F. Katz and C.W. Dirk, "Second-order nonlinear optical properties of donor- and acceptor-substituted aromatic compounds," J. Opt. Soc. Am. B7, 1339.

3. G. Khanarian, R. A. Norwood, D. Haas, B. Feuer, and D. Karim," Phase matched second harmonic generation in a polymeric waveguide," Appl. Phys. Lett. (to be published).

4. H. T. Man, K. Chiang, D. Haas,   C. C. Teng, and H.N. Yoon, "Polymeric Materials for High Speed Electro-Optic Waveguide Modulators", Proceedings of the SPIE 1213, 7.

5. S.J. Lalama and A.F. Garito, "Origin of the nonlinear and second-order optical susceptibilities of organic systems, " Phys. Rev. A. 20,1179.

6. K. D. Singer, M.G. Kuzyk and J.F. Sohn "Second-order nonlinear optical processes in orientationally ordered materials: relationshiop between microscopic and macroscopic properties"; J. Opt. Soc. Am. B4, 968.

7. C. Ye, N. Minami, T. J. Marks, J. Yang, G. K. Wong, Macromolecules 21, 2899.

8. M. Eich, B. Reck, D. Y. Yoon, C. G. Wilson, G. C. Bjorklund, J. Appl. Phys. 66 , 3241.

9. J.A. Armstrong, N. Bloembergen, J. Ducuiny, and P.S. Pershan, "Interactions between lightwaves in a nonlinear dielectric," Phys. Rev. 127, 1918.

10. G. Khanarian and R. A. Norwood "Quasi-phasematched Frequency Dobuling over Several Millimeters in Poled Polymer Waveguides," Postdeadline Paper Integrated Photonics Research Meeting, Hilton Head, SC, March 26-28, 1990.

11. R. A. Norwood and G. Khanarian, "Quasi-phasematched Frequency Doubling over 5 mm in a Periodically Poled Polymer Waveguide," to be submitted to Electronics Letters.

12. H.A. Goldberg, A.J. East, I.L. Kalnin, R.E. Johnson, H.T. Man, R.A. Keosian, and D. Karim, <u>Materials Research Society Symposium 175</u>, 113.

13. J.W. Wu, J.R. Heflin, R.A. Norwood, K.Y. Wong, O.Zamani-Khamin; P. Kalyanaraman, J. Sounik and A.F. Garito, "Nonlinear-optical process in lower-dimensional conjugated structures," <u>J. Opt. Soc. Am. B6</u>, 709.

Poled Polymers for Frequency Doubling of Diode Lasers

G.L.J.A. Rikken, C.J.E. Seppen, S. Nijhuis, and E. Staring

PHILIPS RESEARCH LABORATORIES, P.O. BOX 80000, NL-5600 JA EINDHOVEN,
THE NETHERLANDS

ABSTRACT

We report on the design and characterization of methylmethacrylate copolymers with nonlinear optical 4-alkoxy-4′-alkylsulfone stilbene sidechains, which are transparent down to 410 nm. A fairly stable, polar orientation has been obtained by means of electric field poling, resulting in a reasonably high nonlinearity ($d_{33} \leq$ 9 pm/V). Phase-matched second harmonic generation in planar waveguides of these polymers is achieved by means of spatially periodic photochemical bleaching of the stilbene sidechain.

INTRODUCTION

Aligning polar molecules embedded in a polymer matrix by means of a high electric field, as first demonstrated by Havinga and van Pelt[1] can introduce a significant polar orientation. This makes these materials suitable for second order nonlinear optical (NLO) effects, like electro-optical modulation[2,3] and second harmonic generation[3,4] (SHG). Especially for this last purpose polymeric materials are very promising, as they can easily be applied in optical waveguide geometries, thereby transforming low powers like those emitted by diode lasers into high intensities and consequently giving high conversion efficiencies. This might lead to compact, efficient and inexpensive short wavelength sources that would find wide application in high density optical recording, laser printers etc.. However, up to now no poled polymer has been reported that can be used to double the frequency of the emission of the common AlGaAs-GaAs laser diodes ($\lambda \sim$ 820 nm). Here we will describe polymers that were specifically designed for this purpose and present the linear and nonlinear optical properties that affect their performance in this specific application. Furthermore we will describe preliminary results on achieving phase-matching by means of spatially periodic photobleaching.

FREQUENCY DOUBLING POLYMER

The major requirements for the NLO moiety in a frequency doubling polymer are a high first order hyperpolarizability $\beta(-2\omega;\omega,\omega)$ and complete transparency at both the fundamental and the second harmonic wavelength. As a small separation between the charge transfer (CT) band and the second harmonic gives a large resonant enhancement to β (Ref. 3) a narrow CT band is obviously advantageous. These points are clearly demonstrated in fig. 1[5]. Important secondary requirements are a large dipole moment for electric field poling and a functional group for chemical bonding to a polymer backbone. After having screened a large number of chromophores for their linear and nonlinear optical properties[6], we have selected some stilbene compounds containing an alkylsulfone group as electron acceptor and an alkoxy group as electron donor, as favorable candidates. Sidechain copolymers with methylmethacrylate were synthesized, as described elsewhere[7], with various concentrations of the NLO moiety.

Fig.1 Comparison of absorption and hyperpolarizability determined by EFISH of the two molecules shown.

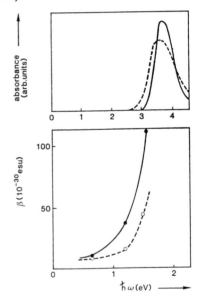

I $R = (CH_2)_5\ CH_3$

II $R = (CH_3)$

The sidechain concentration was limited by the occurrence of semi-crystallinity, above 25 mass % for compound I and 50 mass % for compound II. Figure 2 shows the solid-state absorption spectrum. At 410 nm the absorption coefficient α is 2.8 cm^{-1} and it increases rapidly at shorter wavelengths. At 820 nm, waveguiding losses in thin films of these polymers were below 0.1 dB/cm. The refractive indices of these materials were determined using quasi-waveguide techniques at several wavelengths and fitted to a Sellmeier dispersion formula, as shown in fig. 3. From this, the coherence length $l_c = \lambda/4(n_{2\omega} - n_\omega)$ can be easily determined. This parameter is essential for phase-matching the second harmonic wave to the fundamental one. At 820 nm we find a coherence length of 5.5 μm for the 25% copolymer (II). Part of the wavelength dispersion shown in fig. 3 can be attributed to the polymer backbone (for PMMA, $n_{410nm} - n_{820nm} = 2.10^{2}$), the rest is due to the strong (oscillator strength 0.6) UV CT band of the NLO chromophores.

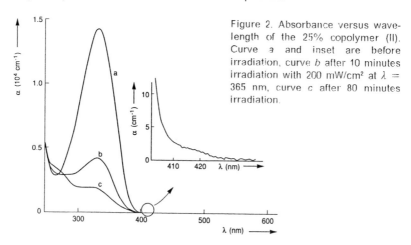

Figure 2. Absorbance versus wavelength of the 25% copolymer (II). Curve *a* and inset are before irradiation, curve *b* after 10 minutes irradiation with 200 mW/cm² at $\lambda = 365$ nm, curve *c* after 80 minutes irradiation.

Polymer films were poled by a corona discharge[8] at a temperature of 370 K, which is about 20 K below the glass transition temperature. Typical fields, as determined by a compensation technique[8] were 1.2 MV/cm. Determination of the nonlinearity was done *in situ* by a Nd:YAG/dye laser system. Equilibrium d_{33} is reached in tens of seconds, after which the sample is cooled down to room temperature in 15 minutes. The nonlinearity shows the expected linear dependence on NLO chromophore concentration (fig. 4). Only at the highest concentration a (positive) deviation is found, which may be indicative of extra orientation due to some collective behavior of the chromophores. The highest value obtained was 9 pm/V at 820 nm for the 45% copolymer shortly after poling. The temporal stability of the nonlinearity was fairly good, as shown in fig. 5, if the sample was stored in a dry environment. Storage at ambient conditions caused a rapid drop in d_{33} after a few days, due to water diffusing into the polymer which acts as a plasticizer. After two weeks, the polymers were found to contain up to 2 weight % water.

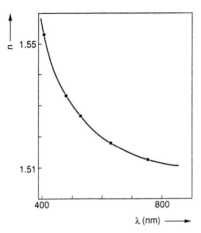

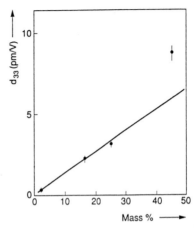

Figure 3. Refractive index versus wavelength of the 25% copolymer (II). The curve is a fit to the Sellmeier formula.

Figure 4. Second order susceptibility versus NLO chromophore concentration at 820 nm fundamental wavelength, shortly after poling at 1.2 MV/cm.

Figure 5. Second order susceptibility versus time at 1064 nm fundamental wavelength of the 25% copolymer (II) after poling at 1.2 MV/cm, when stored in a dry environment. The curve is only meant to guide the eye.

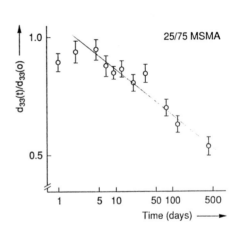

PHASE MATCHING

Several routes are available towards phase-matched second harmonic generation in poled polymers. Amongst others are spatially periodic modulation, anomalous dispersion, modal waveguide dispersion and birefringence. Fairly high birefringence due to the chromophore orientation has been reported for poled polymers[9,10]. By applying a sufficiently thick silicon-dioxide layer as an optical buffer between the ITO electrode and the polymer film, we were able to carry out prism coupling refractive index measurements on poled films. We found that the as deposited films showed a larger refractive index for light polarized parallel to the plane of the film. For the 25% polymer (II) this amounted to $n_\perp - n_\parallel \sim -2.10^{-3}$, probably due to flow induced alignment of the sidechains. On annealing at 370 K for 24 hours, this difference strongly decreased. Electric field poling introduced the expected positive birefringence, albeit too small for phase-matching purposes (for 25% (II) $n_\perp - n_\parallel \sim +1.10^{-3}$ at 413 nm, whereas $n_{410nm} - n_{820nm} = 5.10^{-2}$). An explanation for this low birefringence remains to be established.

So-called quasi-phase-matching may be obtained by periodic spatial modulation of the nonlinearity[11]. This has been demonstrated in polymers by applying a spatially periodic poling field[12]. As such a set-up is difficult to realize on a micrometer length scale, we have chosen a different approach. In figure 2 it is shown that the CT band can be fairly easily bleached by UV irradiation, which results in a very strong loss of nonlinearity, accompanied by a decrease in refractive index. This decrease is wavelength dependent, ranging from 3.10^{-3} at 820 nm to 9.10^{-3} at 410 nm for the 25% polymer (II). Using gratingwise photobleaching with a UV laser beam pattern generator we have accomplished the desired spatial modulation on a poled polymer film. The second harmonic measurement set-up is schematically depicted in fig. 6. Quasi-phase-matching, as witnessed by the strong increase in second harmonic power, is shown in fig. 7 and occurred close to the wavelengths calculated from the linear optical properties. The width of the phase-match peak is much larger than expected, which is most likely due to variations of the effective waveguide propagation constants. Depending on waveguide design, even very small thickness fluctuations may lead to this broadening and therefore a very stringent thickness control of polymer waveguides needs to be developed. As we have also succeeded in defining channel waveguides in these polymers by photobleaching, an entirely 'photographic' definition of a polymeric frequency doubling device seems feasible.

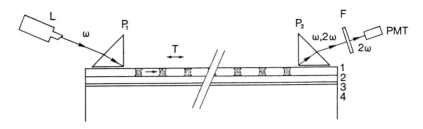

Figure 6. Experimental set-up for determining quasi-phase-matching in periodically bleached polymeric waveguides. The polymer is spin-coated on a silicondioxide (2) - indium-tin oxide (3) covered substrate (4) and bleached with a periodicity T.

Figure 7. Second harmonic power versus fundamental wavelength from two poled polymeric waveguides with bleached periodicities of 11.0 μm (a) and 11.4 μm (b).

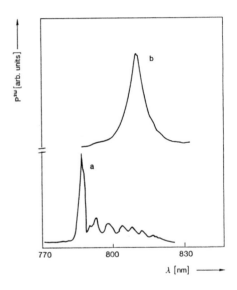

CONCLUSION

In conclusion, we have described the design and characterization of methylmethacrylate copolymers with 4-alkoxy-4'-alkylsulfone stilbene sidechains as nonlinear optical moieties, which are suited for frequency doubling of 820 nm diode lasers. Phase-matching by means of spatially periodic photobleaching was demonstrated.

ACKNOWLEDGEMENTS

We gratefully acknowledge stimulating and helpful discussions with L.T. Cheng (DuPont CR&D), A.H.J. Venhuizen, E.E. Havinga, E.W. Meijer, W. ten Hoeve and H. Wynberg.

REFERENCES

1) E.E. Havinga and P. v. Pelt, Ber. Bunsenges. Phys. Chem. 83, 816 (1979)
2) S.J. Lalama, K.D. Singer and J.D. Sohn, SPIE Proc. 578, 130 (1985).
3) "Nonlinear Optical Properties of Organic Molecules and Crystals", Vol 1 and 2, Chemla, D.S., Zyss, J. Eds.; Academic Press, New York, 1986.
4) K.D. Singer, J.E. Sohn and S.J. Lalama, Appl. Phys. Lett. 49, 248 (1986).
5) E.W. Meijer, W. ten Hoeve, S. Nijhuis, G.L.J.A. Rikken and E.E. Havinga, Dutch patent application PHN 12923 (1989).
6) L.T. Cheng, W. Tam, G.R. Meredith. G.L.J.A. Rikken and E.W. Meijer, SPIE Proc. 1047, 61 (1989).
7) S. Nijhuis, G.L.J.A. Rikken, E.E. Havinga, W. ten Hoeve, H. Wynberg and E.W. Meijer, to be published in J. Chem. Soc., Chem. Comm.
8) R.B. Comizolli, J. ElectroChem. Soc. 134, 425 (1987).
9) J.I. Thackara, G.F. Lipscomb, M.A. Stiller, A.J. Thickner and R. Lytel, Appl. Phys. Lett. 52, 1031 (1988).
10) W.H.G. Horsthuis and G.J.M. Krijnen, Appl. Phys. Lett. 55, 616 (1989).
11) J.A. Armstrong, N. Bloembergen, J. Ducuing and P.S. Pershan, Phys. Rev. 127, 1918 (1962).
12) G. Khanarian and D. Haas, US patent 4.865.406, G. Khanarian, D. Haas, R. Keosian, D. Karim and P. Land, Proc. CLEO'89, THB1.

Non-linear Waveguiding and Degenerate Four Wave Mixing in Rhodamine Doped Epoxy Films

B. Rossi, H.J. Byrne, and W. Blau

DEPARTMENT OF PURE AND APPLIED PHYSICS, TRINITY COLLEGE, DUBLIN 2, IRELAND

1. INTRODUCTION

The critical stage in the development of an all-optical technology is the development and optimisation of suitable nonlinear optical materials. Although many functions and processes have been demonstrated on a prototype level [1,2], exploitation has been limited by the lack of an optimised material. Inherent material requirements include a large nonlinear optical susceptibility, which would allow for switching of low light intensities and for small device dimensions, and a fast response and decay time, enabling high bandwidth operation. Other material properties such as quality of film formation, mechanical strength and chemical, thermal and radiation stability are also critical.Among the materials currently under investigation, organic materials have attracted particular interest. Their ease and low cost of synthesis as well as their processability and the tailorability of their optical and electronic properties are obviously favourable factors. Organic conjugated polymers have been shown to exhibit large nonlinear susceptibilities with ultrafast response times [3,4] and appear to be the most promising organic materials for nonlinear optics. However, studies of the chain length dependence of the polymer nonlinearity indicate that the effective conjugated length of the polymer is limited to about ten repeat units [5]. Although the origins of this behaviour are unclear, a move towards less cumbersome short chain organic molecules seems reasonable.

In this study, the waveguiding properties, both linear and nonlinear, of an epoxy film doped with a commercial laser dye, rhodamine B, are investigated. The wavelength range studied is in close proximity to the resonance of the dye. The temporal behaviour of the nonlinear optical response of the film is investigated as well as the dispersion of the nonlinearity in the region of the resonance. Degenerate forward four mixing of 350psec pulses is observed in a guided mode by coupling out after 1mm to maintain thin grating conditions. The dye/epoxy system should serve as a good model system in which to study the variety of processes observable in organic waveguides.

2. MATERIAL AND LINEAR OPTICAL PROPERTIES.

The film was spun from a solution of one part MV757 epoxy resin, one part epoxy hardener and one part of 2g/l rhodamine B in ethanol solution. A glass microscope slide was used as substrate. A spinning rate of 500 r.p.m. was employed, producing a film of thickness 13.0 $\pm$ 0.5μm. The absorption spectrum of the film, both spectra,is singly peaked at 545nm and compares well with that of rhodamine B in solution [6], indicating that the dye/epoxy film is a monodisperse, quasi-solution, with no indication of aggregation. Assuming a value of $\varepsilon(545) = 1.06 \times 10^5$ lmole^{-1}cm^{-1} for the molar extinction co-efficient of rhodamine B at 545nm, the concentration of dye in the film may be calculated to be 7.7×10^{-3} mole l^{-1}.

3. LINEAR AND NONLINEAR WAVEGUIDING.

Waveguiding experiments were conducted using a nitrogen pumped dye laser. The laser dye employed was rhodamine B, which provided 350 ± 50psec pulses tunable in the range 595-640nm. Typical pulse energies were in the range 40-80µJ per pulse. Coupling to the waveguide was achieved by the prism coupling method. The sample was mounted on an Aerotech ARS-301 rotation stage which enabled control of the coupling angle with 0.01° resolution. The coupling prism employed was an SF6 double prism of base angle $\varepsilon = 60°$, and refractive index at 633nm of 1.806.

At low pulse energies, ($\approx$100nJ per pulse), waveguiding was readily achievable in the rhodamine doped film across the range of the dye and was observable as dark m-lines on the uncoupled beam from the exit face of the prism as well as a visible fluorescence streak in the film. Approximately 20 modes were observable and the zeroth order mode was seen to have a coupling angle $\alpha = 1.05 \pm 0.01°$ with respect to the normal to the input face of the prism. The coupling angle of the highest order mode was found to be $\alpha = -5.81 \pm 0.01°$. Values of the effective indices, β/k, of the modes of the guide may be calculated [7], and are found to lie in the range 1.573 - 1.511. Knowing the effective indices of the modes and the thickness of the guide, the refractive index of the film may be calculated [8] to be 1.5743 ± 0.0001 at 633nm. The total number of modes which may be supported by the guide may be calculated, knowing the refractive index [8], to be 18 for both TE and TM modes. Comparison of this figure with the approximate number of modes observable implies that no distinction between TE and TM modes is observable, indicating the absence of film birefringence.

Using neutral density filters to achieve a variation in intensity, the intensity dependence of the throughput of the guide, as well as the intensity dependence of the coupling angle was monitored at both 616nm and 635nm. The observed dependences are shown in figures 1a(616nm) and b(635nm) for a selection of modes of intermediate mode number. In both cases, there is a clear dependence of the coupling angle on intensity, the angle increasing with increasing intensity at both wavelengths. Estimation of the degree of nonlinearity of the refractive index requires a knowledge of the intensity which is coupled into the film.

At a wavelength of 616nm, there is a dramatic increase in the throughput of the guide with increasing intensities, indicating the presence of an intensity dependent loss process in the guide. Such behaviour is characteristic of an absorption bleaching process, by no means uncommon in organic dye systems [9]. At 635nm, however, the observed trend is inversed. It has been shown that, in the presence of a strong nonlinearity, nonlinear coupling is accompanied by a significant loss of coupling efficiency [10]. In this case, however, the otherwise Gaussian profile of the angle dependence of the coupling efficiency becomes strongly asymmetric. In the present case, however, a reasonable fit of Gaussian profiles to the modes at all intensities can be performed and therefore it may be deduced that a behaviour characteristic of an inverse saturable absorption is observed. Although such a process has also been observed in organic dye systems [11], the problem of understanding the reversal of trends within a small wavelength range must be addressed. With this in mind, the intensity dependence of the transmission of the film, in an unguiding geometry, was monitored at both wavelengths. Figures 2a(616nm) and b(635nm) show the observed

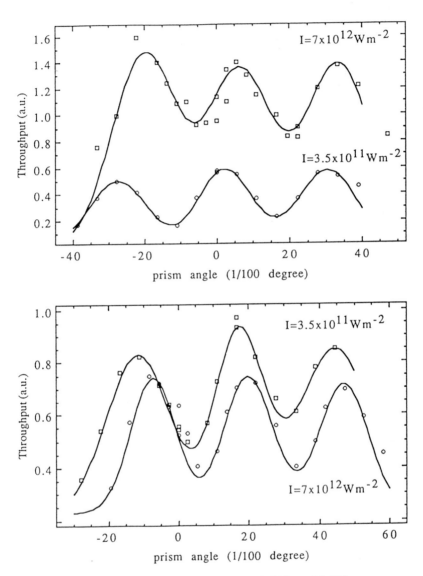

Figure 1; Intensity dependence of coupling angle (a) 616nm, (b) 635nm

dependences of the film transmission on intensity. A similar trend is observed at both wavelengths. The initial decrease of the transmission is indicative of an inverse saturation process, whereas, at higher intensities, a bleaching of the absorption is seen. The difference in the trends observable in the intensity dependence of the guided throughput at the two wavelengths, may, therefore be attributed to a difference in the intensity ranges studied at the two wavelengths (determined by the tuning curve of the lasing dye). The origins of the nonlinear behaviour of the absorption of the rhodamine doped epoxy is as yet unclear and further studies are underway to elucidate the relevant energy level scheme.

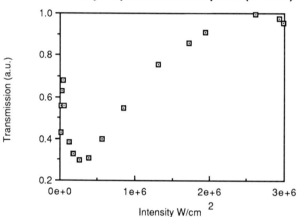

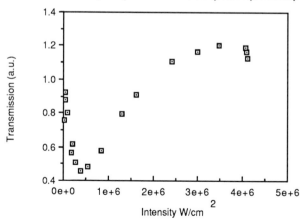

Figure 2; Intensity dependent absorption of rhodamine doped epoxy film at (a) 616nm and (b) 635nm.

The nonlinear throughput of the guide may be used to estimate the coupling efficiency into the guide. This may be achieved by fitting the rate of change of the throughput of the guide as a function of intensity to the nonlinear absorption curves. This procedure assumes that there is no significant loss of coupling efficiency as a result of nonlinear coupling. For the wavelengths 616nm and 635nm, the intensity coupling efficiencies are thus found to be 0.61 and 0.18 respectively. Given a spot size of 250μm radius, these correspond to energy coupling efficiencies of 0.020 and 0.008. It should be noted that the difference in coupling efficiencies at the two wavelengths is not the result of spectral differences in the material properties but rather the day to day variation in the coupling region.

Knowledge of the guided intensity enables an approximate calculation of the material nonlinearity from the intensity dependence of the mode number. Assuming

$$\Delta \beta/k = \beta/k \ (I) - \beta/k \ (I_0) \tag{1}$$

$\chi^{(3)}$ may be calculated from the relationship

$$\chi^{(3)} = \frac{4\varepsilon_0 c n^2 \Delta n}{3I} \tag{2}$$

to be $\chi^{(3)}(616nm) = 1.35 \times 10^{-16} m^2 V^{-2}$ and $\chi^{(3)}(635nm) = 1.0 \times 10^{-16} m^2 V^{-2}$. This calculation is very much an oversimplification to avoid numerical integration of nonlinear coupling equations over the length of the coupling region. To allow for the fact that the nonlinear interaction occurs over this length rather than over the dimensions of the laser spot, a correction factor of 5×10^{-4} should be employed. The nonlinearity of the film at the two wavelengths is therefore estimated to be $\chi^{(3)}(616nm) = 6.75 \times 10^{-20} \ m^2 V^{-2}$ and $\chi^{(3)}(635nm) = 5.0 \times 10^{-20} \ m^2 V^{-2}$. These bulk nonlinearities may be converted to molecular hyperpolarisabilities, assuming a quasi-solution of concentration 7.7×10^{-3} mole l^{-1}, yielding values of $\gamma = 2.7 \times 10^{-42} m^5 V^{-2}$ and $2.0 \times 10^{-42} m^5 V^{-2}$. These values compare rather favourably to off-resonant nonlinearities of organic conjugated polymers [12].

4. SELF-DIFFRACTION FROM LASER INDUCED GRATINGS.

The experimental method of degenerate four wave mixing in the forward direction [13] was employed to investigate the dye doped epoxy system. The dyes used in these experiments were coumarin 485 and rhodamine B, giving tunability in the range 495 - 550nm and 595 - 640nm respectively. The experimental method, which is described in detail elsewhere [14], is based on the formation of a transient grating in the material as a result of the response of the nonlinear refractive index to the interference of two spatially and temporally overlapped beams. Under thin grating conditions [13] satisfied experimentally by keeping the angle between the two beams small, ($< 1°$), an expression relating the diffraction efficiency, η into the first order, to the third order material nonlinearity may be derived;

$$|\chi^{(3)}| = \frac{4\varepsilon_0 c \ n^2 \ \lambda \alpha \ \sqrt{\eta}}{3 \ \pi \ I_0 \ (1-T) \ \sqrt{T}} \tag{3}$$

where c is the speed of light, ε_0 is the permittivity of free space, n is the refractive index of the sample, α is the absorption coefficient, T is the sample transmission and I_0 is the input pulse intensity. In the experiments reported here, n is taken to be the refractive index of the epoxy. For the purpose of investigation of the temporal decay of the grating, a third beam was employed to read the grating in a boxcar geometry. Delay of this beam with respect to the arrival of the other two beams enables resolution of the recovery time of the material nonlinearity.

Waveguiding was achieved again by the prism coupling method. The coupling prism employed was an SF6 prism of base angle $\varepsilon = 45°$, and refractive index at 633nm of 1.806. For simplicity, a two beam geometry was employed for the degenerate four wave mixing in the guided mode. A second, identical prism was used to limit the interaction length within the guide by coupling out the light after 1mm. This was necessary for operation under thin grating conditions. Otherwise a full, three beam geometry would be required.

5. RESULTS AND DISCUSSION.

5.1 Self-diffraction in transmission mode.

In the transmission mode, self-diffraction was carried out using coumarin 485 as the source laser dye. When placed in the interference region, a clear diffraction pattern from the sample was observable over much of the tuning range. The diffraction efficiency into the first order was measured by using silicon photodiodes to monitor the input and diffracted energies. The intensity dependence of the diffraction efficiency was investigated by transmission of both input beams through a variable neutral density filter and figure 4 shows the dependence at 540nm. The plot shows a cubic dependence of the diffracted signal on the

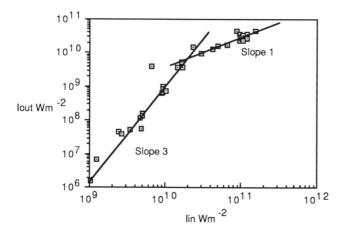

Figure 4 ; Intensity dependence of transmission diffraction efficiency at 540nm.

input intensity at low intensities, characteristic of a third order nonlinear process. At higher intensities, the dependence flattens to a linear behaviour. This may be considered characteristic of the onset of a permanent grating but, as the behaviour is reversible, must be the result of a rapid saturation of the material nonlinearity.

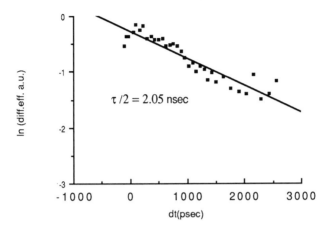

Figure 5 ; Temporal decay of the diffraction efficiency at 540nm.

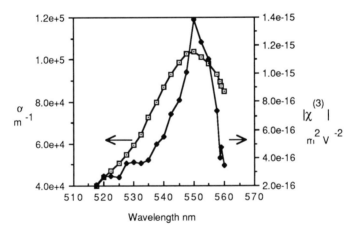

Figure 6 ; Wavelength Dependence of nonlinear susceptibility.

The temporal reponse of the nonlinear process giving rise to the diffraction process was investigated by the introduction of a third "read" beam whose arrival at the sample with respect to the two "writing" beams could be varied via a delay line. Figure 5 shows the decay of the grating at 540nm. A good fit to a single exponential decay is found with a decay time $\tau = 2.05$nsec. This is consistent with the decay of a grating formed by a saturated absorption with a ground state recovery time of 4.1nsec [13]. Similar values have been recorded for fluorescence lifetimes of rhodamine 6G [15].

The dispersion of the nonlinear optical response was monitored by tuning over the wavelength range of the dye. In figure 6 the dispersion of the nonlinearity is plotted as well as the absorption spectrum. The nonlinear response follows closely the linear absorption of the film, indicating that the diffraction process is largely absorptive in nature. This is supported by the observation of a recovery time on the nanosecond timescale.

5.2 Self-diffraction in the guided mode.

Self-diffraction in the guided mode was performed using rhodamine B as the source laser dye, as it afforded a longer absorption depth. The prism coupling apparatus was introduced into the interference region and a second prism was placed on the sample, 1mm from the input coupling region. In the region of 595nm, a diffraction pattern was observable after the sample. The intensity dependence of the diffraction process was monitored by use of a variable neutral density filter. Figure 7 shows the resulting curve. The input intensity has been corrected for the nonlinear throughput of the guide in the absence of self-diffraction, described in previous sections. The behaviour of the nonlinear throughput was

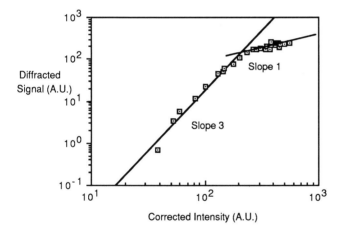

Figure 7 ; Intensity dependence of guided diffraction efficiency at 595nm.

compared to the nonlinear transmission of the sample to yield values of the coupling efficiency, calculated to be 0.61 for intensity and 0.20 for energy. The corrected intensity dependence shows the same behaviour as shown in the transmission mode (figure 4), consisting of an initial cubic dependence followed by a linear dependence at higher intensities, resulting from a saturation of the diffraction process within the laser pulsewidth. A diffraction efficiency of 0.25 at a guided intensity of 2.64×10^{10} Wm^{-2} is observed. The material nonlinearity required to produce this efficiency may be calculated from equation 1 to be 1.5×10^{-19} m^2V^{-2}, which, assuming a quasi-solution of concentration 7.7×10^{-3} molel^{-1}, corresponds to a molecular hyperpolarisability for rhodamine B of 6.0×10^{-42} m^5 V^{-2}. It is worth noting that this value, although considerably resonantly enhanced, compares rather favourably with those measured for organic conjugated polymers [12].

6. CONCLUSIONS

Spun films of a commercial epoxy, doped with a commercial laser dye have been employed to study nonlinear phenomena at visible wavelengths in both transmission and waveguiding modes. Nonlinear coupling and waveguiding of 350psec pulses is observed. Although a strong intensity dependence of both the coupling angle and the throughput of the guide is observed, the Gaussian shape of the angle dependence of the coupling efficiency of the modes is preserved, indicating that there is no significant loss of coupling efficiency. Despite the low energy coupling efficiencies, the high intensities afforded by the guided geometry produce strong nonlinear absorption effects. Coupling nonlinearities have been estimated and, significantly, the values of the nonlinearities are of reasonable size, although considerably resonantly enhanced. Degenerate forward four wave mixing is readily observable in the transmission mode and temporal measurements show the nonlinear mechanism to be largely absorptive in nature. In the guided mode, self-diffraction is observable in the long wavelength tail of the absorption. Material nonlinearities have been calculated and compare favourably to those of the more cumbersome organic conjugated polymers. This indicates the feasibility of a move towards less cumbersome, short chain conjugated systems, in the search for nonlinear optical materials suitable for optical technologies.

ACKNOWLEDGEMENT.
This work has been completed and is presented in loving memory of Barbara Rossi, who died tragically on the 1st of March, 1990.

REFERENCES

[1]   W. Blau, Opt. Commun., 64, 85 (1987)
[2]   P.J. Cullen, W. Blau and J.K. Vij, Opt. Eng., 28, 1276 (1989)
[3]   W.M. Dennis, W. Blau and D.J. Bradley, Appl. Phys. Lett., 47, 200 (1985)
[4]   G.M. Carter, J.V. Hryniewicz, M.K. Thakur, Y.J. Chen and S.E. Meyler, Appl. Phys. Lett., 49, 998 (1986)
[5]   H.J. Byrne, W. Blau, R. Giesa and R.C. Schulz, Chem. Phys. Lett., 167, 484 (1990)
[6]   See for example "Kodak Laser Dyes" Catalogue, Eastman Kodak Company, Rochester, NY 14650.
[7]   R. Ulrich and R. Torge, Appl. Optics, 12, 2901 (1973)
[8]   M.J. Adams, " An Introduction to Optical Waveguides", J. Wiley and Sons (1970)
[9]   A. Penzkofer and W. Blau, Opt. and Quantum Electron., 15, 325 (1983)
[10]   G. Assanto, Seminar in the "International School of Quantum Electronics: Nonlinear Optics and Optical Computing", Erice, Italy, May (1988)
[11]   W. Blau, H. Byrne, W.M. Dennis and J.M. Kelly, Opt. Commun., 56, 25 (1985)
[12]   H.J. Byrne and W. Blau, Synth. Metals, 37, 231 (1990)
[13]   H.J. Eichler, P. Günter and D.W. Pohl, "Laser Induced Dynamic Gratings", Springer Series in Optical Sciences, Springer Verlag (1986)
[14]   P. Horan, W. Blau, H. Byrne and P. Berglund, Appl. Optics, 29, 31 (1990)
[15]   D. Langhans, J. Salk and N. Wiese, Physica, 144C, 411 (1987)

Optical Bistability in Organic Materials

B.S. Wherrett

DEPARTMENT OF PHYSICS, HERIOT-WATT UNIVERSITY, EDINBURGH EH14 4AS, UK

1 INTRODUCTION

There are many opportunities for the use of optics in information processing (optical computing). These range from the exploitation of high communications bandwidth in sequential interconnections down to the chip-to-chip level in conventional electronic computers, through all-optical digital computing and optical implementations of neural networks, to analogue (Fourier) processing. These topics employ data fan-out conditions that can range from one-to-one (data reordering between emitter and detector interfaces with electronics) through to global interconnection from any one pixel on a 2-D input plane to the entire Fourier plane. Optically bistable devices will have their greatest impact at the digital optics level in which integrity of binary data representation is achieved using the two standard (stable) response outputs. Fanning from device to device is kept at a minimum (perhaps below four) in order to maintain accuracy.

There are two regimes in which optics can complement digital electronics. For **ultrafast** switching and logic, beyond the speeds achievable electronically, the necessary high power per switch will limit systems to essentially sequential form (small switching networks, cryptology). Conversely in **highly parallel** systems in which low power per switch is accompanied by a longer timescale the **combination** of 2-D parallelism of active devices with the 2-D interconnect advantage of optics provides new processor architecture possibilities.

To date the greater progress has been made in the low power regime, with the construction of prototype digital optical systems at AT&T Bell Laboratories[1] and at Heriot-Watt University[2]. The latter circuits will be used here in order to illustrate applications of bistable devices and to establish operating criteria (sections 2,3). Investigations of optical bistability in organic materials are reviewed in these proceedings by G. Stegeman[3]; particular studies of low power bistability in nematic liquid crystal (K15) devices and in crystalline polydiacetylene (pTS) are described in section 4 of this article.

Ultrafast experiments in organics have been restricted almost entirely to studies of the third-order nonlinear optical susceptibility ($\chi^{(3)}$), the appropriate spectral component of which can describe the underlying mechanism for fast switching. Brief comments on such studies are included in section 4.

2 COMPONENTS OF AN OPTICAL COMPUTER

The circuit shown in Figure 1 has been designed for processing 2-D images or image representation of 2-D or 3-D physical problems such as may be required in robotic vision, finger print identification tasks, medical image processing, hydrodynamic or aerodynamic studies. Binary images of perhaps 10^4 to 10^6 pixels are input and fall on a 2-D programmable logic unit (F). Simultaneously on each pixel of the unit there is a control laser beam, the beam power is identical on each pixel and has the role of determining the logic function to be undertaken. The architecture is thus SIMD (single-instruction-multiple-datastream) and the array of control beams is produced by multiple fanning from a single beam. In turn this beam power is adjusted using a signal from a host electronic machine so that in effect the cycle rate of the electronics is multiplied by the parallelism of the optical images. Combinatorial logic is achieved between the input and the information that has been cycled around the circuit. After

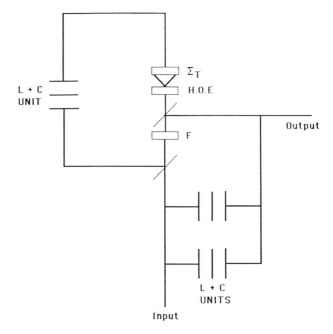

Figure 1. The O-CLIP digital optical circuit.

the processing stage various interconnect options are available. In the upper loop of the circuit schemed a nearest neighbour fan-out/in is achieved using a single holographic optical element (HOE), this gives a regular (space invariant) interconnect pattern usefed to carry out local neighbourhood tasks. A more powerful scheme for tasks such as sorting or discrete Fourier transforms is the 'perfect shuffle' interconnect which is non-local and irregular. The point is that 10^4 or more optical interconnects are achieved simultaneously, between planes of processors; electronic chip-to-chip pin-out interconnects are limited to only a few hundred. The fanned-in power levels are returned to binary levels by the threshold (AND or NAND) plate (Σ_T) in the circuit. Finally in order to synchronise the timing a controllable delay is included, called the lock-and-clock scheme (L&C). Lock-and-clock acts like a parallel shift register and is also usable as cache memory as indicated in the lower right portion of the diagram. This circuit is called the optical cellular logic image processor (O-CLIP) as it is an extended version (in terms of interconnect ability) of the electronic CLIP machine[4,5]. O-CLIP is capable of performing the binary image algebra primitive operations from which parallel processing is built up; it is not conceived of as a general purpose computer architecture.

Three components of the existing, diode-laser driven single channel circuit consist of bistables:- the programmable unit, the threshold plate, and the clocking unit. Each is presently a ZnSe-based nonlinear Fabry-Perot interference filter[6] and could, potentially, be an organic based device. Bistability in a nonlinear filter relies on the refractive index changes of the spacer material brought about by the irradiance level within the structure. The index nonlinearity may be of electronic origin (e.g. in InSb[7]) or of thermal origin as in ZnSe. Either transmission or reflection responses may be used. Figure 2 shows the reflection power output of the circuit filters as a function of the incident power level. Operation as a temporary memory (e.g. in the L&C unit) is achieved by biasing the input to a hold level within the bistable region. Any information input corresponding to one binary value must take the total input beyond the switch down position. If the information is present only temporarily the device output will continue to be switched with a low power output for so long as the bias beam is maintained. The other binary value must fail to take the total input beyond the switch point. Note that inversion occurs for a reflective device.

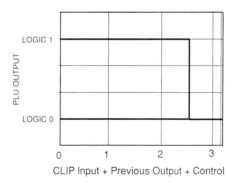

Figure 2. Optical response (schematic) of the latching devices used in the O-CLIP.

To see how optical logic is achieved consider the simultaneous arrival of two information signals. If the power level of either one is sufficient to produce switch-down then the device acts as an optical NOR gate. At a reduced control-beam level then the presence of both information signals may be required for switching - NAND gate operation. Thus one can include an additional signal that sets the effective bias level and thereby programmes the device to be either a NAND or NOR gate. By coupling two such controlled components together all eight, symmetric combinatorial functions have been programmed[8]. Having completed a logic switch in a bistable (latching) device, the control and information beams may be reduced so that standard 0 and 1 logic level powers are determined by the two hold-beam reflection levels. Because logic levels are restored or standardised on each cycle there is no error accumulation over multiple cycles, a requirement of all digital circuits. In the above circuit the use of latching logic and threshold units means that just one further bistable is required in order to complete the clocking.

The single channel circuit has been programmed to carry out recognition of set binary sequences, to compare number sequences and to perform full-addition. Concerning parallelism the status is image transfer between a pair of bistables with 15×15 pixel arrays; the AT&T circuit has 4×8 parallelism and employs a loop of four bistables.

3 CIRCUIT CRITERIA FOR BISTABLE DEVICES

The demands of circuitry such as described above lead to the following active-device requirements.

(i)	Bistability.	With clocking to give logic level restoration.
(ii)	Cascadability.	The output of one device must be usable to address a few further devices; this implies that each has a gain exceeding two.
(iii)	Stability.	Responses must be stable long-term to thermal or photostructural effects.
(iv)	Fabricatability.	Uniform 2-D arrays thermally or electronically structured to prevent pixel cross-talk.
(v)	Compatability.	Use at suitable wavelengths and laser repetition rates.
(vi)	Parallelism.	Power levels should be low enough to allow massive parallelism over perhaps $1\,cm^2$ area with total power consumption of less than the 10 Watts (time averaged) that might easily be dissipated.

CW operation is preferrable such that the electronic host machine can most easily carry out the timing (hold beam modulation) and programming (control modulation). Bit rates, in combination with the interconnect advantages, must be competitive with electronics. Parallelism of 10^4 to 10^6 at timescales from 1 μsec to 1 nsec are long term targets.

Many of the above criteria can be formulated as trade-offs, using the analysis of the dynamic equation of the nonlinear dispersive Fabry-Perot:

$$\frac{d}{dt} \Delta E = \frac{A \alpha I_o}{1 + F \sin^2 (\phi_o + b\Delta E)} - \frac{\Delta E}{\tau} \tag{1}$$

Here ΔE is the degree of spacer excitation (electronic or thermal), α the linear absorption, I_o the incident irradiance, F the Fabry-Perot finesse coefficient and ϕ_o the zero-irradiance spacer phase. The excitation recovery is described by a fixed time-constant τ and the phase (i.e. refractive index) dependence on excitation by the coefficient b. Cavity resonances occur when the total phase is an integer times π.

Steady-state solutions may be pictured by plotting the driving Airy function term as a function of ΔE and the excitation sink term that is a straight line of gradient τ^{-1}. A critical condition is defined where I_o and ϕ_o are such (I_c, ϕ_c) that the line intersects the driving curve tangentially at the point of inflexion. For smaller detunings combined with larger irradiances bistability is possible[9,10]:

$$I_c = \frac{1}{2\pi} \frac{\lambda \alpha}{|n_2|} f_{cav} \tag{2}$$

The coefficient n_2 is the irradiance-dependent refractive coefficient ($\partial n/\partial I$) and f_{cav} is determined by the Fabry-Perot cavity structure (reflectivities R_F, R_B and absorption-length αD). There are three relevant criteria for a cavity optimised for low irradiance operation:

$$\alpha D \approx (2 - R_F - R_B)/4 \quad , \tag{3}$$

$$I_o > I_c = \frac{3\sqrt{3}}{2\pi} \frac{\lambda \alpha}{|n_2|} \alpha D \quad , \tag{4}$$

$$\Delta n_{sat} > \Delta n_c = \frac{\sqrt{3}}{2\pi} \lambda \alpha \quad . \tag{5}$$

For a high finesse cavity the internal irradiance far exceeds the incident, this is manifested in equation (4) by the αD factor. The required internal irradiance and excitation level is similar regardless of finesse, however, and must be such as to produce a change in refractive index $n_2 I_c = \Delta n_c$. Because the nonlinearity will saturate eventually (or because damage will occur at high irradiance) it is necessary that the maximum achievable index change Δn_{sat} exceeds Δn_c. The condition $\Delta n_{sat} > \lambda \alpha$ is common to all bistable switching structures. A further condition, particularly relevant to organics, has been emphasised by Stegeman[3]; the two-photon absorption (βI) must be small compared to α for internal irradiance levels corresponding to bistability:

$$\beta < \frac{|n_2|}{\lambda} \tag{6}$$

The characteristic switching power-time product (energy) has a minimum value that must exceed $\lambda^2 I_c \tau$:

$$\varepsilon_c = \frac{\alpha\tau}{|n_2|}\lambda^3 \alpha D \tag{7}$$

A commonly used figure-of-merit has been $F_m = |n_2|/\alpha\tau$, which can be simplified if the absorption that generates the excitation ΔE dominates over 'parasitic' losses,

$$F_m = \frac{|\sigma_n|}{\hbar\omega} \quad \text{or} \quad \frac{\partial n/\partial T}{\rho C_p} \tag{8}$$

The first option applies to electronic excitation, where σ_n is the refractive cross-section, i.e. F_m the index change per unit absorbed energy density; σ_n depends only on oscillator strengths and on resonance denominators. In the thermal case $\partial n/\partial T$ is the thermo-optic coefficient, ρC_p is the effective density-specific heat product.

There are two trade-offs implied by equations (4,7). Firstly the irradiance or **energy with fabrication tolerance trade-off**; I_c and ε_c reduce with αD but very small αD is only possible for high reflectivities and if parametric losses are minimised. D must exceed $\lambda/2n$ and even near this limit rapid surface recombination can influence τ and decrease the effective n_2 values. Secondly there is a **power-with-speed trade-off**. Switching dynamics are controlled by τ whilst the degree of excitation that can build up is inversely proportional to τ ($n_2 = \alpha\tau\sigma_n/\hbar\omega$ or $n_2^T = \alpha\tau^T\rho^{-1}C_p^{-1}\partial n/\partial T$).

Using the free-electron, λ^2-dependence of σ_n as a guide and writing $\varepsilon_c \approx hc$ $\alpha D/(\sigma_n/\lambda^2)$ we see that only minor variations of ε_c are expected from material to material. Potential advantages in any material will depend upon avoiding parasitic absorptions, fabricating high-finesse cavities, and on selecting excitations of τ value matching the required laser pulse duration or circuit cycle time. Thermal ε_c values tend to be about two orders of magnitude larger than electronic-excitation values, but can still theoretically reach the picojoule region for optimal geometry. If the faster, electronic switching is to be employed then the device cooling rate must be great enough to prevent the dominance of index changes due to thermal build-up. This has proven to be a serious problem in semiconductor bistability and has led to the use of optothermal devices such as the ZnSe filter as test-bed components for optical circuitry.

4 BISTABILITY IN ORGANIC MATERIALS

Given the comments on device fabrication, background absorption and thermal build-up it is not surprising that as yet there has been no clear-cut experimental observation of optical bistability due to nonlinear refraction of electronic origin in organic materials. Following the 1980 experiment of Hermann and Smith in which the transmission of 4 ns pulses through a Fabry-Perot containing the polydiacetylene pTS showed nonlinear behaviour[11] there has been considerable interest in organics for ultrafast switching. Care must be taken over pulsed experiments however. Bistability implies two irradiance or power output levels (in steady-state) for the same input level. It is therefore tempting to produce x-y plots of the transmission of laser power out versus input power during a pulse; any difference between transmission at equal power levels on the rising and trailing edges of the pulse will appear as hysteresis. However if the excitation recovery time is comparable to or exceeds the pulse duration then excitation build-up occurs during the pulse and hysteresis is almost guaranteed. To verify bistability it is necessary to see jumps in the output. For device purposes these jumps must be manifested by an energy-out versus energy-in plot with a clear threshold between high and low transmission regimes; excitation build-up effects will give no such threshold.

There have been many studies of third order nonlinear susceptibilities, in particular in polymer materials, $\chi^{(3)}$ values between 10^{-13} esu and 10^{-6} esu have now been reported, c.f. refs. 12-16. Again caution must be taken in interpretation. For dispersive Fabry-Perot switching the real part of the component at frequency ω is relevant:

$$n_2(\omega) \propto \text{Re} \, \chi^{(3)}(\omega;\omega_3, \omega_2, \omega_1) \qquad , \qquad (9)$$

where two of $\omega_1, \omega_2, \omega_3$ are equal to $+\omega$, and the third equals $-\omega$. n_2 in cm^2/kW corresponds roughly to Re $\chi^{(3)}$ in esu. The imaginary part of $\chi^{(3)}(\omega;\omega,-\omega,\omega)$ corresponding to nonlinear absorption, can also produce bistability. The $\chi^{(3)}$ results, however, include third-harmonic generation (THG), two-photon absorption (TPA), absorption saturation (AS), near-resonance degenerate-four-wave-mixing (DFWM). These measure respectively $\chi^{(3)}(3\omega;\omega,\omega,\omega)$, Im $\chi^{(3)}(\omega;-\omega,\omega,\omega)$, Im $\chi^{(3)}(\omega;\omega,-\omega,\omega)$, $|\chi^{(3)}(\omega;\omega_3,\omega_2,\omega_1)|$. A number of DC electric-field induced measurements have also been made, e.g. EFISH and DC Kerr effect; $\chi^{(3)}(2\omega;0,\omega,\omega), \chi^{(3)}(\omega;0,0,\omega)$. Each mechanism has its own spectral resonances and timescale that depends on the proximity to a resonance. In addition if excitation recombination leads to lattice heating within the pulse duration anomalously large $\chi^{(3)}$ values may be estimated due to the dominance of longer lifetime thermal effects. Many experiments on organics at nanosecond timescales or longer, and some on picosecond scales, have measured thermal nonlinearities.

To date there has been no conclusive observation of optical bistability due to an electronic mechanism in organics. The most likely candidates have been the dyes in which saturable absorption mechanisms produce absorptive bistability[17,18] (in addition many bistable lasers employing saturable absorbers have been described). Contrast between the switched states is, however, poor and no device applications have been made. Optical bistability of thermal refractive origin, however, has been observed for a number of organic samples held within dielectric mirrored Fabry-Perot cavities; these include polydiacetylenes in solution form, liquid crystals, dyes and photochromics[19-23]. Bistability associated with slow, molecular reorientation in liquid crystals has been demonstrated[24]. Switching and bistability in waveguide structures has also been reported[25-29]. An alternative cavity is obtained by replacing one of the dielectric partial mirrors by a thin metallic layer. This relaxes the need for absorption in the spacer as the metal acts both as reflector and absorber. Only the thermo-optic property of the spacer is used. Thermo-optic coefficients of almost all liquid or solid materials are large enough to produce milliwatt bistability in such cavities. Further, any radiation wavelength for which the spacer is relatively transparent may be used, provided that the dielectric mirror is fabricated to reflect at that wavelength and that the metal has a reflectivity of around 90% or higher. Organic materials used for bistability in such cavities include polydiacetylene solutions and even the solvents themselves, alcohols[30] and liquid crystals[31].

Figure 3 shows milliwatt bistability (in transmission) achieved for a pTS single crystal sample grown by solvent evaporation under pressure between a pair of dielectric stacks on glass substrates. Operation was at 710 nm on the edge of the fundamental excitonic absorption of pTS, with mirror reflectivities of 82% and a crystal thickness of order 8 μm. The result indicates a figure-of-merit and thermo-optic coefficient very similar to those for semiconductors. Lower power pTS results had been expected from earlier measurements of the thermo-optic coefficient of bulk material and observations of a novel thermo-optically induced nonlinear beam-splitting phenomenon. In the latter the transmission of a beam of Gaussian spatial profile through a 140 μm thick pTS crystal, was monitored as a function of the beam power. This is a technique that has been used for measuring n_2 coefficients and signs[32]. If a nonlinear phase change is induced within the sample then the radial dependence of the irradiance causes a spatially dependent phase that can produce an annular irradiance profile on propagation to the near-field. The power level at which the beam centre irradiance dips gives a measure of n_2; the phenomenon gives some beam centre power limiting. Surprisingly in the pTS case an anisotropic two-beam profile was observed rather than an annulus[33]. This effect has been explained as a result of the thermo-optic property of pTS in combination with the anisotropy of the thermal conductivity parallel and perpendicular to the polymer chains. Figure 4(a,b) shows an example of the experimental result and the theoretical beam splitting for relative conductivities of 3:1 and for $\partial n/\partial T = -3 \times 10^{-4}$ K^{-1}. The phenomenon has both a limiting and a directional switching potential that is not available for isotropic systems.

One system that has been particularly successful and that could compete with semiconductor etalons for use in optical circuits combines the dielectric-metal cavity structure with a nematic liquid crystal spacer[30]. The significance of such material lies in the extremely high e-ray thermo-optic coefficient occurring at temperatures just below the nematic-to-isotropic phase transitions. Values exceeding 10^{-2} K^{-1} can be accessed, compared to typical values of 10^{-4} K^{-1} in most solid (or liquid) materials. The cyanobyphenyl, given the commercial names K15 or 5CB, has a transition temperature of 35 °C and held at 34.7 °C has enabled bistability with a critical power level of 20 μW[34]. Cavity reflectivities of just 90% were employed, the metal layer consisting of 20 nm of gold. Wavelengths from 514 nm through to 1.3 μm all produce low power bistability in such cavities. The two orders of magnitude power reduction compared to ZnSe bistables is achieved with little loss of speed, which is determined purely by the thermal engineering of the cavity; 50 μs characteristic times have been demonstrated using sapphire substrates. Even lower powers, down to 4 μW, are achieved using a dielectric-dielectric cavity containing K15 with incorporated anthroquinone absorbing dye[35]. In the all-dielectric case the temperature distribution

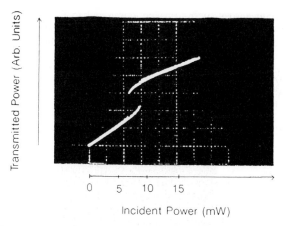

Figure 3. Optical bistability in transmission for a single-crystal pTS film in an etalon.

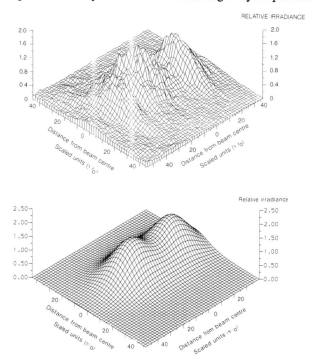

Figure 4. Beam-split output intensity profiles (a) experimental and (b) theoretical, for an incident Gaussian profile at 20 mW power level.

is relatively constant across such a spacer, compared to the gradient distribution peaking at the metal layer. Hence it is possible to operate closer to the transition point, without any portion of the liquid crystal becoming isotropic. Therefore greater $\partial n/\partial T$ values are used and switch powers are lower, Figure 5.

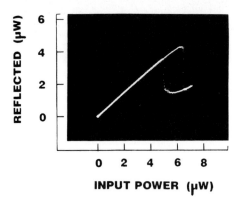

Figure 5. Microwatt bistability in a dye-doped nematic liquid crystal.

For circuitry applications an antireflection coating is incorporated on the metal layer in order that efficient absorption of control/information signals is achieved. Figure 6(a) shows the response to the hold beam, incident through the dielectric partial mirror. If the hold beam is then fixed at the indicated level P_H, the reflected hold as a function of the information signal power is as shown in Figure 6(b).
These results illustrate a number of features of device significance. Firstly, the insertion loss at high reflection is very low, also the contrast can be high because the lower branch of the bistable loop in Figure 6(a) can be brought close to zero reflectivity by appropriate cavity design. Secondly, a high gain can be achieved; in Figure 6(b) information levels of just 5-10 μW are used whereas the reflected hold beam level exceeds 100 μW. In principle cascading from one to ten further devices is achievable for such a gain. There is a catch, however, as illustrated in Figure 7. This shows the switching time as a function of the power level of the information beam. If the latter is weak, corresponding to high fan-out potential then the switch time far exceeds τ. This phenomenon is known as critical slowing down, applies to all bistables, and can be understood simply using the dynamic term in equation (1)[36]. There is thus a **trade-off between device cascadability and speed**, as demonstrated in Figure 7.

To demonstrate that liquid-crystal based devices can be coupled together, and incidentally to attempt to obtain bistability at power levels below critical, a circuit was constructed in which the output of device-1 was used as an information signal to device-2 and vice-versa. This is the Reset-Set Flip-Flop geometry that is used in electronics to produce bistables from mono-stable inverters[10]. A stable response is

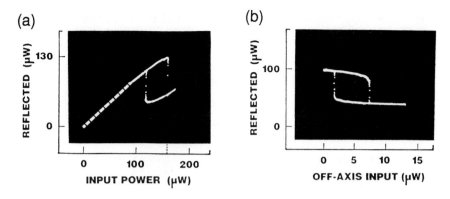

Figure 6. (a) On-axis and (b) off-axis bistable optical responses of a liquid crystal Fabry-Perot.

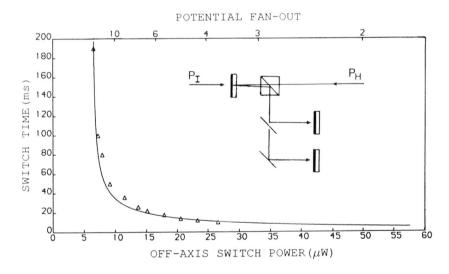

Figure 7. Switching time dependence on overdrive, showing cascadability-speed trade off.

is achieved with device-1 at high reflectivity and device-2 at low reflectivity (identical devices and hold levels are used). A small temporary information signal incident on device-1 is then sufficient to flip the output levels of both devices, as shown in Figure 8(a). The individual devices were not bistable, Figure 8(b), and could therefore be held at power levels below critical. It was found, however, that the coupled circuit would not operate far below critical and indeed it can be confirmed theoretically that the total power (on the two devices) must at least equal the critical power of one bistable device if flip-flop action is to be achieved.

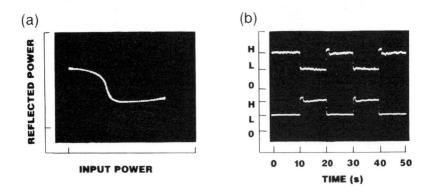

Figure 8. (a) Nonlinear response of a single liquid crystal etalon and (b) flip-flop action of a pair of coupled etalons. H, L, 0 refer to the High and Low output power logic levels, and the zero level.

Finally, it has been shown that pixellated images can be stored on liquid crystal etalons even of uniform 2-D structures (i.e. no physical pixellation). A Dammann, binary phase grating holographic element was used to illuminate a nematic device, with a 15×15 array of beams at power levels equal to within $\pm 2\%$. The input powers could be increased and then reduced such that all 225 elements gave low reflectivity (lower branch of the bistable loop). A brief shuttering of the inputs followed by removal of the shutter brought all elements to the high reflectivity state. Alternatively a mask was introduced to set an image on the device, this was maintained accurately after removal of the mask, demonstrating a degree of 2-D parallelism without cross-talk, comparable with semiconductor devices[37].

5 SUMMARY

As an example of a high performance electronic processor, the AMT-DAP distributed array processor has a potential of 7×10^6 floating point operations per Watt, each operation corresponding to thousands of one-bit switches. Both the 2-D interconnect freedom of optics and the high parallelism must be used to provide greater processing power.

The figure-of-merit F_m over a range of materials and mechanisms from thermal effects in liquid crystals to electronic effects in high-transparency glasses, tends to fall in the region 10^{-2} to 10^2. Implied bit rates for switching, using fairly high finesse cavities, range from 10^{10} to 10^{14} Hz per Watt of available optical power. Possible architecture scenarios therefore in principle could vary from 10^7 pixels per cm^2 operating at 1 ms rates through to single channel operation at sub-picosecond timescales or 10^5 channels using picosecond pulses but only nanosecond repetition rates.

A $\chi^{(3)}$ value as high as 1 esu, with microsecond relaxation and for $\alpha \sim 10^2$ cm^{-1}, is available for the narrow gap semiconductor InSb. Device fabrication and operation difficulties have to date prevented the use of this material in digital optical circuits and indeed, it is fabrication limits that tend to restrict achievable bit rates per Watt for all materials. At the shortest timescales the present inefficiency of optical sources must also be taken into account. A 10^{-7} esu material, picosecond response and with $\alpha < 0.1$ cm^{-1} is a target comparable in merit to InSb.

Organics, acting as thermo-optic materials with Fabry-Perot etalons, give the greatest potential for prototype devices (10^4 parallelism at a 10 μs repetition rate is a present target). Electronic effects in organics will have a role in ultrafast switches only if suitable fabrication figures-of-merit can be achieved, as described above, so that a clear threshold is obtained in the output-versus-input energies, and providing that thermal effects can then be avoided. For parallel arrays one is searching for excitation of longer timescales (> ns).

ACKNOWLEDGEMENTS

Many members of the Heriot-Watt Optoelectronic Devices research group have contributed to the work presented here; particularly A.D. Lloyd and C.H. Wang on the liquid crystal studies; T.G. Harvey, W. Ji and A.K. Kar on pTS, the material itself was supplied by D. Bloor, P. Norman and D. Ando, then at Queen Mary College, London; R. Craig, A.C. Walker and several others are involved in the circuitry work. Support from SERC, MOD and from the Boeing Aerospace and Electronics High Technology Center of Seattle is appreciated.

REFERENCES

1. M.E. Prise et al., Tech. Digest, Optical Computing '90, Kobe, 1990, 114.
2. B.S. Wherrett et al., SPIE, 1990 1215, 264.
3. G.I. Stegeman, this volume.
4. M.J. B. Duff and T.J. Fountain, 'Cellular Logic Image Processing', Academic Press, 1986.
5. B.S. Wherrett, SPIE Milestone Series, 1989, 1942, 185.
6. A.C. Walker, B.S. Wherrett and S.D. Smith, in 'Nonlinear Photonics', ed. H.M. Gibbs et al., Springer, 1990, 30, 91.
7. D.A.B. Miller, S.D. Smith and A. Johnston, Appl. Phys. Lett., 1979, 35, 658.

8.   R.G.A. Craig et al., Appl. Opt., 1990, 29, 2148.
9.   D.A.B. Miller, IEEE J. Quant. Electron., 1981, QE-17, 306.
10. B.S. Wherrrett, IEEE J. Quant. Electron., 1984, QE-20, 646.
11. J.P. Hermann and P.W. Smith, Paper T6, Proc. XI Int. Quant. Electron. Conf., Boston, 1980.
12. T.Y. Chang, Opt. Eng., 1981, 20, 220.
13. D.R. Ulrich, Proc. OMNO '88, Roy. Soc. Chem. Lond., 1989, 241.
14. P.N. Prasad, ibid, p. 264.
15. D.S. Chemla and J. Zyss (eds)., Nonlinear Optical Properties of Organic Molecules & Crystals, Academic Press, 1987, 2.
16. G.T. Boyd, J. Opt. Soc. Am., 1989, 6, 685.
17. Z.F. Zhu and E.M. Garmire, IEEE J. Quant. Electron., 1983, QE-19, 1495.
18. J.W. Wu et al., J. Opt. Soc. Am., 1989, B6, 707.
19. I.C. Khoo and Y.R. Shen, Opt. Eng., 1985, 24, 579.
20. M.C. Rushford et al., Optical Bistability II, ed. C.M. Bowden et al., Plenum Press, NY, 1984, 345.
21. W.M. Dennis, W. Blau and D.J. Bradley, Appl. Phys. Lett., 1985, 47, 200.
22. W. Blau, Opt. Commun., 1987, 64, 85.
23. C.J. Kirby, R. Cush and I. Bennion, Springer Proc. Phys., 1986, 8, 165.
24. M.M. Cheung, S.D. Durbin and Y.R. Shen, Opt. Lett., 1983, 8, 39.
25. J.D. Valera, B. Svensson, C.T. Seaton and G.I. Stegeman, Appl. Phys. Lett., 1986, 48, 573.
26. P.D. Townsend, J.L. Jackel, G.L. Baker, J.A. Shelburne and S. Etemad, Appl. Phys. Lett., 1989, 55, 1829.
27. K. Sasaki, K. Fujii, T. Tomiaka and T. Kinoshita, J. Opt. Soc. Am., 1988, B5, 457.
28. B.P. Singh and P.N. Prasad, J. Opt. Soc. Am., 1988, B5, 453.
29. I.C. Khoo and J.Y. Hou, J. Opt. Soc. Am., 1985, B2, 761.
30. A.D. Lloyd, I. Janossy, H.A. MacKenzie and B.S. Wherrett., Opt. Commun., 1987, 61, 339.
31. A.D. Lloyd and B.S. Wherrett, Appl. Phys. Lett., 1988, 53, 460.
32. D.L. Weaire, B.S. Wherrett, D.A.B. Miller and S.D. Smith, Opt. Lett., 1979, 4, 331.
33. T.G. Harvey, W. Ji, A.K. Kar, B.S. Wherrett, D. Bloor and P. Norman, Inst. Phys. Conf. Series, 1989, 103, 245.
34. A.D. Lloyd and B.S. Wherrett, Proc. OMNO '88, ed. R.A. Hann and D. Bloor, Roy. Soc. Chem. Lond., 1989, 69, 418.
35. A.D. Lloyd, C.H. Wang and B.S. Wherrett, SPIE, 1989, 1127, 143.
36. c.f. B.S. Wherrett and D.C. Hutchings, Springer Series on Wave Phenomenon, 1990, 9, 269.
37. A.D. Lloyd, C.H. Wang and B.S. Wherrett, SPIE , in press.

Index

DATE DUE

DEMCO NO. 38-298